MODERN
HAND
LOADING

Modern Handloading

MAJ. GEORGE C. NONTE, JR.

WINCHESTER PRESS

Library of Congress Catalog Card Number: 73-159427

ISBN 0-87691-046-0

First printing May 1972

Second printing October 1973

Third printing March 1974

Published by Winchester Press
460 Park Avenue, New York 10022

Printed in the United States of America

Contents

CHAPTER 1

History and Development

THE FIRST common men who carried guns into field and battle were, of necessity, handloaders. The fourteenth-century pikeman-cum-musketeer, with his curious but awe-inspiring iron tube on a pole, had to fumble each component (powder, wadding, ball) into his crude and unfamiliar weapon. He then applied a live coal or glowing slow match to "give fire." Coronado's redoubtable arquebusiers of the sixteenth century, seeking the fabled Cibola of the New World, no doubt "rolled their own" in preparation for battle. One can almost see them in corselet and gambion, rolling ball and "corned" powder into twists of paper after the fashion of the day. Bandoleers festooned with similar prepared charges in horn containers appear in contemporary prints.

Some of the earliest hand arms and artillery used separate "loading chambers," which contained all the elements of today's cartridge except a primer. These hollowed blocks of iron or bronze were loaded just as today one recharges a fired case. The chamber was then wedged in place at the rear of the barrel and fired. Once all available chambers were discharged, a fast job of reloading was required if the crew was to avoid being overrun by a charging enemy.

Perhaps that wasn't handloading as we know it today, but to me it is much the same. Progressing farther toward today, we find that in the American Civil War, a fast-firing gun called "Agar Coffee Mill Gun"

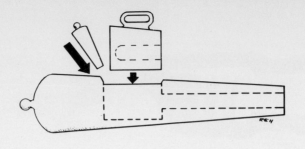

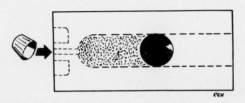

Top: early breech-loading cannon used a mobile reloadable chamber weded at the barrel breech for each shot—the first handloading, and in the field! Bottom: is typical steel reloadable "cartridge" used in early rapid-fire guns such as Agar and Gatling.

actually *had* to be supplied with reloaded cartridges in the field. It was issued with a chest of steel chambers. The chamber had to be charged with powder and ball, then primed with a percussion cap placed on a nipple at its rear. Only then could the magazine be filled and the gun fired. After firing, chambers had to be washed or cleaned of copious black powder fouling and recharged if the gun were to continue in use. The earliest Gatling guns used much the same "reloadable" cartridge, as did some artillery designs as late as the ninteenth century.

The slaughter of the American bison in the 1870's and 80's presented an opportunity to find out how well the newly-developed modern brass case could be reloaded in the field.

Though the hunting was good, the "Buffler Men" trimmed their outfits wherever they could. A lot of space and weight, and no small amount of money, could be saved by carrying lead in pigs and powder in kegs to the hunting ground, along with plenty of caps (primers) and a relatively small supply of cartridges or empty cases. Tool requirements were minimal, the operations simple. The day's generation of fired cases could be cleaned (always advisable in black powder days) and loaded around the evening campfire.

More than one hunter reported that he cleaned the powder fouling from his cases during lulls in the shooting, using a bristle brush and a bottle of water carried for the purpose. Some hunters cast their bullets in the field, others preferred the factory product and carried along an ample supply.

Modern Handloading

Left: Winchester Reloading Tool, 1st Type, listed in catalog for 1875 1st and 2nd edition—to be used with Winchester primers only. *Right:* Winchester Reloading Tool, 2nd Type, listed in 1875 catalog 2nd edition for any type of primer.

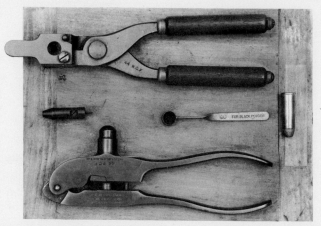

Winchester 1882 loading tool took form of what is now the Lyman No. 310 and still available.

96

TOOLS

FOR

RELOADING CENTRAL FIRE CARTRIDGES,

MANUFACTURED BY THE

WINCHESTER REPEATING ARMS COMPANY,

NEW HAVEN, CONN., U. S. A

In offering these articles to the public, the manufacturers believe that they present the most complete and compact article that can be produced. The aim has been to give in one instrument all the functions possible, without complication.

97

being fired, should be cleaned and washed out carefully with strong soapsuds or soda-water, and dried thoroughly ; otherwise the deposit of burnt powder left on them after firing causes them to oxidize rapidly, and they are soon destroyed. Care should be taken to set the primer well down. The pocket in the shell is always made deep enough to allow the primer to be set below the surface of the head of the shell. Premature explosions and misfires are often caused by failing to attend to this particular.

For powder to be used in rifle cartridges containing more than forty grains, we recommend the following brands and sizes of grains as giving the best results:—

    American Powder Mills' " Rifle Cartridge, No. 3."
    Hazard Powder Company's " Kentucky Rifle, F. G."
    E. I. DuPont, de Nemours & Co.'s " DuPont Rifle, F. G."
    Laflin & Rand Powder Co.'s " Orange Rifle Extra, F. G."

In reloading the .32, .38 and .44 W. C. F. cartridges used in Winchester Model 1873 and Model 1892 rifles and all center-fire pistol cartridges, use either of the following brands and sizes :—

    American Powder Mills' " Rifle Cartridge, No. 4."
    Hazard Powder Company's " Kentucky Rifle, F. F. G."
    E. I. DuPont, de Nemours & Co.'s " DuPont Rifle, F. F. G."
    Laflin & Rand Powder Co.'s " Orange Rifle Extra, F. F. G."

The American Powder Mills' " Rifle Cartridge Powder," as its name implies, is made especially for use in rifle cartridges.

In rifle cartridges none of the high grades of powder should be used ; we refer to such brands as Hazard's Electric, DuPont's Diamond Grain, etc. These powders (most excellent for use in shot-guns) owe their quick burning properties to their peculiar manufacture ; they are not hard pressed powders, and, when compressed in a cartridge shell, they cake behind the bullet more than the harder pressed brands, and give high initial pressure and very irregular shooting, without greatly increased velocity.

In casting bullets, keep the mold and lead very hot, and use the proportions of lead and tin given in the pamphlet for each bullet.

If using naked bullets, see that the grooves are filled with lubricating material ; beef tallow or Japan wax is best for this purpose. Wipe off all surplus grease before loading. When patched bullets are used, place a lubricating disc of wax with a card-board wad both above and below it, between the ball and the powder. No wads or lubricating discs are required in reloading any of the various cartridges adapted to Winchester Rifles, or in .45 caliber, 70 grains, Government cartridge.

Our reloading apparatus is accompanied with bullet-molds for either smooth or grooved balls, as may be ordered ; but all hand-made balls cast in this manner are comparatively imperfect, and will seldom give satisfactory results. All the bullets we use in the manufacture of cartridges are made by very heavy and perfect machinery. Balls thus made give the most accurate and satisfactory results in firing, and are constantly kept in stock.

Winchester 1894 catalog listed these loading tools and provided instructions for use.

History and Development

3

Such activities proved the quality of properly reloaded ammunition. Reloading became sufficiently popular that nearly all gun manufacturers produced simple tools for the purpose. A Winchester Model 1886 tool in .45/70 caliber cost only a few dollars which were easily regained in reduced ammunition costs.

Tools of the day were mostly variations of the pliers-like tong-tool that survives today in the Lyman 310 tool. Some carried integral molds, some used interchangeable dies. Crude, perhaps, by today's standards, but they turned out perfectly good ammunition.

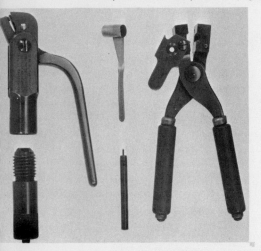

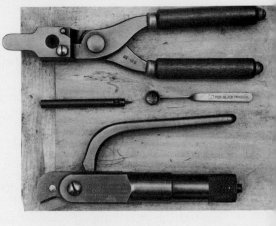

98

WINCHESTER
MODEL 1894 RELOADING TOOL.

This tool has been devised to make more easy the reloading and resizing of large sporting ammunition.

By a strong lever, a small motion (not more than .03 of an inch) is imparted to the slide. The cartridge is contained in the die A, as shown in the cut. The die screws into the frame. The shell, with its charge and bullet, is put together by hand and put into the die A. The die is screwed into the frame as far as it will go readily. A motion of the lever toward the die will force the cartridge into the die through a short distance. The backward motion of the lever releases the pressure on the cartridge and the die can then be screwed up through a part of one turn. The repeated motion of the lever, and the continued screwing up of the die bring the cartridge its full length into the die, insert the bullet to the right distance, crimp the cartridge around the bullet, and reduce the shell to its original size, so that it will go freely into the gun.

With this tool it will be found possible to *easily* reload the largest cartridges, compressing the powder, putting the bullet to place, and reducing the shell on the outside to its original form. By the reverse motion, that is, by lifting the handle of the lever away from the die, the slide is lifted, and the cartridge, by means of the extractor, is drawn a slight distance out of the die. When the handle is returned to the die, the latter can be unscrewed a corresponding distance. Another motion of the lever draws the cartridge still further out of the die, and with a few motions it becomes so loosened in the die that the latter may be easily unscrewed. As soon as the cartridge is loosened in the die, the extractor loses its grip, and the die and cartridge are taken from the frame together. This tool, new in principle, permits, with the use of *little force*, the most exact reloading, including the resizing of the shell.

This model will only be made for the following sizes of cartridges:—

32-40.	40-65 W. C. F.	45-70-500 U. S. G.
38-55.	40-70 W. C. F.	45-90 W. C. F.
38-56 W. C. F.	40-82 W. C. F.	50-110 Ex. W. C. F.
38-70 W. C. F.	45-70-405 U. S. G.	

Price per set, for Solid Ball,...............\$3.00
Price per set, for Express Cartridges, ... 3.50

99

TOOLS CAN BE FURNISHED FOR THE FOLLOWING SIZES OF CARTRIDGES.

22 W. C. F.	40-50 R. N. Patched.*	44 Webley.
25-20.	40-50 S. S. Patched.†	44 S. & W. Russian.
32 W. C. F.	40-70 B N. Patched.	44 S. & W. American.
32 S. & W.	40-70 S. S. Patched.	44-77 Patched.
32 Colt's.	40-70 S. S. Grooved.	45-90 W. C. F.‡
32 Short.	40-70 Ballard.	45-125 Express.
32 Long.	40-60 Marlin.	45-75 W. C. F.
32 Extra Long.	40-60 W. C. F.	45-60 W. C. F.
32-40.‡	40-65 W. C. F.‡	45 Sharp's Patched.
38 S. & W.	40-70 W. C. F.‡	45-70 W. C. F., Mod. 86.‡
38 Short.	40-82 W. C. F.‡	45-70 U. S. Gov't, 405 grs.‡
38 Long.	40-90 B. N. Patched.	45-70 U. S. Gov't, 500 grs.‡
38 W. C. F.	40-90 S. S. Patched.	45-70 Marlin.
38-55.‡	40-90 Ballard.	50 Carbine.
38-56 W. C. F.‡	40-110 Express.	50-70 U. S. Government.
38-70 W. C. F.‡	41 Long Colt's D. A.	50-95 Express.‡
38 Express.	44 W. C. F.	50-110 Express.‡

* R. N., Bottle Neck.   † S. S., Straight Shell or Sharp's Straight.   ‡ The Model 1894 Reloading Tool (page 98) furnished for these Cartridges.

REDUCED PRICE LIST OF RELOADING TOOLS.

For Reloading the Following.	Reloader.	Bullet-Mold.	Charge-Cup.	Per Set.
All Pistol sizes, and .32, .38, and .44 Winchester, Model 1873,..........	\$1.50	\$1.10	\$0.10	\$2.50
All Sporting and Military Cartridges, ...	2.00	1.10	.10	3.00
For Express Cartridges, ...	2.00	1.70	.10	3.50

Extractor Plugs for Reloaders (any size),....................\$0.25
Swages, .38 to .44 Caliber, for smooth bullets, ... 3.00
Swages, .45 to .58 Caliber, for smooth bullets, ... 4.00
Shell-reducing dies, ... 2.00

In ordering, state what particular cartridge is to be reloaded, as only one size can be reloaded with a single set.

The majority of cartridges require no wads in reloading. For such cartridges as require wads, a wad-cutter will be added to the set at an addition of 50 cents to above prices.

Top, left: new Winchester 1894 patent tool resized case after bullet was seated; *top, right:* Model 1894 Reloading Tool; *bottom:* Nov. 1894 catalog.

Modern Handloading

The epitome of lead bullet handloading was reached by the many devoted Schuetzen Rifle shooters of the late 1800's and early 1900's. Those capable and devoted gentlemen went to great lengths in determining precisely what combination gave maximum accuracy in a particular rifle. A confirmed Schuetzen addict might work several years with a single rifle, winding up with slightly different loads that produced maximum accuracy during the various seasons of the year. It was not unusual for a single cartridge case to be used literally thousands of times. Special "Everlasting" cases were machined from solid brass rod for the purist who reloaded a *single* case on the range for all his shots.

The tenacity and inventiveness of the Schuetzen rifleman who "stood up on his hind legs and shot like a man" lives on today in the bench-rest fraternity. These boys shoot sitting down. but they use *only* handloaded ammunition and shoot ridiculously small groups. They have been responsible for many new developments that increase accuracy.

The advent of smokeless powder and jacketed bullets driven at high velocities had a definite impact on reloading. Ammunition became more costly and loading became more complicated. Certainly the change did not reduce interest in reloading. Many an old-timer distrusted the new-fangled powder and "hard bullets," but the majority knew a good thing when they saw it, and clamored for the goods to brew their own. Always looking for ways to reduce costs, the U.S. National Guard units loaded their own so-called "mid-range" lead bullet ammunition for practice and social shooting. For a short time, Frankford Arsenal turned out decapping tools and re-sizing presses for the Guard and the old Ideal Company catered to its needs with "Armory" reloading presses.

The new high-velocity cartridges demanded more sophisticated loading methods and were a challenge to many a man. Outstanding among researchers of the day was the late Dr. Mann. His classic *The Bullet's Flight, From Powder to Target* is more in demand today than ever before. It is a "must" for any serious handloader, and a new printing is currently available. Dr. Mann was the first of the "wildcatters" with his high velocity experiments.

The reloadable center-fire metallic cartridge and reloadable paper/brass shotshell of the middle-late nineteenth century naturally created a demand for their components. Manufacturers were glad to supply these items. Maintaining a constant and adequate flow of loaded cartridges to the frontiers was a chancey business at best and certainly no trapper or Indian fighter would buy a new "cartridge gun" unless assured of a reliable resupply of ammunition in one form or another.

The powder mills continued to supply black powder for the muzzle loading trade. Naturally, it was available for reloading. As the muzzle loading demand dried out, cartridge-gun owners required more powder for their reloading, so the mills saw no reason to change their habits, though by then the bulk of their sales had shifted from hunters and fighters to the cartridge plants.

When the new smokeless powder loads came along, the manufacturers continued their policy of selling components as well as loaded ammunition. The powder makers did likewise, offering new propellants for individual sale as they were developed and proven.

It is interesting to note that if U.S. makers had, at any point in history, stopped individual sales of components, reloading would have died a-borning. Bullets and primers could have been produced independently on a small scale with relatively little capital, but powder, particularly smokeless, is a different breed of cat.

I've long believed that the virtual non-existence of reloading in Europe and England can be attributed to a calculated effort by the manufacturers to keep components away from the peasants, so they will *have* to buy factory loads. Such a state of affairs *may* increase manufacturers' profits, but I rather doubt it.

The period between the two great wars saw slow but steady improvement in reloading. The old Pacific Gun Sight Company introduced the now-familiar C-type reloading press and the now-standard ⅞ x 14-thread, two-die set. During that period, more than ever before, the *American Rifleman* (journal of the National Rifle Association) acted as a sort of clearing house for airing of experiments and developments of the recognized experts. Such names as Keith, Whelen, Wotkyns, Roberts and a host of others appeared regularly over the results of their experiences. Reloading tool manufacturers published handbooks which enabled the neophyte to get started, even if he didn't have a good teacher handy.

Reloading experiments by the late Philip B. Sharpe and others produced the sensational new cartridges of the day: the .357 Magnum, the .257 Roberts, the wildcat .22/250 Varminter, and others. Reloading was slowly becoming an end unto itself rather than a simple necessity or convenience as in the past.

Those were the lean years — a wealth of ideas, but scarce little money with which to prove them. Many shooters made their own special loading tools. The old home-brew bottle capper became a loading press (without being altered so much it couldn't still effectively seal Dad's fine hop-bearing product) and cut-off, worn-out barrels became bullet seaters. More than one fellow made bullet-pullers and decapping tools from strap hinges. Others converted tobacco cutters (how many of you ever even *saw* one of those?) into resizing presses. The boys even made their own jacketed bullets from fired cases and shot black powder in .30/06's (it was cheaper than smokeless). Improvization was the order of the day and methinks the clan was all the better for it. Rare indeed was the devoted shooter who couldn't cobble up workable reloading tools from household discards and farmyard scrap.

Left: Pacific Gun Sight pioneered the C-type reloading press back in the 1920's. Today's version here differs hardly at all from the original. *Right:* Out of all the die types offered, this ⅞ x 14 T.P.I. design is now basic and interchangeable.

Many new tools appeared on the market, but the economic climate was rough. Today, only a few survive, the Pacific C-Press, the Lyman tong-tool, the old Belding & Mull, and a few fortunate others.

During these trying times the ODCM (Office of the Director of Civilian Marksmanship) and the NRA were beacons in the darkness. If a fellow was an NRA member—and what reloader wasn't—he could purchase primers and powder, as well as occasionally some .30 caliber bullets and .30/06 cases, from the ODCM at prices that seem today like a license to steal. Of course, the bullets were military full-jacket type, the powder may have been in storage since 1918, and the primers were a type that left a corrosive residue that required repeated laborious cleaning with soap and water — but a fellow could keep shooting at little cost. That was the important thing.

Reloaders of the 1920's and 30's may have been short on cash, but not on ambition or influence. They made their experiences known and their wants were felt. For the first time, self-styled experimenters produced wildcat cartridges of such superior performance that they achieved standardization and factory production. The .22 Hornet and the .257 Roberts were born that way. The hollow base lead wadcutter bullet for pistol cartridges was thus developed by a reloading addict and is now *the* standard factory bullet for .38 caliber match ammunition.

During the above-mentioned period, handloaders (who seem to be strictly an American type) tried virtually every type and shape of case and bullet, powder combination, and velocity imaginable. Velocities in excess of 4,000 fps were achieved and unheard-of accuracy was produced. Then, war clouds boiled up and bullets and powder and primers became more valuable for other purposes. Reloading came to a near standstill. Many a gun-nut/reloader took his skills to war. Keith and Ackley, for example, labored mightily in the arsenals. Sharps studied enemy arms and ammunition for our Ordnance Corps. Others carried their own products into combat and did themselves proud. Not all of them came back.

During the war, primers and reloading tools were literally worth their weight in gold. Factory loaded ammo was practically unobtainable. Many a group pooled its gasoline coupons and spent weekends touring all the stores in small towns, searching for ammunition of any kind. Some pretty odd stuff came to light that way, but the boys usually found a way to use it. One fellow I know happily received a windfall of .30 carbine cartridges in those days. He didn't have a carbine, but he did love his S.A. Colt .357 Magnum. He spent hours carefully splitting those carbine case heads with a jeweler's saw so as to remove the primers undamaged. Primers in hand, he then developed a pretty fair .357 load for the carbine powder. That case of carbine ammo kept his .357 talking all during the war.

Many lessons of leaner years were re-learned, and more than one shooter swaged his own jacketed bullets and made cases from military brass bootlegged past the M.P.'s. Some tried making their own powder, and a few even reloaded fired primers. Dangerous, of course, but a born shooter just *has* to shoot—and damn the consequences.

When the smoke and flame of Nagasaki and Hiroshima cleared away, the resurgence or reloading interest that has since created an industry all its own, reared its head. Many of the old manufacturers of equipment had either not survived the war, or had gone on to other things. Soon there was no lack of equipment, though, for plenty of new makers sprang up in garages and small shops. They brought forth lots of good ideas and many of them have grown tremendously. RCBS, Inc., for example, grew from Fred Huntington's Oroville, California, garage and is today a dominating figure among makers of reloading presses and dies and accessories. The late Charlie Heckman did much the same with his C-H Die Company, as did other handloaders-turned-manufacturer.

While tools were available in plenty, components initially were not. One could always buy factory ammunition to get the fired cases and cast his own bullets. Powder was no problem, especially since millions of pounds of surplus was available, but the giants of industry seemed not at all interested in supplying handloaders with bullets and primers. Oh, they could be bought occasionally, but prices seemed exorbitant and deliveries terrible. Though denied repeatedly by industry spokesmen, there seemed to be an unwritten policy that reloading was to be discouraged. The idea being that the industry feared a loss in sales of factory ammunition if too many people loaded their own.

For nearly a decade this attitude prevailed, and perhaps we are fortunate that it did. During those years a couple of enterprising young fellows named Speer (Vernon and Dick) got into the bullet and primer business out in Idaho. A Nebraskan named Hornady, and also a trio out in California started producing bullets, the latter under the Sierra name. Federal Cartridge Corp. decided to release primers to reloaders, and shotgun champion Homer Clark started making shotshell wads and importing powders and primers. An outfit called Herter's Inc., out in Minnesota, got into the act from all angles—tools *and* components. Literally dozens of other small producers of components, tools, dies and accessories sprang up. Many of them are still with us today.

Demand grew, and so did production. One day in the middle fifties it dawned on the big manufacturers that millions of dollars were being spent for components annually, and they were getting very little of it. The reason was fairly obvious. The independent producers were pricing their goods to sell, yet still making a comfortable profit. The big boys in their ivory towers were still thinking they could force people to shoot only factory loaded ammo, and as a result, were losing a very substantial market.

Here is an interesting sidelight, About this time I saw one lot of popular pistol caliber cartridges loaded in cases that appeared to be designed to be unsuitable for reloading. They were cut partially through at a cannelure and would frequently pull in half when resized! One make of shotshells of the same period was so thin-walled that most cases blew through at the base, so were unfit for reloading. The manufacturers, when queried, denied that such characteristics were intentional. I wondered then. . . .

Then suddenly, the big outfits announced price reductions, making their products competitive with the independents. Obviously the ivory tower boys had changed their minds. They even managed to make deliveries. Since then, reloading has boomed even more than ever before. The big makers now even design their products for maximum utility by the handloader. They readily admit that a primary consideration in development of a new shotshell is its reloading life. Now that he has money to spend, and lots of friends, the reloader is ardently courted by big business!

Aside from the simple desire to shoot as much as possible for as little money as practicable, the massive importation of surplus military weapons probably did more to boost interest in reloading than any other factor. As the battle-fields were cleared and nations began to replace their war-weary arms, they suddenly discovered that virtually any gun that would shoot (and some that wouldn't) could be sold for cash in the United States. As Mausers, Enfields, Springfields, Lebels, Arisakas, and even older, long-obsolete black powder guns became available from the world's arsenals, sharp business men brought them to this country by the hundreds of thousands. More than a few not-inconsiderable fortunes were made in this fashion. With these guns came literally millions of rounds of old and new military ammunition which could be bought by the shooter for mere pennies per round, only a small fraction of its original cost.

All those cheap rifles and even cheaper ammunition caused a resurgence of interest in shooting just for fun. Many a man or boy spent his fifteen or twenty bucks for a good, shootable military rifle and a hundred rounds of ammo. He shot and liked it, then went back for more and more cheap ammo. Eventually he became accustomed to that sort of cheap shooting, and when surplus ammo became scarce, he naturally turned to reloading to continue shooting at a price he could afford. Even as this is written, the surplus market still thrives, though much less extensive than a few years back and destined to oblivion by Federal legislation. It continues to produce converts to reloading, and it is quite likely that a high percentage of you who read these pages got started just that way.

Progress continues in all areas. More variety and quantity than ever before is available in both equipment and components. Literally dozens of different loading tools are available, from the venerable tong-tool to a $5,000 powered rig that cranks out over 3,000 rounds per hour. You can choose from seven or eight makes of primers, at least a dozen bullet makers, seven or more in metallic cases and probably that many in shotshells.

As far as material is concerned, we are in the best shape ever. The big fly in the ointment is legislation and regulation, and we'll get to that in detail farther along.

Handloading has grown from a pure necessity to a thriving money-saving and performance-improving hobby in hardly a century. It's bound to be with us for many a year yet to come, certainly all of our lifetimes.

CHAPTER 2

Your Handloading Shop

WHILE IT is possible to do a pretty decent job of loading a few rounds of ammunition on the kitchen table, I've not yet met a single person who would put up with that sort of a setup any longer than absolutely necessary. I know of nothing more likely to produce a torrent of new cusswords than to have a C-clamped press slip off the edge of mama's table just as you near the end of the handle stroke in resizing a case. Don't laugh. It's happened to me, and it will to you unless you beg, borrow, steal or buy a good solid bench early in the game. In the instance referred to above, I wound up flat on my face on the floor, with C-press draped unbecomingly across the back of my neck, where it accommodatingly removed a sizeable strip of skin—which I suppose I didn't need anyway.

Consider first a permanent or semi-permanent setup in the average household. That seems to be the most common. Then we'll get around to apartment dwellers and others who are restricted in space or have to move around the country a lot in this highly mobile society we've built.

No great amount of space is needed for your reloading operation, and if the setup is well planned, everything you need now or later can be installed on or housed right in the bench you are about to build. Pick out a corner of basement (must be dry and well-ventilated), garage, spare room or attic. Personally, I don't go much for the latter because most I've seen become uninhabitable during the hot summer months. If dry, ventilated quarters aren't available, play it smart; begin by installing a cheap window or exhaust fan to get rid of heat, and a dehumidifier to keep moisture down. Their cost will be a profitable investment.

Lay out a space six to eight feet long, and perhaps a bit less in width, parallel to the wall. Clear all the family heirlooms, broken bicycles, old picture frames and similar debris away so you will have a little freedom of movement. Defend this space with your life, and if necessary, threats of divorce. The work you'll be doing there is personal and rewarding and demands your full attention while in progress. You won't want a stream of neighborhood children, your wife (or anyone else's wife), or the neighborhood bore drifting through while you're at work. If visitors must be admitted, restrict them to fellow handloaders who have sense enough to keep their mouths shut and their hands in their pockets while you are attending to the important matters we know so well. It usually isn't practical to wall in this space, so rig a drape or curtain hanging from wires or rods so you can close off the entire area.

By constructing a sturdy bench along the wall of your new inner sanctum, and putting in storage shelves or cabinets and a chair or stool, you'll be able to work in private comfort. And no matter what your wife says, never admit that sometimes you go there just to hide in a nice, peaceful atmosphere.

If you've only a simple layout and but a few calibers to load for, a bench four feet in length will provide plenty of space. An eight foot bench will handle almost anyone's layout. Make it with a top about eighteen or twenty inches in width, and roughly in accordance with the accompanying sketch. It need not be fancy, but it must be solid, and weight contributes to solidity. If you'll be doing heavy work with up-stroke tools, better tie at least the front legs to the floor with lag screws set in solid lead anchors.

Make the top first, from ten or twelve lengths of 2 x 4 lumber. Used wood is as good as any for this purpose, so long as the pieces are straight and clean. A few sizable knots won't hurt a thing. Spread a goodly layer of Elmer's Glue on one side of the first piece, slap another length up against the glue-coated surface, then spike it solidly in place with 20-penny nails. Glue will squirt out and run all over both you and the floor, but it'll wash off easily enough. With these first two pieces laid down on a spot of fairly flat and level floor, glue and spike the rest of your 2 x 4's in place. When laid up to the width you've decided upon, wipe off all the excess glue with a wet cloth and let the top lie for a day or two to dry and cure. The long nails and glue will make it plenty strong for any reasonable use, but if you're one of those fellows who wants everything twice as strong as it really needs to be, add threaded steel rods as shown. An electrician's drill bit will make those deep holes, and the job will be neater in appearance if you counterbore both ends so the nuts and washers are set below the surface of the edges.

When the top is dry, cut bench legs from 4 x 4 lumber and attach from the bottom with heavy angle brackets and long screws as shown. Drill pilot holes for the screws and smear a bit of soap on the threads and they'll drive a lot easier. A power screw driver helps a lot. Align legs carefully with a big carpenter's square and tack the 1 x 4 braces in place with small nails.

If you live near a good-sized hardware store, you might save lots of time by picking up a set of *heavy duty* steel bench legs. They'll cost little more than wood and do a fine job with lots less work.

You should now be able to tip the bench over onto its feet. Drill pilot holes through the top into the legs for 8" lag screws. Counterbore deep enough to place the lag screw heads below the surface and run

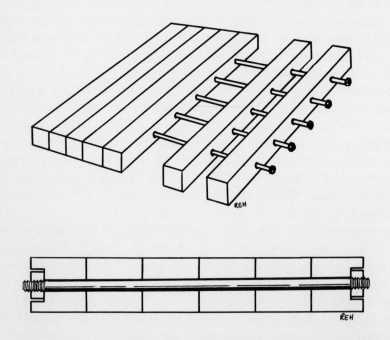

A solid bench top is essential and is easily made by gluing up 2x4 timbers as shown. Addition of tie rods increases rigidity and makes glue and nails unnecessary.

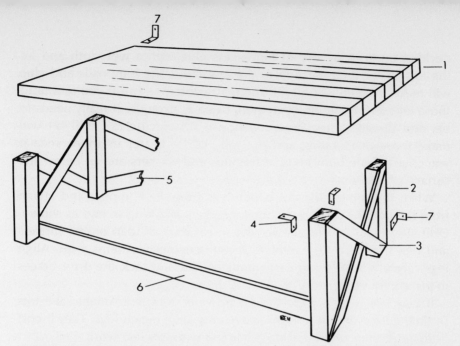

Heavy timber bench legs and frame are essential for stability. Don't leave out braces 3, 5, 6, or wobbles will develop. Top, 1, is tied to legs with brackets, 4, and to wall, 7.

them in tight — tight — with whatever trusty wrench you happen to have. Now go back and use 2½" or 3" woodscrews to attach the bracing permanently to the legs. Level the bench by whatever means suits you best.

You now have a bench that will do the job well enough, but a few improvements won't hurt it a bit. Square off the ends and round off the corners a bit to avoid damaging your carcass should you cut a corner a bit too fine. The top isn't too flat, so borrow the biggest block plane you can find and use it to take off the humps. A sheet of cheap ⅛" Masonite, smooth face up, will make a much nicer work surface. Spread a layer of glue over the planed top and lay the Masonite in place. It needn't be cut to a perfect fit. You can even up the edges later. Just make sure it covers all of the top. Stack plenty of heavy objects (boxes of bullets, your lead supply, case of shotshells, etc.) on the Masonite to hold it down while the glue dries, at least overnight, longer if possible. When the glue is completely dry, take saw and rasp in hand and trim the Masonite flush with all four edges of the timber top. Should you wish to be a bit more fancy, and can stand the cost, use Formica instead of Masonite, but pick a dark matte finish that won't bounce light up into your eyes.

Shove the bench into place along the wall. Once you are satisfied with its location, it's a good idea to secure it solidly in place. You'll be applying a good bit of force to it in the months or years to come, and you don't want it skittering around over the floor. About four heavy angle brackets will do the job. Attach one to each rear leg and a pair to the underside of the top with 2" or longer screws. Run the same size screws into the wall studs and that bench is there to stay. If you happen to be up against a concrete or block wall in basement or garage, use any of the better screw anchors for this part of the job.

Now is the time to get a press fastened down so you can get some use out of the new bench. Caution: don't bolt the press down in the first location that looks good to you. Set up all the tools and equipment that you'll normally be using. Use C-clamps to hold the items in place and work with them a while until you're certain you have the arrangement that is best for you. To go ahead and drill a mess of holes and bolt everything down right off simply means you're going to have a lot of extra holes once you later move them around to where they are convenient.

Once every tool position has been determined, drill holes and cinch the press down solidly with 4" lag screws with large washers under their heads. If, though, it's likely that you'll be removing the press often, either to use it elsewhere or to make room for some other operation, then you'd better use six-inch ⅜" machine bolts and large wing-nuts for the job. The lag screws won't hold well after they are removed and replaced a few times. Fasten all your other pieces of equipment in place with long woodscrews or small lag screws. Even a lubricator-sizer machine will be held solidly enough for any normal use with 2" woodscrews.

The basic bench is there now. It is solid and strong enough for the heaviest resizing or bullet swaging operation, and if you calculated correctly in the beginning, it's big enough for any new job you plan. Storage space for hand tools, dies and components is the next consideration.

A visit to a large hardware store or a look in a good mail order catalog will locate an assortment of ready-made plastic or metal drawers and bins. Get some deep drawers that you can attach under the top of your bench. These will do for keeping your hammers, screwdrivers and other common hand tools out of the way and also safe from pilferage by the rest of the family, if you have the presence of mind to fit a lock of some sort. Of course, you can build perfectly good drawers out of scrap lumber, but that's more trouble than it's worth, in most instances. It is another way to save a little powder money, though.

Personally I prefer to keep all of my reloading dies covered except when they are actually in use. For this reason I don't care for the many types of spring clips and storage pegs used by some people. Leaving a highly-polished resizing die out in the open simply allows it to collect dirt, dust and grit from the atmosphere. And this is a sure way for it to become scratched—or even ruined—the next time it is used. With common hardware store shelf brackets, attach three or four 1 x 6 shelves on the wall above your bench. Make them as long as the bench and use plenty of brackets. You'll eventually have quite a bit of weight on them, and it's no fun to be buried under a cascade of dies when you least expect it. On these shelves stack your die sets, still in the heavy cardboard or plastic boxes in which they were packed. Unfortunately, one or two die makers pack their products in fancy see-through boxes that fall apart in no time at all. If you are stuck with some of those, pick up some cheap plastic boxes with tight-fitting lids for the job. Set aside a few extra boxes for the extra expander rods, decapping pins, lock rings, etc., you'll accumulate.

Primers and powder can also be stacked up on the shelves over the bench, but I don't favor this unless your reloading hideaway can be locked securely—to insure that no inquisitive child, neighbor, or guest of friend wife can get in. At no time can you afford to have anyone not under your direct supervision handling those two sensitive items. My favorite storage spot is a heavy fibre footlocker under the bench. Not ony can it be securely locked, it can be chained to the wall or bench to keep anyone from making off with the whole lot. Another reason for this location is simply that air temperature is normally coolest down at floor level. This aids in preventing exposure to unusually high temperatures that occasionally occur up near the ceiling, especially in an attic location. Should you be so unfortunate as to be forced into a shop location that does become unusually warm, better pick another part of the house to store your powder and primers.

The National Association For Fire Prevention recommends powder be stored in a wood chest or cabinet constructed of lumber not less than 1" thick throughout and fitted with a tight lid or door. This thickness of wood prevents heat from being transferred through to the powder in case of a fire. Ignition of powder during a fire will be delayed more by wood than metal, and when ignition does take place, the wood will give way easier, eliminating dangerous fragments a metal container could produce. Keep this in mind when selecting a storage place for powder and primers.

Naturally you'll need good light to work by, and the odds are against the available location having it. I find a couple of tightly-stretched wires over the bench ideal for hanging a pair of droplights. On simple wire hooks, the lights can be moved wherever a little extra illumination is needed. There was even a time when I had over twenty feet of bench space, all with parallel fluorescent tubes which could be raised and lowered at will. That was an ideal situation, but pretty much too costly for the average handloader. For that very reason I don't have such facilities at my present location. A long-corded trouble light will come in handy when you drop your last spare decapping pin and it rolls into the dark under the bench. By whatever means, be sure you have good light.

But to get back on the subject of your handloading bench and work/storage area. It is essential that you have a system for storing and locating items. Every handloader I've known has been an incurable packrat. They all collect small lots of fired cases, handfuls of bullets, old tools and dies, empty cartridge boxes, gun parts, and God only knows what all else. One fellow I know keeps a collection of miniature whiskey bottles up on the same shelves with his powder and primers— full, blast his mangy hide! I once used cigar boxes for this type of miscellaneous storage, but over the years was able to gather up quite a number of empty .30 and .50 caliber machinegun ammunition boxes. Many of you know these items well. They are of heavy-gauge stamped steel, have carrying handles that fold up, hinged lids that are easily removed, and even gaskets that exclude air and water when properly closed. In addition to all that, they stack very nicely against a wall. The .30 boxes are ideal for small lots of goodies, while the .50 size will hold large amounts of almost anything. As this is written, about two hundred cans of this type are stacked against the wall of my storage area. Each has had one end sprayed with flat white enamel, and on this is written the contents with a black grease pencil (usually sold as "china marking pencil"). If you make shelves of rough (and cheap) lumber spaced to hold a single row of these boxes, you'll never have any trouble finding and retrieving anything you have stored away. The boxes can usually be found at so-called Army surplus stores at exorbitant prices, but a little shopping around will produce some you can afford to buy or for trade. Another very useful military box is the heavy steel 20mm and 40mm ammo cans often found among Navy surplus items.

Protection of the equipment bolted to your bench should be given at least a little bit of consideration. Some fellows are partial to up-ended plastic food bags, but they only encourage condensation and promote rust when the humidity is high. Plain cloth covers are better. If you can't talk mama into sewing some up for you, look over her

Modern Handloading

rag bag for worn out trousers. Chop off 12" or 18" pieces of leg and sew or staple one end shut. Presto, ready-made covers to drop right over press, powder measure, etc. When you've finished loading for the evening, simply wipe tools off lightly with an oiled rag and drop a pants leg in place. Everything will be bright and clean for the next session.

A while back something was said about the poor apartment dweller and the fellow who has to travel around a bit. Should they be deprived of the pleasures and economies of reloading simply because space is in short supply? Certainly not. Various mechanical-type magazines (*Popular Science, Science & Mechanics*, etc.) have carried numerous articles on how to build small, compact workshop setups for the apartment or mobile-home dweller. Almost any of these designs can be used for a reloading setup with only minor modification. One good friend of mine simply has drilled tool-mounting holes around the edge of his desk in his den. Tools and components are stored in a handy closet. When he feels the urge to concoct a few loads, out comes a compartmented box with all the necessaries in it. In a couple of minutes all the tools have been installed on the desk by means of wing-bolts, and he's ready to go. With a primer catcher on the press and a little care taken to avoid spilling powder on the carpet, his wife doesn't get offended and everyone is happy. I must admit, though, that an occasional guest will finally surrender to his own curiosity and ask about that half-dozen holes in the desk top.

Another apartment rig I've seen, consisted of a bench top hinged to the inside of a closet door. Open door, swing down bench top, attach two quick-release legs to fittings in floor and on bench, attach tools with wingnuts, all in less than five minutes, and our friend was ready to load. Another man I once knew picked up an old rolltop desk at a used furniture auction. With a little reinforcement and modification here and there, he was able to fit all of his equipment and supplies into it in such a manner that closing the rolltop hid them from sight. Refinished to match the landloard's furniture it never once interfered with the smooth running of the household, and being fitted with a good lock, it was never tampered with by wife, child or guest.

One of the neatest apartment-size loading shops I've ever seen was built by Texas writer and gunning aficionado, John Wootters. In converting the spare bedroom to combination office/gun room, John blocked off half the double closet with heavy plywood. Then, he built in carefully designed and dimensioned shelves, compartments and drawers and a solid, well-balanced bench top. There he loads many thousands of rounds of both metallic and shotshell ammunition each year. Yet, when it's dinner or cocktail time, or guests drop in, he simply slides the closet door shut, and everything disappears from view. It wouldn't be hard to duplicate this setup in any reasonably similar space with a little work and some slight carpentry skill.

At one time I owned a medium-size wooden chest that looked pretty much like a standard footlocker. Whether in a tent, in the field, or in my small room in bachelor officers quarters, I could set that chest up for reloading in but a few minutes. Inside, it was neatly compartmented to hold a RCBS Junior press, several sets of dies, powder measure, and scale, and components for two rifle and two pistol calibers. Standing on end with the lid swung open, the press swung out and up to clamp in place at bench height. C-clamps held it steady against table, dresser, door-frame or jeep bumper. Many thousands of rounds were loaded in woodsy bivouacs with that outfit.

It is impossible in the space allotted here to illustrate enough apartment-size reloading setups to meet everyone's needs. The ideas and examples given above will, I think, suffice to enable anyone reading this tome to come up with a reasonably effective setup with which he can enjoy reloading, no matter how restricted his living space might be.

While preparing for the 1964 *Handloader's Digest,* John Amber, of The Gun Digest Association, conducted a contest for reloading bench designs. While the winning designs bear no great similarity to the

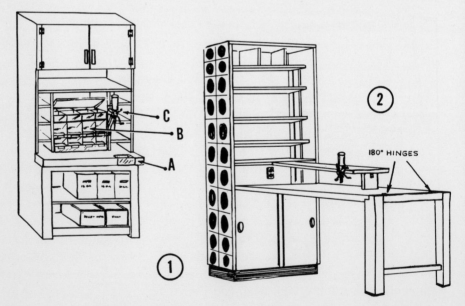

Loading bench designs from *Gun Digest* contest.

Modern Handloading

bench presented in this chapter, they are no doubt better suited to the needs of many handloaders. For this reason, I persuaded John to allow those winning designs to be reproduced here. For this, John, I thank you sincerely.

One last note on security and safety. You are the handloading buff of the house. You know what is safe and what isn't. The rest of the family and occasional guests *don't* know. Small children especially, with their insatiable curiosity, are particularly attracted to any hand-loading setup. The actual hazards to casual investigators are discussed in detail elsewhere. It should be sufficient to say that you must take all steps possible to keep other people clear of your gear and supplies when you aren't around. A separate locked room is best, but lacking that, locked chests or cabinets are a must for components and ammunition and for easily-damaged tools. Don't leave an accident lying around to happen.

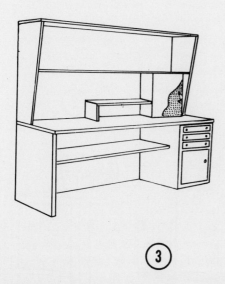

③

CHAPTER 3

Characteristics and Operation of Basic Tools

HANDLOADING DOESN'T really require a great deal of equipment. The basic setup consists of a loading press or tool; dies to suit the caliber concerned; a shell holder; and a means of weighing or measuring powder charge. Normally the press will contain provisions for seating the new primer, but, if not, a separate priming tool is required.

While designs of the various makes and models of all those items differ a good deal, most function in essentially the same manner and perform the same operations. Here we will describe the basic and variant designs of the *essential* tools and equipment. The many other items are covered farther along in this volume.

LOADING PRESS

The most common form is the "C-type" sold by several makers but originally introduced over forty years ago by Pacific. At the top of the "C" frame we have a hole threaded ⅞ x 14 TPI (threads per inch) into which loading dies and accessories are screwed. At the base of the "C" is a horizontal or slightly angled flange, drilled for bolts or screws to attach the press to a bench or work stand. A massive tubular housing extends downward from the "C" roughly aligned with the die hole at the top. This extension is carefully drilled and reamed

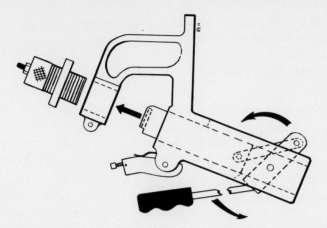

The classic "C-type" loading press introduced by Pacific in the 1920's is the simplest form of bench tool and an accepted standard today.

full length, the hole being accurately aligned with the die hole. The frame is generally cast of iron or steel, sometimes an aluminum alloy.

A ram fits closely in the hole and is free to move vertically. Its upper end is machined to accept a shell holder head to grip the cartridge case. Some older presses had the shell holding surfaces machined directly into the upper end of the ram, but this expensive practice has ceased. An egress hole for fired primers is drilled in the ram, some directing primers out to the front, others to the rear.

The lower end of the ram is slotted and drilled to receive a toggle arm, usually of flat steel stock. This arm is pinned to a heavy toggle which is in turn pinned to the frame inside appropriate recesses. The toggle is further drilled to accept an operating handle. The toggle and arm form a "toggle-joint" or "knee joint" which functions exactly as the human knee. It is actuated by the handle to raise and lower the ram with considerable force.

When the toggle and arm are in line (all three pivot points falling in the same line) they support the ram at the upper limit of its travel. As the toggle joint is "broken" by handle movement, the ram is drawn downward. Reversing toggle movement, straightening the joint, moves the ram back upward with great force. The toggle joint exerts progressively greater force as it approaches the closed (straight) position, making it ideal for forcing oversize cartridge cases into resizing dies.

Characteristics and Operation of Basic Tools 23

Removable shell holder heads such as these made by Bonanza slip into T-slots in ram. They interchange quickly to allow caliber changes.

Most modern presses "break" the toggle away from the operator on the handle upstroke and use the more powerful downstroke for closing the joint and forcing cases into dies. Some presses are designed so that the toggle may be assembled to close on *either* the up or down stroke to suit the operator. Early models closed on the upstroke which resulted in many an exasperated handloader lifting tool and bench off the floor on a heavy resizing job.

A detachable shell holder head fits into grooves in the head of the ram. It is machined so that it encloses both front and rear of the case rim in such a manner it can both press the case into a die and draw it out with great force without deforming the extraction rim. The holder is drilled at the center for passage of fired primers and a primer seating punch. Holders may be held in the ram by setscrews or spring clips, and there have also been screw-in and lock-nut variations.

At the lower end of the "C" the press is fitted with a swinging arm carrying in its upper end a primer seating punch. The punch is usually adjustable in height, and several interchangeable diameters and shapes are available to fit different primers. The top of the punch is surrounded by a spring-loaded sleeve to align and keep the primer in place.

A slot to accept the arm is machined in the ram. When the ram is at or near the top of its stroke the arm is manually pressed forward against its supporting spring into the slot. Dimensions and placement of slot, arm, and punch combine to align a primer in the sleeve with the primer pocket of a case held properly in the shell holder. Then, with the arm held forward, the ram is moved downward, pressing the case over the pimer to seat the latter. Some designs provide a mechanical stop to control depth of primer entry into case; others depend on individual operator "feel."

Raising the ram slightly then allows the priming arm to be snapped clear of the ram by its spring, freeing the ram for further movement.

The length of the ram stroke must be as great as the longest cartridge to be loaded; this requires the opening in the "C" to about ½" more. Nearly all modern designs are made sufficiently large to accept the long H & H Magnum cartridges.

Modern Handloading

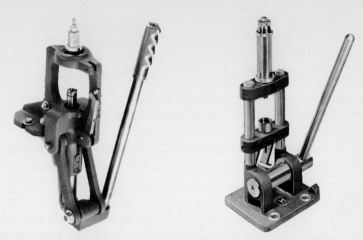

Left: typical O-type press is this RCBS A-2 model, now replaced by a smaller and lighter version, by RCBS. *Right:* H-type press is simple and effective. It resembles in miniature many big industrial presses.

The "O" type press is a variation of the "C". The open side is simply closed to form a more rigid and distortion-free link between ram and dies. One side of the "O" is usually offset or the entire area rotated somewhat to provide more finger room for handling cases. In other forms, the "O" is rotated a full 90° for the same purpose.

A classic example of the "O" press is the RCBS Rock Chucker and its predecessors, the A-series, which introduced a form of compound ram linkage in place of the simple toggle already described. In it the lower pivot point of the toggle is supported on swinging arms and is not fixed. Consequently, greater power may be generated by the same force application; in other words, the design produces a greater mechanical advantage. This feature was patented by Fred Huntington of RCBS, and it is found only on a few other makes and models produced under license. Heavy cast "O" presses are the least subject to distortion and are best for heavy work such as swaging bullets.

The "O" press is found in another variation called the "H" type. It consists of a heavy base and die head joined by two (occasionally three) vertical rods. A plate-like ram or riser bar carries the shell holder and rides on the rods between base and head. A large shaft passes through the base, carrying a toggle arm on each end. They are attached to a second pair of arms pivoted on the ends of the ram. All this forms a pair of typical toggle joints operated by a handle attached directly to the shaft in the base. Unless the die head and ram are unusually thick in section, the "H" press uses a fixed vertical primer punch beneath the ram. Some makes and models provide a cam-operated shield to keep falling primer debris from clogging the

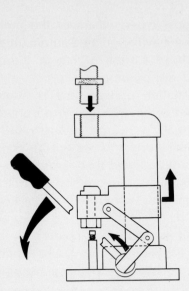

Left: post-type press is not common, produced mainly by Hollywood. *Right:* unusual Bonanza CO-AX press combines features of several types; is most advanced design on the market.

sleeve. Those that do not require the primer punch and sleeve be blown or wiped clean before inserting each primer. Not very convenient.

There is a third variation of the "C" press, the "Post" type typified by the old Hollywood Super. In it a heavy steel column carries a rigid die head and is attached to a large base. In between, a ram *encircles* the post and is free to move vertically upon it. The ram is driven by a toggle joint: a large diameter shaft passing through the base, carrying arms on both ends. They connect with arms pinned to the ram. The handle is attached directly to the large shaft.

A threaded shell holder is attached through a hole in the ram, aligned with the die hole. Directly beneath the holder is situated a fixed (usually) vertical primer seating punch. This places the open primer sleeve directly in the path of expelled primer debris which interferes with re-priming.

Numerous minor variations of the foregoing presses will be encountered and rather exaggerated claims are made for same. Nevertheless, they are all based on the C, O, or H types, differing only in detail and/or envelope. One useful variation is found in the Bonanza Co-Ax press with its cam-operated, multiple-caliber shell holder and slotted rather than threaded die seat. Another is the Bair Grizzly press with

Modern Handloading

its horizontally-sliding bar semi-automatic priming system mounted on the left side of the "O."

Most presses, except H and post types, whatever the model or make, are provided with a seat for a gravity-operated automatic primer feed. This device automatically drops a fresh primer in the cup of the punch as the arm moves to its rest position.

In addition, we find multiple-station presses; that is fitted with holes for more than one die. Most common is the turret press, such as the Lyman Spar-T. This is essentially a "C" press, but with the top arm of the "C" formed by a circular turret rotating on the upright of the "C." A number of die holes are arranged in a circle. Some form of detent is provided to insure alignment of successive die holes with the ram. The number of die stations ranges from four in the Redding-Hunter to twelve in the Hollywood. A few "H" presses are produced with two or three die stations and as many shell holders on the bar aligned with them. One other variation is the discontinued C-H "Slide-O-Matic" press with four die holes in a sliding bar riding in a T-slot at the top of the frame.

Though available for over seventy-five years in the Lyman/Ideal

Redding-Hunter #26 is typical turret press. Contains one or more sets of dies and powder measure in a rotatable head.

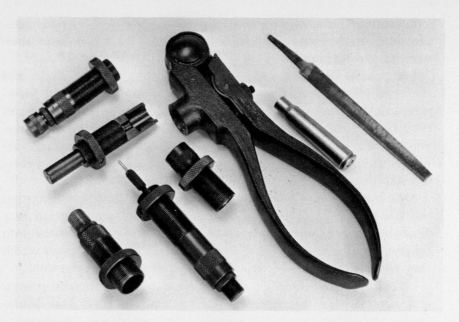

Old Lyman tong tool (No. 310) is simple and efficient, derives from 19th century tools supplied by Winchester and other gun makers

"Tong-Tool," hand-type tools have become widely popular only in the last decade or so. Three basic types are offered; those utilizing a mallet or bench vise (even a large C-clamp) for power, those in the form of hinged tongs or a lever hinged to a body or handle, and those where a screw supplies power. The first is typified by the Lee Loader in which resizing, decapping, repriming, and bullet seating are accomplished by driving a punch into a die or a die over the case with a mallet, The second type is represented by the Lyman 310 Tong-Tool consisting of a pair of hinged, pliers-type handles. Short dies and tools for decapping, repriming, resizing, and bullet seating are screwed into the lower handle, while the upper handle forces cases into and out of the dies. The third type is presently available only from Pacific and uses a coarse-thread, T-handle screw to force cases in and out of simple dies.

None of the hand tools permit full-length resizing. Full length dies are available to accompany them; the case being forced into the die in a vise or large C-clamp, then driven out with a punch and mallet.

Hand-type loaders are generally used with simple charge cups for measuring powder and one, the Lee, is supplied with a set of plastic, graduated cups for this purpose.

Modern Handloading

LOADING DIES

Reloading dies are the heart of the game. They perform the necessary work on the case, with the press furnishing the power and framework to make it possible.

The resizing die screws into the press to meet the shell holder and is shaped inside just like a barrel chamber of the same caliber. However, inside it is a bit smaller in diameter throughout, so that the soft brass fired case will be reduced in size when it is pressed into the die. This cavity is surface-hardened and smoothly polished to reduce friction. All resizing dies function in exactly this same manner, regardless of make or design.

A decapping pin is provided in all bottle-neck caliber resizing dies and some others. It is situated on the end of the expander rod or a separate decapping rod, positioned so that as the case is pressed into the die, it enters the flash hole and forces out the fired primer which then falls clear.

An expander button is attached to a rod passing down the center of the die cavity. The expanded case neck passes freely over it into the die and is then reduced. The neck is then pulled over the expander on its way out of the die, thus brought up to proper and uniform inside diameter. Straight-case caliber dies carry the expander rod and decapping pin in a separate die.

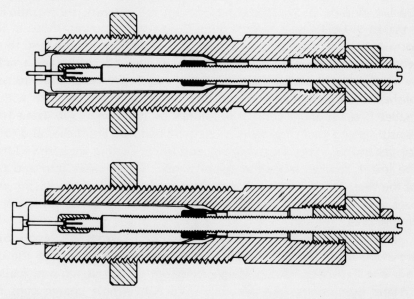

Typical resizing die in section. Most have the expander plug rigidly installed on lower end of decapping rod. In this die by Bonanza, plug is threaded and may be moved vertically on stem to suit individual preference.

Typical set contains 2 dies; this one has 3 for straight pistol calibers.

The bullet-seating die body is shaped just like the resizing die, but with the cavity large enough to freely admit a new resized case. Yet, the cavity fits the case closely enough to guide it with reasonable accuracy. At the top of the cavity an adjustable plug, profiled to match the bullet point, is fitted. The case, with bullet set on or in its mouth, is pressed into the die; the bullet meets the plug and is halted; the case is forced on over it, seating the bullet in the case to the proper depth.

Some seating dies, notably those in revolver calibers, are further fitted with an internal shoulder which turns the case mouth into the bullet or into a cannelure (groove) on the bullet to form a crimp.

POWDER MEASURE

Powder measures are essentially quite simple, with the accuracy and uniformity they produce dependent more upon precision of manufacture and operation than sophisticated design.

The typical measure consists of a metal housing bored through hori-

A 4-die set for maximum uniformity in loading pistol calibers.

This RCBS Uniflo powder measure typifies the conventional rotating drum type.

zontally to accept a close-fitting, rotating drum. This drum has a cavity machined in it, and a movable plug forms the bottom of the cavity. The plug is adjusted by a threaded stem and nut (or, in one or two instances, a sliding vernier) passing through the drum. The cavity may also be of fixed volume, simply a hole drilled in the drum.

The drum is rotated by a handle to position the cavity under a reservoir, from which powder flows by gravity. With the cavity filled, the drum is again rotated, sharp edges of cavity and reservoir mouth shearing off the column of powder, to empty the cavity through a drop tube into the case.

Other measures use a horizontal sliding member containing a cavity, or a separate hand-held chamber, but the principle is the same: cavity

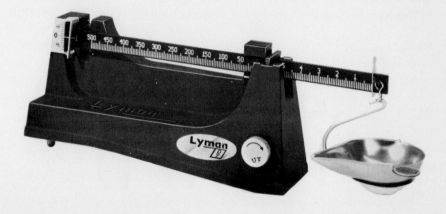

Typical beam balance powder and bullet scale.

filled to overflowing by gravity, surplus sheared off by movement, volume of cavity varied by moving a plug in it, or by insertion of different bushings.

POWDER AND BULLET SCALE

Scales used by handloaders to weigh powder and other items are nothing more than simple beam balances, most with weights sliding on the beam. Most have weight markings on the beam so that there is no need for check weights except when zeroing or when damage or error is suspected. Generally they are accurate to 1/10 grain, but this depends considerably on the skill and attention of the operator.

PRIMING TOOL

Separate priming tools are more common than in years gone by. Those several currently available consist of a housing containing an operating lever and a primer seating punch. An interchangeable shell holder of one sort or another fits over the punch and holds the case in proper alignment. Some hold the case horizontally in order to use a gravity-fed primer magazine, others hold the case vertically and primers must be individually placed on the punch. Either works well. These units are usually bench-mounted, and downward pressure on the handle cams the punch toward the case head to seat the primer.

There are also hand-type priming tools. They generally consist of a small body housing a primer punch and a shell holder. The punch is linked to a small lever pivoted on the housing. Simply squeezing the lever against the body moves the punch to seat a primer in the case. Such units come in a variety of sizes and designs, the simplest and smallest and most economical being the Lee, which can be completely enclosed in one's hand.

Actually, separate priming tools are normally a convenience rather than a necessity, inasmuch as most loading presses incorporate an entirely satisfactory priming device.

SHOTSHELL LOADER

Though shotshells were once loaded on equipment of the type just described, a more or less standard form of tool has developed since

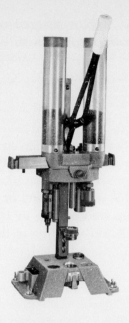

Typical modern shotshell loader.

WWII and is now used almost universally. It consists of a single multiple-stage tool, assembled and adjusted to perform all operations in loading a fired shotshell without any change or auxiliary equipment.

The most common form consists of a base and one or more vertical columns, with a die head sliding on the columns. Powder and shot reservoirs are mounted on or above the die head. It is fitted there with a sliding or rotating charge bar (volumetric measure) containing fixed-charge, interchangeable bushing cavities to transfer shot and powder charges to the cases through a long drop tube. The die head contains resizing die, wad seating punch, a die for starting the case crimp, another for finishing the crimp, and a rod for ejecting the completed round from the sizing die. Because shotshell cases are relatively weak, final crimping is done with the case in the resizing die.

A shell holder or seat is located on the base beneath *each* die station, and a priming punch is provided at one of those points, and another contains a spring and plunger or other means of determining the pressure exerted on wads. Completing this typical loader we have a handle and linkage to provide for movement of the die head, and some form of guide to facilitate seating wads.

Differently from metallics, a shotshell case is successively processed through each station of the loader, emerging as a completed shell, without change, adjustment, or use of any other item. The limited number of basic loads and gauges makes this type loader the most practical.

The basic form appears in many variants. Increased production rate is obtained by adding automatic feeding of primers, shot, and wads, and by utilizing an automatically indexed turntable to carry the cases between die stations. Where basic tools can produce 50 to 100 rounds per hour, the more massive and sophisticated models offer as much as 500 to 600 rounds per hour, with one exceeding 1200.

Shotshell loaders are generally assembled and adjusted at the factory for a single load. Conversion to another load in the same gauge requires exchange of fixed-charge powder and shot bushings and, perhaps, an adjustment of wad pressure. A change of gauge requires all that, plus a complete change of five or six dies and shell holders, and extensive minor adjustments. In this respect the shotshell loader is far less flexible than the metallic press. This is no disadvantage because relatively few "standard" shotshell loads (only three or four in each of six gauges) suffice to meet 99 per cent of all shotgunning needs. Thousands of loads are necessary to cover the metallic field as well.

One or two makers, notably Hollywood, still produce shotshell loading dies and accessories after the metallic type for use in conventional single-stage loading presses. This equipment offers the ultimate in versatility and is quite useful to the experimenter, though normally quite slow of operation.

There are, of course, numerous other accessories and tools useful to handloading. They are covered in subsequent chapters concurrently with their applications.

CHAPTER 4

Cartridge Cases and Their Manufacture

THE CARTRIDGE case as we know it today provides both the means and the motive for reloading. The means by virtue of ite reusability; the motive because it alone represents more than half the cost of the self-contained center-fire metallic cartridge.

While knowledge of the history and development of the case may be of little value in producing minute-of-angle loads, it is nevertheless a fascinating subject. Familiarity with the subject will help you to appreciate the toil, tears and money that have gone into what looks such a simple brass cup. It will also give you meat for many an interesting bragging session with your shooting compatriots.

Hazy, indeed, is the historical development of small arms ammunition. Even today you can stir up a right good fuss over just where, when, and by whom "gunpowder" was first used to propel a projectile. In fact, much of the early development of firearms is shrouded in rumor, fancy and fiction, with just a smattering of fact. It seems more than passing strange that a development of such great importance to civilization should be so dimly chronicled when minute details of lesser, even earlier events are so readily available.

The origin of the cartridge is just as hazy. There is reason to believe the good King Gustavus Adolphus of Sweden started the ball rolling. He is alleged to have ordered his crack troops to carry their powder and ball together in prepared rolls of paper, twisted together at the ends. The leaden ball reposed in one end of the paper cylinder and a measured charge of powder was poured in on top of it. This was in the early 1600's. Gus was evidently no slouch elsewhere in the weapons field. There is evidence that his artillery men were the fastest-firing and best trained in all of Europe.

Even before Gustavus there appears to have been cannon which used a separate chamber. It was loaded beforehand with powder and shot, then wedged into place in a recess, hopefully in alignment with the gun's bore. This device would seem to be the most direct antecedent of today's breech-loading cartridge, even though not adapted to use with more personal arms.

Getting back to Gustavus' paper cartridge, the musketeer bit or tore open the end containing the powder. He then poured the powder down the cavernous muzzle of his piece, then dropped in the undersize ball (paper and all) and wacked it a few times with his ramrod. A few thumps would compact the paper to give it purchase on the bore which helped to hold the ball in place. It would be embarrassing while leveling one's thunderstick at an enemy, to have the ball roll sedately from the muzzle just before the trigger was pulled. Regardless of the many attempts made to combine powder and ball for ease of handling, a practical cartridge had to wait on the development of percussion ignition. Rev. Alexander John Forsyth, of the Scotch clergy, first applied a percussion compound to firearms ignition, on which you'll find more in the primer chapter.

The first completely self-contained cartridge, thus the first case, to achieve any significant degree of success was the Dreyse, adopted by the Prussian Army in 1842. It was variously of paper, linen or silk, containing a substantial charge of powder and a conical lead bullet in the usual arrangement. A pellet of detonating compound (fulminate of mercury) was placed centrally against the base of the bullet. In this position it was struck by a long, slender "needle" firing pin. To do so, the pin had first to pierce the head of the case, then penetrate completely through the powder charge. Quite a task. When the mechanism was new and perfectly clean, it worked, after a fashion. After corrosion induced by powder fouling, aided and abetted by the gummy residue of several firings, it worked poorly in the field. In terms of ability to continue shooting under poor conditions, the caplock musket with paper cartridge was superior to the Dreyse Zundnadelgewehr (needle gun).

The same general principle was soon applied to cartridges for both pistols and shotguns, a pin within the cartridge replacing the delicate

Modern Handloading

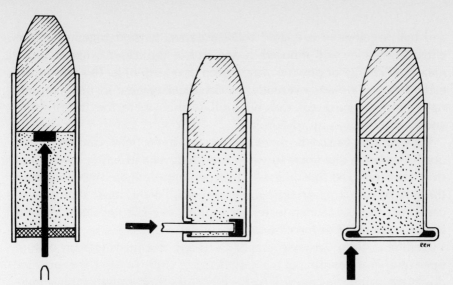

Left: is basic Dreyse cartridge with primer at base of bullet, detonated by a needle-like firing pin which pierced the powder charge. *Center* is pinfire where a transverse firing pin (a part of the cartridge) was struck by a hammer to detonate a cap inside the case. *Right,* the original rimfire cartridge, still with us today, almost unchanged.

An early attempt at a re-useable cartridge in this Roper (shotgun) intended to be loaded by the shooter with loose powder, wads, and shot, primed with a percussion cap on an outside nipple.

needle. We call this development the "pinfire" cartridge. A transverse pin entered the case to rest on a pellet of detonating compound cemented to the wall of the case. The pin was located as near the case head as possible and protruded a fraction of an inch. With the cartridge chambered, the pin stuck out through a notch in the barrel breech. There it could be struck by a conventional hammer driven down on the detonating compound, which exploded to ignite the main charge. The pinfire case, with its metal head and paper body, sealed the chamber well and tied all components together.

There is little doubt that the pinfire was the first practical self-contained cartridge, though most authorities choose to ignore it and jump from paper to the rimfires, perhaps because of its European origin. As late as the early 1950's, I saw new pinfire pistol ammunition for sale in Europe and even encountered one fellow who regularly reloaded that type of shotshells.

Cartridge Cases and Their Manufacture

The first practical cartridges developed in this country were the combustible type used with percussion caps. The Sharps is typical, using a linen tube glued to the base of the bullet to hold the powder. The flash from the cap penetrated the linen (which might be nitrated) to ignite the cartridge. It had its disadvantages, the major one being that there was no positive means to seal powder gases in the chamber. The case was, of course, consumed in firing.

Next came the Maynard & Burnside cartridges with drawn brass case which effectively sealed powder gases where they belonged. Still, they used a standard percussion cap on a separate nipple for ignition. Its flash was directed to the powder charge through a small hole in the base of the case. A step in the right direction, and a major one, since obturation was positive, but still not good enough, still not fully self-contained.

Then came the rimfire cartridge—fully self-contained, including primer. Though suitable only for low power cartridges, the rimfire case was a tremendous improvement. Following immediately came many methods of inside central priming to strengthen and adapt the case to more powerful calibers. Some of these were reasonably successful in .44/40, .50/70 and .45/70 calibers.

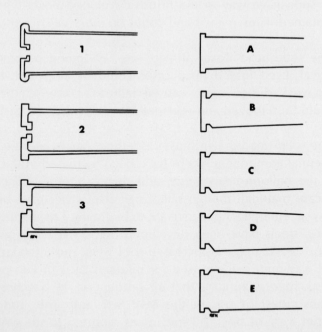

(1) Folded-head case   (2) Balloon-head case (once known as solid-head)
(3) Modern solid-head case.
(A) Rimmed   (B) Semi-rimmed   (C) Rimless   (D) Rebated rimless   (E) Belted rimless

Modern Handloading

The major inherent weakness of the rimfire case was carried over into inside-primed C.F. cases. The rim was formed by a simple fold of the thin brass or copper case. The sharp folds further weakened an already marginally strong basic design. This was referred to as folded head construction. Numerous attempts were made to improve upon it, but until the advent of the external C.F. primer, the cases had to be made of very thin, soft metal.

The external primer was inserted into a pocket in the head of the case from outside. Early cases of this type continued to use the folded head for a short time. However, the new system made no demand that case metal be thin so the "solid head" case quickly evolved. It was immeasurably superior to the F. H. type. So much so, in fact, the letters SH were used in the headstamp of Union Metallic Cartridge Co. cases during the transition period.

Today this type case is referred to as "balloon head" since metal surrounding the primer pocket bulges or balloons into the powder space. The term solid head is now reserved for those cases having web thickness considerably greater than primer pocket depth.

The early U.S.A. outside C.F. cases used Berdan-type primers. Therefore, the primer pocket contained a conical anvil rising from its floor. This form soon gave way to the British-developed Boxer-type primer which contained its own anvil and could be more easily removed and replaced.

With the solid-head Boxer-primed case, reloading became simple and practical. Even folded head cases could be reloaded, but their inherently weak design often caused failures. Likewise, the Berdan primer could be removed and replaced, but not nearly so easily as the Boxer.

Since the solid head drawn brass case, there has been little advance in this essential component. Webs have been increased in thickness to eliminate the balloon-head bulge, and that is about it. Brass is still principal case material, though billions of steel cases have been produced in war emergencies. Once thought totally unsuitable for cartridge cases, steels have been developed that do the job quite well. Their major disadvantage is increased tool wear and attendant higher cost. Steel was quite widely used for military cases in Europe during WWII. Also, reports indicate that all production of Czechoslovakian 7.62 x 45mm cases for use in the M52 SHE auto rifle and M-1952 machine gun were of steel. All specimens reported to the writer were coated with a dark green lacquer. Most Soviet rifle and machine gun ammunition is loaded in steel cases.

Aluminum and plastic cases (as well as combinations of the two) have been tested and found lacking. One Ordance officer tells how aluminum 7.62mm NATO cases functioned fine, but occasionally would generate a thermite reaction in machine guns. This means simply that the aluminum caught fire, burning with sufficient heat to nearly melt the gun in half!

Plastic cases are used successfully for some purposes. Several firms produce them in pistol calibers for short range shooting with plastic bullets. Little or no powder is used, the primer alone often providing the propelling force. Wax bullet practice loads (no powder) are sold by Colt's in plastic cases. Plastic case pistol blanks and shotshells have been produced by the Alcan Co., and some European firms produce varieties of plastic case blanks. Thus far, though, plastic cases have not been successful in full charge rifle or pistol cartridges.

Caseless cartridges have been developed for military use. but to date none have been good enough to achieve adoption. Almost invariably they consist of a propellant mass held together by a combustible binder or shell. The bullet is fitted to one end of this mass and a pellet of priming compound to the other. Several examples are shown in the accompanying illustrations.

Thus far no such cartridge has been developed which can successfully withstand the forces applied to them in automatic guns. Since their greatest value lies in military usage, this has kept them out of the running. Such cartridges have no valid sporting use and would drive a reloader out of his mind.

So it would seem that the drawn brass case is destined to be with us for a long, long time yet in spite of being nearly 120 years old. The entire arms industry is based on it, and even if something better were to be developed, a changeover would be a long, costly process. At the risk of being proven wrong, I'll prognosticate that anyone who is a reloader as this is written will load his last round in the same tried and trusted brass case we've known all these years.

Manufacture of the shiny, modern, solid head, externally primed brass case that makes reloading possible is fascinating. I've spent many days watching closely the various automatic presses and other machines as they clatter away and spit out streams of hot cases in dozens of calibers.

Any drawn, as opposed to extruded, case begins life as a roll of high-grade rolled sheet brass. "Cartridge brass" is a specific alloy developed over the years and is so named in the metals trade. Olin (owner of Winchester-Western) is probably the largest producer of this metal and supplies it to other cartridge makers.

Typical of cartridge brass formulas is this one intended for .30/06

cases; 68-71 per cent copper; 29-32 per cent zinc; .05 per cent (max.) iron; .07 per cent (max.) lead. Traces, no more, of other impurities are allowed so long as the brass meets all physical property requirements. Most case-drawing plants conduct their own analytical and physical tests on new lots of brass received, even though the material may be "certified" by the supplier. Since this is the most important single material in the cartridge, its quality must be assured, at any cost. The formula mentioned above, while typical, is best for only *some* calibers. A formula suitable for .30/06 is perfectly satisfactory for .270 cases, but would not necessarily be used for small rifle or short pistol cases. Different formulas are specified for different cases, dependent to a great extent on the depth and number of draws.

The brass sheet ranges from 3/32" to ¼" in thickness. Since drawing cannot increase thickness, the stock must be at least a bit thicker than the greatest thickness of the case web.

Brass is fed into automatic punch presses equipped with multiple dies. Circular discs are punched out, varying from dime-size up to 2⅛" diameter for the hefty cal. .50 M.G. cartridge. As they come from the sheet, discs (slugs) are forced on through a second and smaller portion of the die and drawn into shallow cups. This first operation is known as "blanking and cupping," and the piece produced is called a "cup" until it becomes a finished case. Cups are ofter purchased direct from the brass source by smaller case-making plants.

Punches and dies are arranged so that the cups are punched out in a staggered layout to hold brass scrap to a minimum. The perforated waste is automatically cut to manageable size and sold as scrap. In times of metal shortages it is usually routed right back to the sheet producer to go into his next melt of cartridge brass.

Brass becomes hard and brittle as it is worked. For a graphic example of this, compare the neck of a case that has been resized and reloaded many times with one that is new. The former will be quite hard and springy, while the latter is soft and easily deformed. This phenomenon is called "work hardening" and is actually used to contribute to the strength of the completed case. Work hardening makes it necessary to anneal (soften) the cup at several stages of manufacture. At one time it was necessary to anneal after every draw, but techniques have reduced that requirement somewhat.

Annealing is accomplished by feeding the caps through a gas fired furnace, then quenching them abruptly in a mixture of soap and water or drawing lubricant and water. Prior to every drawing operation the cups must be lubricated.

After cupping, the cups go to "first draw." Here they are forced through a slightly smaller die which reduces their diameter and deepens the cup, at the same time reducing the wall thickness. The thickness of the bottom of the cup remains unchanged. The drawing process may be repeated as many as four or five times, depending on the case to be produced. Each draw reduces diameter and increases length, at the same time reducing wall thickness and giving the interior case profile needed.

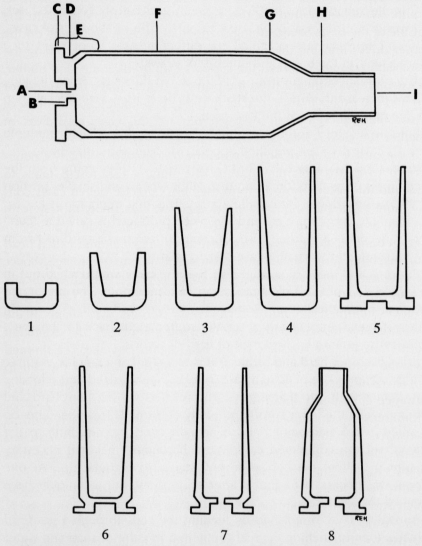

(A) Flash hole (B) Primer pocket (C) Rim (flange) (D) Extraction cannelure
((E) Head (base) (F) Body (G) Shoulder (H) Neck (I) Mouth
(1) Cup (2) First draw (3) Second draw (4) Third draw (5) Head & Trim
(6) Head turn (7) Taper (8) Neck.

Modern Handloading

The mysteries of the drawing process are disclosed far more clearly in the accompanying drawings than they could be by another page of print here.

When the cup is long enough, the bottom is "bumped" nearly flat, at the same time forming a rough primer pocket. The uneven mouth is trimmed back to clean it up. At this point the cup is a wee bit longer than a finished case and is ready for forming operations.

The cup is pressed into a die by a close fitting rod that enters its mouth. While supported by this rod, a "bunter" die slams into the outside of the base. Several things happen at once. A protrusion on the bunter drives into the thick brass of the cup base (now becoming the case head) trapped between the support rod, holding die, and bunter, all excess flows outward from the primer pocket. Space is allowed for this and the excess brass is placed in the cup deliberately. This flow of excess brass produces an oversize rim on the case, regardless of whether the case is intended to be rimmed or rimless.

If the case is to be of belted type, the holding die contains a counterbore to match belt dimensions. The bunter also forces brass to flow into this counterbore to form the belt. Cups for belted cases contain more brass than others to insure there will be plenty to fully form the belt, as well as the very thick web characteristic of the type.

From bunting, the cases go to a head turning (sometimes called head trim) machine, sort of an automatic lathe. An air-actuated punch shoves the case mouth first into a chuck or collet which closes automatically to grip the case. The chuck spins rapidly and a formed cutting tool advances automatically into the case head, turning it to shape. If the case is rimless, the excess brass rim is cut off completely, the extraction groove cut, and the rear face of the head lightly beveled. If a rimmed case, the cutting tool is shaped so that it merely trues up the rim to proper cross section, thickness and diameter, then perhaps cuts a clearance groove just ahead of the rim.

Rim thickness of rimmed cases is a critical dimension since it controls headspace. For this reason, the head turning operation is watched closely. The brass now begins to look more like a cartridge case than milady's lipstick container, which it has resembled to this point.

From head turn, the case goes to tapering. It is forced into a die which squeezes the thin brass walls down to establish correct body taper. Aside from its outside dimensions, this die is much like those you use to resize fired cases—and it functions in exactly the same manner.

The case goes next to a necking and shouldering operation. Proper dies shape the shoulder and neck to final dimensions. Some large

Cartridge Cases and Their Manufacture

diameter, small caliber cases with a sharp shoulder (the .264 Winchester is an excellent example) require more that one operation to accomplish this. Just as in drawing the cups, only so much movement of brass can be done at one pass.

The case is now complete except for minor, but necessary, details. It must receive its neck anneal. Necking and shouldering hardened the brass. It must be softened so that it can later grip a bullet securely without cracking or splitting from the strain. Annealing produces a bluish-brown discoloration of the surface. This is polished off commercial cases, but left on those for military use.

It must be pointed out that previous working and annealing have been regulated so that the final bunting operation produced the required hardness in the case head. After neck anneal the case graduates from very hard (for brass) at the head to quite soft at the neck. Crystalline structure at the head is small, dense and compact. At the neck it is large and irregular, with uniform transition in between.

The case is still just a trifle too long, so the surplus is removed from the mouth in another automatic trimming machine. The case is now complete, ready for delivery to the priming machine. You may have noticed that we've not mentioned punching the primer flash hole yet. Virtually all U.S. produced cases have this hole pierced on the priming machine.

Cases are fed mouth down into a track that holds them by their rims. Fingers grip the case head and a rod rises inside from below to support the web. A thin steel punch plunges through the web to form the flash hole. The case moves on down the track where a probe drops into the flash hole. If, for some reason, the hole isn't there, the probe strikes the web and stops the machine. If the probe drops through the hole, the case continues on.

A swinging arm picks up a primer from a feed device and carries it over into position directly over the primer pocket. As the case is held by fingers, a punch presses the primer part way into its pocket and the arm swings back for another primer. The case then passes on to where another punch seats the primer fully. If the case is part of a military order, it then meets a hollow punch which forces brass in over the curvature of the primer to secure it tightly in place, then a drop of lacquer is applied to completely water and oil proof the joint. These last two operations are not normally performed on commercial cases.

You fellows who wonder why it is difficult to obtain unprimed cases should now understand the problem. It is possible to leave the primers out, but the machines are geared primarily to primed cases and any change interferes with a smooth work flow.

So there it is—a complete, primed case in all its virginal brightness, ready for delivery to the loading machines or for packing and shipment to reloaders.

During all those operations cases are subjected to continuous gauging and visual inspection and any not meeting rigid specifications are sent to the scrap barrel. A percentage sample of each lot of cases is subjected to strenuous firing tests, including reloading. Until it passes these tests, that lot cannot be released for either loading or sale. The manufacturers protect their name and your life by all these extensive testing and inspection procedures.

If you're a confirmed wildcatter you may have had the idea that you'd like to have one of the factories make up a special case to your design. Can it be done? Well, yes, it can. But unless you're willing to wait months, or even years, and spend a hell of a lot of money, forget it. I once had such ideas, though the case I wanted was the old .45/120/550 Sharps in 3¼" length. No one in this country would consider the job unless I could buy a few hundred thousand cases outright, cash on the barrel head. So I went to Europe. There I presuaded the Kynoch branch of Imperial Chemical Industries to do the job. I visited the Birmingham, England plant and the gentlemen there agreed, for a reasonable sum, to produce 10,000 (the minimum order they'd accept), of said cases with my name on the headstamp.

Somewhat over a year later I received a pre-production sample of the cases. Minor corrections were made and before long I had my cases, all 10,000 of them. They cost me right at 16 cents each when cost, transport, duty, taxes, commissions, etc., were all totalled up. Most of them were sold to foot the bill so I could end up with a few hundred cases for my own use. Had I not wanted a particular case other people would be willing to pay for, I'd have had to shell out $1,600 to get what I wanted. No one I know can afford *that* for his hobby. If you run across one of those big, rimmed cases head-stamped "Nonte-Taylor," pick it up. Today it's a collector's item.

CHAPTER 5

Primers: History and Development

SMALL ARMS centerfire cartridge primers exist today in extensive variety and in three basic types. Each type was developed to meet a specific need and set of circumstances. Present ammunition requirements could be met with a single mechanical design if certain changes in case design were to be made. The three types are firmly entrenched, though, and no sudden change is likely. Certainly the cheapness of the unreloadable rimfire type will keep it in favor for many a decade.

The first two basic types are "Berdan" and "Boxer" utilized in metallic cartridges, the latter primarily in the United States and Canada, the former for the same purpose in most other countries of the world. Third is the "Battery Cup" type of primer used in shotshell cases, be they plastic or paper. It is also used to some extent in metallic shotshell cases. As a matter of interest, Europe (once completely standardized on Berdan) now produces quite a lot of Boxer-primed ammunition.

Ignition of powder in a barrel or chamber has been accomplished down through history by weird and wondrous means. First came the application of a glowing coal, hot wire or slow-match to a touch hole leading to the main charge, the cannon lock and matchlock, which was eventually refined to the point where a trigger moved a slow-match into contact with priming powder.

The intricately-wrought wheellock followed. It spun a serrated

Top, left: Boxer primer pocket, flat bottom, single central flash hole. *Right:* Berdan pocket, conical bottom, multiple off-center flash holes. *Bottom, left:* Boxer case head viewed from inside case body. *Right:* Berdan, ditto.

wheel against which a piece of pyrites or flint-like stone was pressed. Sparks were generated to ignite priming powder. It is doubtful, in this scribe's opinion, that any firearms form since (or before, for that matter) has even produced such skilled artisans. Those who produced ornate, delicately chiseled and sculptured wheellocks during the sixteenth and seventeenth centuries have gone unmatched in skill ever since.

Eventually the very expensive wheellock gave way to new forms where a flint was struck against an angled, roughened steel plate to throw a shower of sparks into a pan of priming powder.

This type evolved into the true flintlock familiar to us in the American Kentucky and Pennsylvania rifles. In its most highly developed form, the flintlock ruled the world for nearly two centuries, but still left much to be desired. Untold numbers of thinkers strove to improve upon it.

Primers: History and Development 47

Chemical compounds such as the fulminates of gold and mercury had been known since the mid-seventeenth century. However, no practical use for them had been found. Their characteristic of exploding at the slightest blow made them useless as "working" explosives.

The Reverend Alexander John Forsyth, Scotch clergyman, had interests other than the mere saving of Presbyterian souls. He saw in the dangerous fulminates a way to rid gunners of dependence upon flint and steel. Forsyth's development was a fulminate that could be handled with a fair degree of safety, yet would explode violently when struck sharply. In 1805, he announced his "Detonating Powder" as admirably suited to ignition of firearms.

Detonating powder was used in a number of ways. Loose, it was placed in a magazine which discharged a measured amount as required into a shallow cavity of the gun lock. When struck by the hammer, it exploded and flame flashed through a hole to ignite the black powder propelling charge. Some locks of this type were called "scent bottle" locks because the pivoted powder magazine resembled perfume flagons of the period. Forsyth built this type of lock, but it was none too successful, what with the possibility (and none too remote, at that) of the detonating powder container exploding grenade-like in a shooter's face. Even the most free-thinking gunner values his physiognomy rather highly.

Forsyth also encased detonating powder in soft metal tubes (for waterproofing and ease of handling) which could be struck by the hammer. At least one form used a long tube of powder from which a knife-edge on the descending hammer sheared off short lengths. Just how one could expect the hammer to shear off a short length without exploding the entire tube is not clear. Only the severed section was intended to detonate and fire the main charge. This lock type was short-lived, and not widely used.

The Pill Lock utilized small pellets of detonating powder. The powder was mixed with a binder (some form of gum) and rolled into small roundish "pills." These fiery (and touchy-dispositioned) pellets were placed in a cavity from which a flash hole led to the propelling charge. A protruding nose on the hammer entered the cavity to crush and detonate the pellet, which in turn flashed through to ignite the black powder in the barrel. Pill Locks could be water-proofed by closing the cavity (over the properly seated pill) with wax. This most practical use of Forsyth's powder was the development of an American named Guthrie.

A logical development of the detonating pill was to encase the powder in sealed metal, paper, or foil "caps" which could be handled

Modern Handloading

more easily. Countless inventors worked with the idea. Pauly, of Paris, made paper caps, Dreyse made metal ones in 1824, and later used them in the celebrated "Needle Gun" (Zundnadel Gewehr) cartridges.

Of the many methods for encasing detonating powder, none were particularly efficient or long-lived until an Englishman-turned-American artist and sportsman entered the picture. Joshua Shaw (1776-1860) arrived in Bordenton, N.J. in 1814, and soon moved to Philadelphia, continuing to reside there the balance of his life. Shaw is reported to have gotten the basic percussion cap idea in England, but kept it to himself until arriving in New Jersey.

Like most worthwhile ideas, Shaw's was simple. Make a small, thin metal cup and place detonating powder (mixed with a gum to make it cohesive) in the bottom of the cup. A nipple (cone) was supplied over which the inverted cup (cap) was pressed. The fit was tight enough to insure the cap would not fall off in normal handling of the arm. The nipple was drilled through its length, the passage communicating with the propelling charge. When the gun was fired, the heavy hammer crushed the soft cap against the nipple, igniting the detonating powder. Flame flashed through the nipple to ignite the main charge. As an added, and certainly valuable, refinement, Shaw covered the pellet of powder with a disc of thin metal foil, shellacked or varnished in place. This not only held the pellet securely, but water-proofed the cap.

Shaw's first caps were made of thin steel in 1814, in essentially the same form the percussion cap still carries today. They were not satisfactory, so he switched to pewter, and finally again in 1816, to the ideal material, the soft copper used ever since. The compound used changed as he went along, too. Potassium chlorate used in early caps was too touchy. The standardized compound evolved as a mixture of fulminate of mercury, chlorate of potash, and ground glass, which served admirably for nearly the entire percussion period.

The malleable copper cup which would cling tightly to the nipple and not shatter when fired, made Shaw's percussion cup the safest and surest form of ignition firearms had yet known. They were extremely simple to use, not expensive to manufacture, were relatively free of danger of accidental explosion, and were sure-fire, even in rain.

Hundreds of other inventors got into the act, and dozens in Europe and England claimed to have invented the cap before Shaw. It was a time of much experimentation and some of the claimants may have had an edge on Shaw, but it seems unlikely. In any event, only a few years were required for the percussion cap to become the standard form of ignition over virtually all of the civilized world.

Other types stemmed from Shaw's cap. Not because of particular

Primers: History and Development

deficiencies in design or function, but in attempts to eliminate capping the nipple as a separate loading operation. Best known today are the Lawrence Disc-Primer and the Maynard Tape Primer. Both allowed a quantity of primers to be stored in a compartment in the lock. They were then fed over the nipple by action of the hammer.

The Lawrence Disc was catapulted forward out of its magazine as the hammer fell. The object was for the flying copper wafer to be caught in mid-air by the falling hammer and smashed against the top of the nipple. It worked, after a fashion, but was less certain than standard caps.

Maynard's system was more certain. Individual pellets of compound were cemented between two long, narrow strips strips of foil or waterproofed paper. The pellets were uniformly spaced so that upon cocking the hammer a pawl would advance the strip to position a pellet over the nipple. Dropping the hammer would then detonate the pellet and fire the charge.

Maynard's tape primers were rolled up compactly and advanced singly from a small lock plate magazine by means of a feed pawl. The strip of primers resembled nothing so much as the "roll caps" sold for use today in toy cap pistols. Some say the Maynard tape primer worked well, others say, "No." Perhaps the best evidence is that the U.S. Army soon dropped it, and that it never received wide commercial acceptance. Shaw's cap reigned supreme until the advent of the practical self-contained metallic cartridge.

The rimfire cartridge sounded the death-knell of the percussion cap in 1858, though one of the bloodiest wars of history (the American Civil War) would yet be fought with caps. But that war gave tremendous impetus to the development of improved small arms ammunition. Numerous "inside" primers came along as the drawn copper or brass cartridge case was developed. Best known are the Martin, Hotchkiss, Laidley, and Benet. All consisted essentially of an internal support which held a pellet of priming compound centered against the *inside* of the case head. The head was thin and soft enough to be dented by a firing pin, crushing the compound against whatever anvil or support was being used.

None of the inside priming systems were eminently successful because of the limitations they placed on case strength. A case head soft and thin enough to be so indented could hardly be very strong. In spite of these shortcomings, it remained in use by the U.S. Goverment for over twenty years — long after better types were common in sporting ammunition.

The "outside" centerfire primer is what we use today, and it grew

Modern Handloading

naturally from the unsuccessful inside type. Between the Civil War and the late 1870's, a tremendous number of priming systems were devised. Of them all, the U.S.-developed "Berdan" type and the British "Boxer" type survived to come into wide use in the 1880's.

The Berdan Primer is named after Col. Hiram Berdan of the U.S. Army and most references give him full credit for developing it. Yet, the late Phil Sharpe, in his extensive *Complete Guide to Handloading* states that the primer had already been designed and produced at Frankford Arsenal where the Colonel first encountered it. Perhaps the good Colonel merely "adopted" the design. Berdan primers were widely used in this country, but after a few short years, the Boxer type achieved domination of the field.

Berdan primers consist of an open cup into which a pellet of compound is pressed. The anvil against which the pellet is crushed to detonate it is contained in the cartridge case. A conical projection rises from the floor of the primer pocket. From one to three small-diameter flash holes are drilled through the bottom of the pocket, usually spaced around the projecting anvil.

In use, the Berdan primer is seated in the pocket so that the pellet is pressed with some force (not much) against the top of the anvil. The tight fit of the cup in the pocket holds the primer securely in the position. When a firing pin strikes, it dents the cup, crushing the pellet against the anvil. Detonation of the pellet follows, and flame passes through the flash holes to ignite the propelling charge.

The Boxer primer is constructed quite differently. It utilizes the same type cup and pellet of priming compound, but there the similarity ends. A conical anvil stamped from sheet metal is pressed tightly into the mouth of the cup, its peak bearing snugly against the priming pellet. Thus, anvil, cup, and pellet are formed into a single complete assembly.

The Boxer case contains a simple, flat-bottomed primer pocket, pierced centrally by a fairly large-diameter (usually about .080", or a bit less) flash hole. When the primer is seated, the "feet" of the anvil must bear on the bottom of the pocket. If they do not, part of the force of the firing pin blow will be absorbed in driving the entire primer forward before the pellet is crushed and ignited.

Generally speaking, the Boxer primer is used in the United States and Canada, while the Berdan-type is used by the rest of the world. However, as U.S. equipment and personnel are furnished to various nations for production of military ammunition, Boxer primers became more widely used. For example, Israeli-produced military ammunition is Boxer primed. RWS is also now producing military and sporting cartridges of the same type. Famed DWM is also using Boxer primers in a portion of its output.

In America, we find it hard to understand why anyone would *ever* use Berdan-type cases and primers. Remember, though, that widespread reloading of metallic cartridges is virtually unknown in most other countries of the world. Except to the handloader, the differences between the two are insignificant.

Due to the location and size of the flash holes, it is virtually impossible to remove fired Berdan primers by means of a punch through the case mouth. The most common method is to pierce the cup with a slender, pointed instrument from the outside, then pry it out. During black powder days, a tool called a "cap awl" was used for this purpose. European tools for the purpose look like a three-handled pliers and combine the piercing and prying into a single motion. Regardless of the type tool used, the point must be flat (horizontal) and chisel-shaped A conical point such as on an ice pick will tear through the thin cup, leaving it hopelessly stuck in the pocket. Remembering this simple requirement can save you much profanity.

Boxer primers are a cinch to remove. The large central flash hole allows a punch to be inserted from the mouth of the case to push the expended primer out of its pocket. It is just that simple. It has long puzzled me. Did we develop as a handloading nation because of the simply-used Boxer primer? Or did our grandfathers adopt the Boxer primer because they were handloaders?

Either type of primer is equally simple to insert in the case. They are simply pressed into the proper depth, and the same types of tools will handle either type so long as diameters do not differ greatly. Odd though it may seem, by the late 1880's and early 1890's the U.S. Berdan primer was solidly entrenched in England and the rest of the world. And the British Boxer was standard in virtually all U.S. production ammunition.

The U.S. handloader is inclined to sneer at the Berdan primer and loudly proclaim the Boxer far superior. Well, for reloading with our tools, the latter is certainly the most convenient. As for being superior, an objective analysis indicates otherwise. From a purely technical viewpoint, there is little doubt in this scribe's mind that the Berdan primer delivers equal results.

From a mechanical viewpoint, the interposing of a light-weight anvil between pellet and case is bound to inject ignition variations. The anvil may or may not be seated solidly on the bottom to the primer pocket. Assembly of anvil to cup can easily vary from the ideal. Pellets (particularly the dry-mix type) pre-stressed by the anvil are more

Modern Handloading

likely to crack or flake in storage, handling, and loading. All of the foregoing are potential troubles that do not exist in the Berdan-type case-primer combination. In addition, the Berdan primer is simpler and cheaper than the Boxer to produce. At the same time, production of Berdan cases is extremely little (if any) more costly than Boxer types.

All of this convinces me that purely from the viewpoint of doing the best job of igniting small arms ammunition at lowest cost, the Berdan primer might even be superior to the Boxer. Do not rise in wild indignation at this statement. I'd howl louder than the rest of you if the Boxer primer were replaced. Cold chills run up my spine just to think of the headaches that would have developed if even a part of the 2,000,000 plus primers I've used had been Berdan. When it comes to reloading, Col. Boxer's baby has no peer.

All early priming compounds contained materials that had a very bad effect on both barrels and cartridge cases. Potassium chlorate was the heart of the compound, responsible for long storage life and unusually fine stability under almost any climatic conditions. Yet, it left a residue in the bore that promoted very rapid rusting. The fulminate of mercury in the priming compound made the cartridge case so brittle that, under some circumstance, it could not be reloaded.

Some explanation is in order. A product of combustion of potassium chlorate is potassium chloride — similar to common table salt. The potassium chloride was throughly distributed throughout the bore of the weapon, even being driven into the pores of the metal. There it exhibited its hygroscopic properties. In short, it absorbed moisture from the atmosphere, promoting rapid development of rust. The higher the humidity, the faster the rust developed. During a damp, warm spell, a rifle left uncleaned overnight will develop a coat of rust on its bore, and two or three days would ruin it completely. I know — it's happened to me.

Mere oiling of the bore does little good. The primer residue is *not* soluble in oil, so it mostly just works away under the oil film, forming rust like mad. Potassium chloride is, however, soluble in plain water. Hot, soapy water cleans it out in a hurry. Bores properly cleaned with hot, soapy water soom after being fired with so-called "chlorate" primers, will suffer no ill effects from the residue, even in very damp climates.

Military small arms instruction manuals of today still specify "cleaning on each of three sucessive days following firing." Even during WWII, the rifleman or machine gunner who forgot those simple instructions could easily wind up signing a statement of charges for his ruined barrel. U.S. manuals further recommended hot, soapy water; plain hot water; cold water and soda ash; and plain cold water, in that order of effectiveness and desirability. Soldiers were instructed to pump hot water through the bore until the barrel was "hot to the touch," then wipe dry and oil while still hot. Such extensive treatment will generally remove all the chloride residue in one cleaning, but troops were required to do it three days running just in case. The wrath of a platoon sergeant upon catching a recruit skipping one of the three cleanings is something that is awesome to behold. As a former company commander, I can recall taking a rather dim view of same, too.

For many years, riflemen blamed the powder used for the havoc wrought by chorate primers. Strange and wondrous claims were made for a multitude of "powder" and "nitro" solvents sold for the purpose of preventing ruined bores. Eventually, someone got bright enough to run controlled tests and that innocent-looking primer was properly identified as the culprit. From that day on, chlorate primers have been known as "corrosive primers."

Through WWII, nearly all military small arms ammunition was loaded with chlorate primers. Not until the early 1950's did the U.S. military establishment switch over exclusively to a "noncorrosive" type. Details of the changeover are shown below. Other nations made the change a bit earlier or later, the exact dates being difficult to determine.

This simply means that the bulk of the hundreds of millions of rounds of surplus military ammunition sold in this country is "corrosive"-primed. Over the past ten years, I've examined dozens of rifles ruined by lack of proper cleaning. That old devil, potassium chloride, will do it every time. If you use surplus military ammunition and don't know about the primers — clean, brother, clean — and clean again.

The effect of early primers on brass cartridge cases has been mentioned. Early in the game, ammunition makers determined that the combustion of fulminate of mercury released minute particles of free mercury. This was forced into contact with the entire inner surface of the case by the explosion. Mercury attacks brass, amalgamating with it to destroy its characteristic strength and malleability. In short, the mercury residue makes the brass so brittle it can often be crushed between thumb and finger, breaking into ragged chunks.

Large diameter straight cases using large charges of black powder were less affected than small bottle-neck types. For this reason, black powder era reloaders didn't object much. Their cases held up pretty well in spite of the mercury.

Much of the free mercury was absorbed by the solid products of black powder combustion, leaving less to attack the brass. Then, too,

Modern Handloading

the pressures involved in those early cartridges were quite low, never about 25,000 psi. Higher pressures drive more finely-divided mercury into the brass. Consequently, modern bottle-neck cases operating near 50,000 psi speeded up and increased the mercury's effect on the brass.

Also, it was found that smokeless powders were much harder to ignite than black. Where almost any old warm spark would set off black powder, the new types required a hotter, longer and more violent primer flame. This meant more compound in the primer, consequently, more free mercury driven into the brass case.

The action of mercury is irreversible once the cartridge is fired. It completes its action in a very short time, and no amount of washing or other treatment can prevent its completion. I have fired mercuric-primed .303 military ammunition in the afternoon, and by the next morning the cases would break in the resizing die, usually when the neck was pulled over the expanded plug.

Short, straight cases of pistol type are much less affected by mercuric primers and can often be reloaded many times before failing. Primarily, this is due to low chamber pressures and the fact that pistol primers contain less mercury than the larger rifle varieties.

The advent of the high-pressure bottle-neck cartridge caused the primer boys to replace the fulminate of mercury with a less vicious compound. Commonly used was a mix of potassium chlorate, antimony sulphide, lead sulfocyanate, and black powder. The latter later being replaced with granular TNT or similar high explosive. Even so, mercuric primers continued in wide use for sporting ammunition well into the 1930's. U.S. military ammunition has not been loaded with mercuric primers since before the turn of the century. Some foreign calibers were loaded with mercuric primers as late as the early 1950's.

The chlorate primer's deleterious effect on barrels has already been covered in detail. It is obvious, of course, that shooters the world over objected strenuously. Much effort was devoted to development of a primer that produced no harmful residue. Competition among companies was fierce because, naturally, the first one to offer a "non-corrosive" primer could expect to gain tremendously in sales.

The earliest non-corrosive primer was produced in Germany around 1910. It was designed only for .22 rimfire cartridges and contained 55 per cent fulminate of mercury. Ammunition thus loaded was quite popular since, for the first time, a fellow need not be overly concerned that his .22 barrel would be eaten away in a short time. This new primer replaced potassium chlorate with barium peroxide. While its residue was not corrosion-inducing, it was highly abrasive. It eroded barrels significantly, and the mercury couldn't be tolerated in center fires, so the ideal primer was still far in the future.

Switzerland began using a non-corrosive primer about 1911 and much development work took place in Europe during the first quarter of the century. RWS Sinoxid priming was the first commercially successful development and by the late 1920's, Remington had introduced its "Kleanbore" rimfire priming based on European developments.

Even so, early U.S. non-corrosive primers weren't all they should have been and it wasn't until the 1930's that truly successful non-corrosive/non-mercuric types became standard in sporting ammunition. Another twenty years was to pass before non-corrosive primers reached a sufficient state of development to be adopted by the major powers for military use. Military requirements for storage life and stability were (are) much more exacting than for sporting ammunition. Actually, nearly all sporting primers today will meet those military specifications.

Today U.S. military and commercial primers are all non-corrosive and non-mercuric and are essentially alike — the only differences being that some commercial numbers are "hotter" than others, and that so-called "magnum" types are available from some makers. All the compounds are built around lead styphenate which during post-WWII years was determined to be superior to most other practical formulas.

Also during the post-war years, the differences between Remington and Winchester primers and cases were resolved. Originally, Winchester pockets were radiused at the bottom and primers were designed to seat there. Remington, on the other hand, used a square-bottomed pocket and primer to match. For this reason one sees much in older writings about matching case and primer makes. The rounded bottom of the Winchester pocket would "pinch-in" the bottom of a Remington primer, while a Winchester primer would be poorly supported in a Remington case.

Today all primers and pockets are made after the Remington design, and this is no longer a problem. Until relatively few years ago, primers were made in two shapes. The *flat* type is standard now, meaning the top of the primer cup is flat, at right angles to the sides, joined to the sides by a small radius. The *round* type had an arched top and was made at one time or another by most commercial producers.

Both types worked equally well, but required different priming punches to fit their individual profiles. For this reason, all older loading tool brochures specify that one should order the correct punch to match the primers being used. When a flat punch was used to seat round primers, the arch or dome was flattened at its crest, often damaging the priming mix pellet beneath. Use of a rounded punch on

Modern Handloading

flat primers left the center of the cup unsupported and seating pressure often caused the anvil to raise a dimple in the center of the cup. This also caused pellet damage.

There are actually only two sizes of Boxer primers standardized through out the World — *large* (.210" nominal dia.) and *small* (.175" nominal diameter). Each size is divided into both rifle and pistol types, thus "Large Pistol," "Large Rifle," etc. In addition, there are "Magnum" and special purpose types within each size. There are some larger sizes, but they exist only for military use in heavy automatic weapons ammunition.

Pistol and rifle primers differ only in that the former generally contain less priming compound (being required to ignite smaller powder charges) and are assembled in thinner, softer cups. Pistol firing pins and hammers can't produce a great deal of energy, so the cup must be relatively weak to be easily dented. Fortunately, pistol cartridges develop relatively low pressures which can be confined by a weak cup. On the other hand, the heavy, powerful firing mechanisms of rifles can easily indent the thick, stiff rifle primer cups.

Shotshell primers have also been standardized dimensionally in the past decade. Before that there were two major sizes and a couple more not so prominent. Now one size, that previously referred to as "209-size," is used by all, in both paper and plastic cases.

The battery cup primer is enclosed in a steel reinforcing cup because of the inherent weakness of paper shotshell construction. The cup confines the primer proper during firing so that it does not expand and destroy its seat in the case.   in the primer, a separate anvil, is sandwiched between the bottom of the battery cup and the primer cup or "cap," as it is often called. When the cap is indented by a firing pin, the pellet is crushed against the anvil and ignites, flashing flame through a hole(s) in the bottom of the battery cup to ignite the propellent charge.

Primer manufacture is a meticulous business at best and a great deal of attention must be paid to reducing explosion hazards. The priming mix is very sensitive and must be kept wet to reduce handling risks. Only very small quantities are mixed and delivered to the charging line as needed. The explosion of only a few ounces can be devastating.

Cups and anvils are formed on automatic presses from metal strip, then pickled, washed, annealed, plated, etc., as required. Cups are placed mouth-up in a series of holes in a plate to receive the mix.

Mix is spread over a perforated plate, then squeegeed into the precise holes, and the plate wiped clean. Each hole will accept only the amount of mix required for one primer of a particular type. The priming mix plate (charge plate) is then mated with the cup plate and placed in a press. Punches force the mix from the charge plate into the cups, compressing the individual pellets slightly in the process.

This is followed by "foiling," where discs of paper or foil are punched from a strip directly into the cups against the pellet. A drop of waterproofing compound is then applied, the anvil is seated, and the primers are sent to drying rooms where excess moisture is evaporated from the still-wet pellet.

During primer manufacture floors in work areas are kept wet and are frequently scrubbed or flushed to remove all traces of priming mix. Any accumulation of mix that becomes dry is an explosion waiting to happen. Naturally, non-sparking tools, shoes, and electrical devices are mandatory, as are good ventilation and high humidity. Once assembled and dried, primers are quite stable and will withstand rough handling, but during manufacture, the mix *is* touchy.

Some primers are assembled by dry-mix methods as opposed to the wet method already described. All the same operations must be performed, but each is more hazardous because the mix is much more sensitive and it is exposed to greater pressure in seating and foiling. As much as possible of the dry-mix process is performed by remote control with operators working behind barricades for safety.

When finished, primers are packed in the small grooved plastic (formerly wood) trays we know so well, ten per groove, packed side by side. This type of package is prescribed by Federal regulation and minimizes explosion danger in case primers are exposed to fire or impact. For these reasons, primers should always be kept in their original packaging until used.

Shotshell primer manufacture remains the same through the foiling step. At that point, anvils are placed in the battery cups, and the caps are seated in the cup to the required depth and pressure on the anvil.

Primer quality is judged more by sensitivity than any other feature. Obviously, they must function correctly in a wide variety of guns and be sensitive enough to be ignited by an old soft-springed junker or a crisp new model. Without going into great detail, we can describe the process of drop testing. A test fixture contains the primer (in a case), a firing pin, and a free-falling weight. The weight is dropped on the firing pin to ignite the primer.

The height from which the ball must be dropped to fire *all* the primers of a sample is determined; then the height to fire 50 per cent; then the height at which *none* will fire. From this a measure of sensitivity is determined to learn if the primers will be ignited in most guns but yet are not so sensitive as to be dangerous. Unusual variations or

Modern Handloading

erratic behavior will also be discovered. The *all fire/none fire* spread must not be too great, and the *none fire* height must be enough to indicate the primers won't be subject to inadvertent ignition by floating firing pins, priming tools, etc.

Every production lot of primers must be so tested before it can be released for plant use or sale to handloaders. Modern primers withstand storage well. Unless exposed to extremes of humidity and temperature, they remain perfectly serviceable for many years.

Yes, the primer is the spark plug of any cartridge. As such it deserves special attention, especially those points mentioned elsewhere in this volume under Priming.

NON-CORROSIVE PRIMERS, U.S. MILITARY AMMUNITION
.30-06 and .45 caliber

Headstamp	Maker	Changeover date	Safe Headstamps
FA	Frankford Arsenal	Oct. 1951 (.30)	FA 52, 53, 54, etc.
		July 1954 (.45)	FA 55
FCC	Federal Cart. Corp.	Nov. 1953 (.45)	FCC 54
DAQ	Dominion Arsenal (Canada)	All non-corrosive	
VC	Verdun Arsenal (Canada)	All non-corrosive	
WRA	Winchester	Aug. 1951 (.30 ball)	WRA 52
		June 1954 (.30 AP)	WRA 55
		Nov. 1951 (.45)	WRA 52
WCC	Western	June 1951 (.30)	WCC 52
		Nov. 1952 (.45)	WCC 53
TW	Twin Cities Arsenal	Dec. 1950 (.30 Ball)	TW 51
		Feb. 1952 (.30 AP)	TW 53
SL	St. Louis Ord. Plant	May 1952 (.30 Ball)	SL 53
		July 1952 (.30 AP)	SL 53
RA	Remington	Nov. 1951 (.30 Ball)	RA 52
		Sept. 1952 (.45)	RA 53
LC	Lake City Arsenal	June 1951 (.30 Ball)	LC 52
		April 1952 (.30 AP)	LC 53

All .30 Carbine, 9mm Luger, .38 Special with U.S. headstamps was loaded with standard commercial non-corrosive primers. All 7.62mm (.308) and 5.56mm (.223) is non-corrosive.

Primers: History and Development

59

CHAPTER 6

Smokeless Powder Manufacture

THE HANDLOADER has a wider variety of powders available today than at any time in history. Not including the lesser-known imported powders that are not widely available, there are over fifty available. Three large domestic manufacturers (DuPont, Hercules and Winchester-Western) two reliable importers (Alcan and Norma) and one supplier of surplus powders (Hodgdon) concentrate considerable effort to insure the reloader of a continuous supply of powders for his various needs. Naturally, all of the firms just mentioned compete with each other for the reloading market and this eventually results in better products.

We classify all small arms propellant powders in two groups: "straight" nitrocellulose, *single base,* and those containing also a substantial percentage of nitroglycerine, *double base.* For many years the basic formulae for smokeless powder have changed very little. As a fuel, which it is, in reality, common garden variety smokeless powder is pretty efficient. In the gun (essentially a single-cylinder internal combustion engine) it is considerably more efficient than gasoline in your auto engine. So please don't knock it simply because it is made from basic formulae many years old. While it may be possible for new small arms propellants to come along as a result of the vast amount of rocket propellant research, you can plan on using the current crop for many years to come. On the other hand, manufacturing methods have undergone much change over the years. These changes have resulted primarily in greater stability (freedom from deterioration due to age or climactic conditions) and uniformity and reduced production costs.

Acids are extremely vital to powder manufacture, yet, are conversely powder's greatest enemy. Unless all traces of acids are removed in the final stages of manufacture, powder deteriorates rather rapidly with age. During the early 1920's, millions of pounds of powder left over from World War I had to be destroyed because it deteriorated in storage magazines. Other large quantities destroyed themselves simply by spontaneous fire. These incidents were brought about directly or indirectly by residual acids which crash, wartime production methods failed to remove.

In addition to the vast quantities of bulk powder destroyed due to deterioration, much loaded ammunition was destined for the same fate. During the years following the two Great Wars, Ordnance inspectors tested samples of all lots of military ammunition in storage at frequent intervals. When deterioration was discovered, the ammunition was condemned and often effectively destroyed by dumping it in deep water at sea. It was also sometimes destroyed by burning, but that is a slow, laborious and expensive way to do the job.

With such problems in the background, the military establishment was greatly interested in a product better suited to storage. DuPont and Hercules, primary WWI powder producers, developed and perfected new production methods before WWII thundered into being. The War Department's Picatinny Arsenal has also conducted considerable research into better methods. Fortunately, these efforts set the stage for WWII powder production that staggers the imagination, yet today, much powder produced during the painful early 1940's is still perfectly safe and usable. Magazines fires and explosions like those of the 1920's are not unknown, but are rare.

Naturally, production facilities were not adequate to the task at the start of WWII. The quantities produced aren't important for our purposes here. It is sufficient to say that in the peak year (1943) of U.S. small arms ammunition production, the entire U.S. powder production of WWI could have been matched in slightly over *one week!*

One plant produced 206,700,000 rounds of .30/06 ammunition per month at its peak rate. This required 1,476,430 pounds of IMR 4895 powder — one full freight car (50,000 lbs.) — each day. Another plant (the smallest in operation) produced 1,200,000 caliber .50 M.G. cartridges daily. This consumed nearly 43,000 lbs. of powder each day. And, it was the smallest in operation.

Since the WWII production methods were eminently successful and are still in general use, we'll describe them in some detail, digressing only where significant changes have taken place.

For convenience sake, we'll break production into two stages — that of processing the nitrocellulose, and that of converting the nitrocellulose to powder. High grade nitrocellulose (actually cellulose-nitrate) is required for all powders, so up to the point of delivery of the nitrocellulose to the powder mill, processes are the same.

PREPARING NITROCELLULOSE

Powder begins with the production of nitrocellulose of a specific type. It is produced from cellulose or vegetable fibre, the most common material being cotton "linters" and wood pulp. Linters are the extremely short, fine cotton fibres removed from cotton seed after all fibres useful in fabric making have been salvaged. Linters are essentially a by-product of no value to the fabric industry, so are disposed of to powder and plastic producers.

Baled linters received at the powder mill have been carefully washed and purified to remove all natural oils, so are ready to be made directly into nitrocellulose.

Each bale is sampled and tested to insure no contamination is present. Bales are broken and the compressed material is fed through special machines which fluff it up and remove any foreign material. Cellulose is highly hygroscopic (moisture-absorbing) so it is necessary to dry the fluffed linters by means of heated air.

Wood pulp may also be used to make the basic powder nitrocellulose and its use was highly developed during WWII when demand for powder exceeded linters supply. Wood pulp is much more widely available and during the war its cost was less than half of that of linters.

The pulp used by Hercules during the war was known as "Purayonier," of very high purity, containing a minimum of 90 per cent virgin (?) cellulose of a high alpha content. This pulp was supplied to the powder mill in the form of heavy paper, about .050" thick, much like rough blotter stock. It was shipped in 550-560 lb. rolls 30-32" wide. As received it had a moisture content of just about 6 per cent so was run through a continuous feed drying furnace. This reduced moisture content to approximately 0.5 per cent. Naturally, extensive sampling and testing verified its purity first.

After drying the pulp was shredded in special machines into a fine

Modern Handloading

fluff, very much resembling the linters. At this point it was ready for the nitrating processes which are identical to those used with linters, except for individual charge weights.

The DuPont mechanical dipper method of nitrating cellulose is the most commonly used in this country. Other methods are Pot; Nitrating Centrifugal; and Thompson Displacement, but we need not go into them here.

A nitrating unit at the Radford plant (operated by Hercules during the war) consisted of four stainless steel pots or kettles with their tops at floor level to facilitate charging. Each nitrating house contained eight units, a total of thirty-two pots.

At this point a brief explanation of the functions of the required acids will be necessary to follow the mechanical nitrating processes. Sulphuric and nitric are the basic acids used. Nitrogen from the nitric combines with the cellulose to form nitrocellulose. Sulphuric has the job of combining with water released by the nitric, thus preventing dilution of the nitric during the nitration process.

Nitration of one pound of cellulose requires one pound of nitric acid and a half-pound of sulphuric. Approximately 1.5 pounds (dry weight) of nitrocellulose is produced depending on the degree of nitration desired. Because of the high cost of the acids and the fact that they cannot be simply pumped into a sewer when spent, they must be recovered after the nitration process. An extensive (and expensive) recovery system salvages the spent acids and most of their vapors. After collection they are separated into nitric, sulphuric, and water. The acids are re-used, but only after reconcentration to orginal strength and purity. The nitric is run through some of its original concentration processes for this purpose, while the sulphuric is brought up to strength by adding "fuming sulphuric," or "Oleum."

Naturally, the continuous handling and storage of such large quantities of acid present a major problem. Often nitric acid is manufactured in a companion plant next to the powder mill. Both acids are used at strengths (nitric 98-99 per cent, sulphuric 68 per cent) which are most injurious to materials and equipment. Miles of stainless steel and other special alloy piping, tanks and plumbing are required. I once examined a defunct wartime mill where hundreds of tons of special stainless plumbing lay around like huge piles of cordwood. Yet, being heavily contaminated, it had no value, in spite of costing over a dollar per pound originally.

Two basic nitrocelluloses are required for smokeless powder production. They differ primarily in the degree of nitration and are known as guncotton, or high grade, designated S1, and pyrocotton (low grade), S7. While quality is the same, high grade S1 is approximately 13.4 per cent nitrated and S7 only 12.7 per cent. The two grades are blended to produce a blend of 13.15 per cent nitrogen content for single base powders and 13.45 per cent for double base types.

Why blend two grades to get these percentages when they could be obtained in the initial nitrating? Solubility is the main reason, though others we'll not get into do exist. Low grade at 12.6 per cent is 99 per cent soluble in ether-alcohol solvent to be used later, while high grade 13.4 per cent is practically insoluble in this solution.

When the colloid (more on this later) we call "smokeless powder" is formed, the soluble low grade dissolves and acts as a vehicle carrying the insoluble high grade in suspension. The latter, with its greater power, producing a large volume of gas at high pressure in your gun, is diffused in a fibrous state throughout the low grade and acts as an igniter for the low grade. A volume this size would be required to elaborate on that particular phase of powder making. Suffice it to say that the development of this blending of two grades, soluble and insoluble, has been found essential in producing the many superb powders we have today. Only this nitrating time differs in producing the two grades.

Nitrating consists of stirring the cellulose (linters or pulp) with a mixture of nitric and sulphuric acids in the proper proportions. Agitation of the mixture and control of its temperature are also important, as is the time interval.

The acids are extremely dangerous to life, so they are handled through extensive remote-controlled plumbing and all vapors are continually drawn off by vacuum lines to the recovery system.

The stainless steel kettle is charged with cellulose (handled in a fibre bucket) while the acids are being pumped in. The resulting soupy mass is called "slurry" and is constantly stirred by an agitator consisting of sets of paddles rotating in opposite directions. The slurry is stirred in this manner for (at Radford) fourteen to twenty-two minutes with the pot covered. The four-pot unit mentioned earlier allows a single operator to keep them all going, number I being ready to dump and recharge when he finishes charging number 4. Signal lights and other devices keep the operator on his toes and assure uniform quality and proper timing.

Normally the pots are located high enough in the building to allow the slurry to flow by gravity into a high-speed centrifugal wringer below. The wringer contains a perforated drum rotating at high speed. The nitrated fibres are held by the drum as the spent acids and some water is hurled out by centrifugal force into recovery system drains.

Modern Handloading

Wringing is a danger point in the process. The acid-damp fibre can flash spontaneously upon exposure to the atmosphere. Nitrator fires occur fairly often, so every conceivably safety measure is taken. Escape routes are plentiful and strategically located, and always open. When any unit flashes, the operator calls one word, "Fire," and all personnel drop everything and leave rapidly but in orderly fashion — no mad scramble. Emergency buttons within reach of everyone (some systems are fully automatic) allow immediate functioning of exhaust fans to clear out the smoke.

Nitrator fires burn only a few minutes and cause little interruption in the routine. Everyone gets right back to work and a clean-up crew starts in on the pot that flared.

The dangerous wringing process takes only three to four minutes per charge. The charge must immediately be forced through a hole in the bottom of the wringer into a "drown" tank below. It still contains about 25 per cent acid and will flare very easily. The operator opens the wringer and uses a non-sparking fork to scrape the pulp from the drum surface and force it through the bottom of the water below. From this point on, the nitrocellulose is hairy stuff to handle and must be kept wet at all times.

Washing and rinsing operations follow to remove the bulk of the remaining acids from the slurry. It is piped, mixed with water, through complex plumbing to boiling tanks in separate buildings. These were (at Radford) wood tubs constructed of heavy planks, bound tightly by metal bands such as seen on old silos and water tanks. The twenty foot diameter tubs held 12,000 lbs. of nitrated linters or 18,000 lbs. of wood pulp.

Steam fed into the tubs from below was used to bring the slurry to a boil. A network of perforated pipes supplied the low pressure steam which bubbled up through the slurry. Boiling removes the water-soluble acids, carrying them away in water vapor to the acid recovery system.

After fifty hours of continuous boiling, the water was drained off and replaced with fresh water. This was followed by an additional ten hours of boiling, after which water was drained. Two fresh water rinses followed, under agitation, to remove final traces of acid.

After being dumped (still mixed with water) the slurry is pumped to another location for beating. This consists of feeding it into specially built machines which chop the individual fibers into even smaller pieces. Beating is a continuous process, the nitrocellulose passing from one machine to another to be chopped successively finer. Continuous inspection insures that the required degree of fineness is achieved.

In addition to reducing individual fibre size, the beaters perform another vital function, that of releasing minute residual quantities of acids trapped in the fibres where washing cannot release them. Once released, these acids are taken up by the water in the slurry.

Unless the nitrocellulose is entirely free from acid when delivered to the powder mill, the end product will deteriorate with time. The degree and speed of deterioration is dependent on the amount of acid left in the nitrocellulose. Beating cannot be slighted.

Once properly beaten, the slurry is pumped to de-watering pots where free water drains away. The smooth, soup-like slurry flows by gravity over sloping screens. The screens will pass water, but restrain most of the nitrocellulose fibers. Even so, some of the very fine (and very dangerous, now that its water content is being lowered) pulp passes through the screens. A very sophisticated recovery system returns 100 per cent of this wayward pulp to its parent mass.

After de-watering the thickened slurry is pumped to yet another building for "poaching." In this process a tub filled with the pulpy mass has soda ash and water added to neutralize any remaining acidity. The mixture is boiled by means of steam as before, agitated mechanically all the while. The nitrocellulose is then rinsed to remove all traces of the alkaline soda ash and piped directly to blenders nearby.

At this point, low grade S7 and high grade S1 are mixed to produce the precise blend required. (You'll recall the earlier explanation of the necessity for a blend of two separate grades of nitrocellulose rather than a single grade.) Blending is done by continuous pumping of the two grades into each other, and goes on for several hours until the proper mechanical mixture is obtained. Continuous sampling and testing goes on and, as often as not, additional nitrocellulose of one grade or the other is added to achieve the proper blend. When additional material is added, blending begins anew, so the process can be time-consuming.

At this point, the pulpy mass must be stripped of most of its water down to about 30 per cent content. To go below this percentage at this stage would create unacceptably great danger of spontaneous ignition and flash fires. Wringers are loaded with several hundred pounds of slurry and rotated at successively higher speeds. Centrifugal force throws excess water through the drum screen, so the water goes to a recovery system which picks it up.

When wringing is finished, a non-sparking "plow" scrapes the caked nitrocellulose from the screen. The charge is then dumped into fibre buckets for transfer to the dehydrating press. Its moisture content must be maintained to avoid spontaneous ignition. Among powder men, it

Modern Handloading

is well known for its "mean" disposition and commands much respect. Despite the most stringent controls it can, and sometimes does, catch fire.

At this point, nitrocellulose is complete and ready to be transformed into powder. It requires only pressing into a block and complete removal of its remaining water. This operation is actually the beginning of powder manufacture. Previous steps were the production of nitrocellulose proper.

The dehy press forms the nitrocellulose into a cylindrical block, and also removes the water, replacing it with alcohol to keep the material wet enough to prevent spontaneous ignition.

MAKING POWDER

Just as it comes from the wringer, the soft, irregular chunks of nitrocellulose are placed in the cylinder of the dehy press. The press has two opposed vertical rams operating in a single cylinder. The nitrocellulose is placed in the cylinder with both rams retracted. Hydraulic pressure is applied and the rams move toward each other until stopped by the mass of nitrocellulose. Then, alcohol (pure grain alcohol of 180 proof, not the poisonous denatured variety) is forced into the nitrocellulose through the top ram. As alcohol feeds into the cake of nitrocellulose, water surrounding the individual fibres is displaced by the alcohol. Being heavier than alcohol, the water flows downward through the screened head of the lower ram into a tank. Later, stray alcohol is pulled out of the water by an efficient recovery system.

Once the water is displaced, high pressure (around 3,000 psi) is applied to compact the nitrocellulose. Excess alcohol is squeezed out in the process and returned to supply. The dehydrated cake (containing about 49 lbs. dry nitrocellulose and 13.5 lbs. alcohol) is removed from the press and broken up into chunks by means of non-sparking tools. At this point it becomes known as "powder."

This is as good a time as any to point out that all of the foregoing operations are continuous and interdependent. Once the dry linters or wood pulp begin nitration, they must pass rapidly and smoothly from one operation to the next. No hold-overs can be allowed at any point. All operations are conducted to meet the nitrocellulose requirements of the powder mill. This is done by varying the output of the nitating plant, not by stock piling. Because of this requirement for continuous operation, nitrating plants operate around the clock or not at all. At no time can an operator or shift cease operation until replacements are on hand to insure continuous operation. Also, the completed nitrocellulose must be delivered immediately directly to the first operation of the powder mill before any significant evaporation can take place. If it should sit idle for any great time, a fire is certain.

From the dehy house, the broken cakes are transferred in fiber buckets direct to the mixers. No delay is permissible because the highly volatile alcohol evaporates rapidly. Any great amount of evaporation will produce a flash fire. This can be a real problem in hot, dry weather.

In the mixing house, individual mixers are housed in heavily-barricaded rooms or compartments.

The broken cakes are dumped into mixers virtually identical to heavy-duty bread-dough mixing machines found in large commercial bakeries. The charge is carefully weighed and any alcohol lost by evaporation in transit is replaced. The alcohol is readily drawn from a nearby tap to meet this requirement. While I've never seen a tin cup hanging alongside the spigot, I've often wondered whether some shrinkage might not be accounted for by thirsty mixer tenders. I recall my own younger days of enlisted service when assigned duties kept a number of us in close contact with large supplies of grain alcohol. With a little experimentation, we developed some pretty fair-tasting substitutes for store-bought hootch. One loveable old sergeant (rest his pickled soul) took his straight, filling an eight-ounce pocket flask each day at work call and recall.

To the alcohol-wet nitrocellulose is added a precisely-calculated amount of previously blended ether-alcohol solvent. Ether cannot be used straight for this purpose. The usual mixture is 65 to 67 per cent ether and 33 to 35 per cent alcohol.

Other ingredients required to control the burning characteristics of the powder are now added to the nitrocellulose-solvent mix. These will vary considerably, according to the type of powder being produced and the individual specifications for that powder.

Most important of these is the stabilizer which functions to prevent decomposition of the powder from this point onward. Pretty much the standard stabilizer in this country since 1909 is diphenylamine (DPA). It is an oily, pale yellow, granular material. Should there be infinitesimal amounts of acids left in the nitrocellulose, DPA counteracts their effect by combining with liberated oxides to form stable compounds. This effectively prevents deterioration with as little as 1 per cent DPA.

Potassium sulphate may also be added at this time to reduce muzzle flash if the powder is intended for military small arms use. If for artillery use, then DPB might be added for the same purpose. Artillery powders would also require addition of a deterrent at this point to slow down burning. Dinitrotolulene (DNT) is normally used for this purpose. It is highly hygroscopic so must be broken up and screened just prior to addition to the mix.

Once all secondary ingredients have been added, the mixer is closed and started. The operator leaves the building or enters a barricaded safety area while the mixer is running. Mixing is done by two counter-rotating Sigma blades, two-blade paddles resembling the letter S, rotating in a horizontal plane.

After a short time, the operator shuts down the mixer and uses a non-sparking tool to break loose any mix that has gathered where the blades could not reach it. Solvent mix is also added at this time if appears necessary to the operator. Experience is invaluable in determining whether the mixture is too dry.

Mixing then proceeds to completion which is determined to a degree by elapsed time, but also by the operator's experience. Considerable quantities of alcohol-ether vapors are generated by mixing and must be drawn off into the recovery system.

Flash fires were once fairly common in this operation, but in recent years they have become rare. Experience and sophisticated safety measures, plus the use of highly responsible personnel, paid off in safer operation.

It is worthwhile to mention here that scrap powder generated in the various processes is reworked by either using it (plus solvent) to charge the mixer, or by adding it to the nitro-cellulose charge. The scrap consists generally of broken powder strings, defective grains, "heels" from the various processes. Nothing is wasted, as is clearly shown by the vast and expensive recovery systems.

Coming from the mixer, the powder is ivory to yellow in color, depending on the mix, and looks somewhat like damp brown sugar. It is partially colloided and the process must be continued in a machine called a macerator.

The powder is transferred directly to the macerator where gear-like blades and teeth knead the mixture under tremendous pressure. The blades are closely meshed and rotate in opposite directions. Solvent content of the powder is very important at this stage, neither too wet nor too dry. No more solvent can be added during maceration, yet the powder will ignite if it becomes too dry while the machine is running. Considerable heat is generated by the kneading process, so the macerator is fitted with a cooling jacket through which refrigerated alcohol is circulated. By this means, powder temperature is held at about 68°-72°F.

The macerator continues the colloiding operation, converting the powder to a rubbery mass that can be pulled apart by the fingers. It is removed in irregular chunks and transferred immediately to the preliminary blocking (pre-block) stage.

In pre-block, the rubbery chunks are charged into the cylinder of a large vertical press. With the press closed, ram pressure is applied very slowly to force excessive solvent from individual chunks. Pressure is slowly increased to compact the colloided powder into a solid block. If pressure is applied too fast, an attendant temperature increase may cause the powder to flash and burn — not conducive to the press operator's peace of mind. When the compressed cake is fully formed, it is ejected from the press cylinder and immediately transferred to the "macaroni" press in the next barricaded compartment.

The macaroni press performs two functions — that of extruding the powder into thin strands (thus its name), then re-compressing them into a solid block. The pre-block is placed in the chamber of the macaroni press. The ram applies force to the top of the block, forcing the plastic powder to flow through a number of individual dies set in a plate in the bottom of the cylinder. It emerges as long, round strands which fall directly into the cylinder of a floor-level blocking press with its ram retracted.

This extrusion continues the colloiding action begun by the mixer and carried on by the macerator. Screens placed between the powder blocks and the die plate, catch and hold un-colloided lumps of nitrocellulose or other free ingredients that still exist. Great pressure is required to accomplish the extrusion, being built up slowly to just over 5,000 psi.

When extrusion of a cake is complete, the macaroni press is stopped and the powder strands broken off and dropped into the blocking press. This press is then capped and its ram rises to compact the mass into a twelve inch diameter cake fifteen to eighteen inches long. The cake or "block" is then ready for the graining presses.

When quality control of mixing and macerating is of high enough caliber, powder may go directly from pre-block to graining, eliminating the macaroni press and floor-blocking stages. Powders were produced both ways during WWII and passed all acceptance tests in good order.

Most small arms propellent powders are formed into cylindrical grains or kernels. I prefer the latter term since neophyte reloaders seem to be continually confused between a powder grain (kernel) and

Modern Handloading

powder weight in grains avoirdupois. These kernels usually contain a single round perforation lengthwise through their centers. Powder men call this "mono-perforation," as opposed to the seven-hole "multi-perforation" used in artillery powders.

The wall thickness (web) of the tubular kernel must be uniform since it controls the burning rate. This is equally true with any size kernel.

While the graining operation looks simple and proceeds with great rapidity, this is only an illusion brought about by highly developed machinery and skilled operators.

First the blocks (usually two) are charged into a vertical press. In the bottom of the cylinder is a die plate carrying a large number (often forty-nine) of individual extrusion dies, each forming a perforated strand as powder is forced through them. Between the die plate and powder block are positioned straining screens like those mentioned earlier. They prevent any previously undiscovered uncolloided lumps from reaching the dies.

The extrusion dies consist of two basic parts. One is the "nozzle" or agate containing a hole the same diameter as the finished "green" powder strand is to have. The lower end of this hole is cylindrical for a short distance, then flares out to a conical shape at the upper end. This agate fits into a "pin plate", from the center of which extends a small diameter pin. This pin extends, carefully centered, into the hole in the agate and ends flush with the lower end of the hole. Holes drilled in the upper end of the pin plate allow powder (still fairly soft and plastic) to flow into the agate, around the pin, and out the nozzle. Under great pressure, the powder flows through these pin plate holes in thin strands, then consolidates into a homogenous mass again. As it flows on out through the agate, the pin produces the single perforation in the center of the strand.

When the graining press ram exerts about 5,500 psi on the powder, each extrusion die in the die plate produces a continuous perforated strand of powder. Still green, the strands are quite flexible and fall to coil naturally in an individual fibre container under each die. Since green powder shrinks approximately 30 per cent in drying, the fresh strands are larger in diameter than finished kernels.

The powder strands are rushed directly to a cutting machine to avoid air drying. There individual strands are fed through holes in a plate and into a pair of precisely adjusted horizontal feed rollers. The rollers must feed the strands into the cutter at a precise, uniform rate, yet not mark or distort them. They do. The plate contains extra holes so that if one becomes plugged, the operator can switch to a spare, avoiding a shutdown for cleaning.

As strands come from the feed rollers they are guided into openings in the cutting die. As they emerge from the face of the die, a series of rotating knives neatly clip off the protruding ends. Proper relationship between the rate of feed (by the rollers) and the rotational velocity of the knives determines the length to which the individual kernels are cut.

The knives strike the face of the cutting die lightly and this contact must be carefully adjusted. Too hard and the knife life is shortened, while too little contact and the kernels will not be cut cleanly. They'll have those ragged "tails" occasionally noticed by reloaders.

A water-alcohol solution is fed in small amount to the face of the cutting die. It serves several purposes — that of keeping blades and die cool, preventing kernels from sticking to the metal, and prevents early dulling of blades.

The cutters are high-speed machines. Production rate will vary according to the powder being cut, but as much as 30,000 lbs. of .50 M.G. powder can be processed by a single machine in one twenty-four hour period. As powder comes from the cutter it is still green, but otherwise finished. It pours from the canvas-covered cutter chute into buckets, to be transported to the solvent recovery building. It is still mighty dangerous stuff, but there it will be tamed by drying it — under water!

Drying actually consists of removing the remaining solvent from the green powder kernels. Called "solvent recovery," this operation is carried out in a barricaded building surrounded by a large earthen dam. The dam's purpose is to deflect the force of any explosions skyward. Even the lighting for the solvent recovery building is remote. Highly explosive solvent vapors (with a low flash point) are continually exuded from the powder kernels. One spark and — eternity.

Each visit to the foreboding confines of the dam gives me a worms-in-belly feeling. Solvent recovery may not be the most dangerous stage in powder manufacture, but it would make by far the biggest bang if that 30 tons of green powder and solvent did let go.

Powder is conveyed in carts to the recovery building. The carts are metal lined and sealed to prevent evaporation and are fitted with hydraulic dump mechanisms, and rubber tires. The loaded cart is lifted to the upper level of the building where the powder is dumped into large copper recovery tank. A seal closes the tank to prevent any loss of solvent vapors. The water also serves to absorb pressure surges within the tank.

Inert carbon dioxide gas is used as the solvent recovery medium. It is pumped into the loaded tank where it fills the air space and, being heavier than air, settles to surround each powder kernel. The gas is circulated under pressure, picking up alcohol and other vapors. The vapors are liquified in a refrigerated condenser as the gas leaves the tank. Liquid alcohol and ether are then returned to storage tanks.

Solvent recovery must proceed slowly or the powder kernels will warp and split as they shrink some 30 per cent in size. The carbon dioxide must be warmed before entering the tank, since it is cold as it comes from its generator. The tanks and their contents tend to heat up during the recovery process. Water spray on the outside of the tanks retards temperature rise. The amount of cooling required depends considerably on the weather.

Recovery removes all but a residual 3 to 5 per cent of solvent from the powder. The powder is then transferred to the Water Dry House where the last traces are removed. There the powder is dumped into huge wood tanks. It is then covered by water heated to 65°F. This water circulates through the powder for a long time — six days, in the case of Radford wartime production.

At the end of the prescribed time, the powder is drained, rinsed repeatedly, then tested for freedom from solvent. If it passes the tests, the powder is drained and moved on to the next operation. If not, water, drying continues.

The moist powder is bagged (normally 50 lbs. each) and transported to the coating building. Here Dinitrotolulene (DNT) will be applied in the form of a thin external coating to control burning rate. It is applied while the powder is still moist from water drying. Even so, coating is still a dangerous process.

The coating building is barricaded. Large copper barrels shaped like a flattened ball are used to apply the coating by tumbling powder, DNT and water together. The barrel is charged with (in one instance) 1112½ lbs. of powder, 96 lbs. water and 65 lbs. DNT. The drum is sealed and operators move out of the barricaded area. The drum rotates about 20 rpm for approximately one hour.

At the end of the tumbling, the DNT will have uniformly coated each kernel. Only the outside of the kernel is coated. Though the DNT may have built up over the perforations of some kernels, sealing them, it burns off quickly and does not affect the general burning characteristics of the powder. The powder is then transferred to a wash house where it is hosed down with water and screened to remove lumps and kernels stuck together. It is then air dried. Warm air is circulated through it in tanks. Several hours may be required for this process.

Once dry, the powder requires only "glazing" or coating with graphite. This coating has no effect upon the powder's performance.

Indeed, it could be used without it. However, the graphite prevents build up of static electricity which could fire the powder. It also makes the powder flow freely in packing and loading machines. The distinctive gray/black color of cannister powders does not appear until the graphite is applied. Prior to that point the kernels are relatively light in color.

The powder is placed in a drum similar to those used for coating and tumbled with a very small amount of graphite. As an example, 1½ lbs. of graphite will treat 5,000 lbs. of powder. After about an hour of tumbling, the charge is dumped and placed in 50 lb. bags for movement to the blending machine. There, bags from different batches of powder are emptied into a barrel to make up a "pre-blend" charge. Rotation of the drum thoroughly blends the powder together and it is again bagged. A second blending takes place in the same way, only this time, bags from different pre-blend batches are mixed.

Upon completion of the final blending, the powder is packaged with the assurance that ballistic qualities will vary very little from lot to lot. At this point, perhaps you'll wonder just what constitutes a "lot" of powder. Lots generally consist of a quantity that can be easily and conveniently handled. During the wars, lots consisted of 50-55,000 lbs. simply because that was the capacity of the railroad cars available for shipment. Thus each carload was a separate lot and there were no portions of lots left over after loading. In peace time, powders for the handloading trade are produced in smaller lots. For example, if the eastimated market requirements for a given period is 10,000 (or 20,000) lbs. of a particular powder, then a production order for that amount of powder will be issued. The powder will be then produced, packaged as a single lot and placed in the magazine for filling future orders. If, on the other hand, Lake City Arsenal is (as this is written) producing 7.62mm cartridges at a high rate, requiring continuous deliveries of powder, it will be shipped in the traditional individual carload lots.

It goes without saying that continuous testing takes place throughout the production processes. Once the powder is completed and blended and a lot number assigned, samples are fired extensively for pressure and velocity to insure that it meets the proper specifications. Powders for the reloader must be very uniform from lot to lot. Powders for commercial or military purposes can vary without causing trouble, since the users can adjust loading data to produce standard velocities and pressures. For example, each lot received at a loading plant is tested and charge weights are calculated and proven by actual firing before that lot is used.

DOUBLE BASE POWDER

Nitroglycerine is a major ingredient in double base powders, as mentioned earlier. Double base powders do have slightly higher energy content, which means they can be used in smaller quantities, than single base types. The major producer of double base powders for the reloader is Hercules Powder Company at its Kenville, N.J. plant.

Nitrocellulose constitutes from 65 to 82 per cent of the weight of double base powders. A common military powder formula contains 75.50 per cent nitrocellulose, 20.00 per cent nitroglycerine, 1.00 per cent potassium nitrate, 1.50 per cent barium nitrate, 0.75 per cent diphenylamine, 1.00 per cent DNT (the deterrent), and 0.25 per cent graphite. This formula is nearly identical to that of the discontinued Hercules Hi-Vel #2. Some special purpose military powders contain as much as 45 per cent nitroglycerine. The nitrocellulose used is identical to that described earlier in single base powders. Nitrocellulose may be diverted to either single or double base powder lines (both of which often are operating simultaneously at the same mill) at any point before final blending of the high and low grades. You'll recall a 13.15 per cent blend is prepared for single base use, 13.45 per cent for double base.

The latter is insoluble in ether-alcohol solvent used in single base manufacture, but readily dissolves in the acetone-nitroglycerine double base solvent. Acetone not only serves as a solvent when mixed with nitroglycerine, but somewhat desensitizes the latter to make for safer handling.

Naturally, the process of manufacturing nitrocellulose remains the same, clear through the dehydration stage where water is displaced by alcohol and mix is compressed into a cake.

Nitroglycerine must, of course, be manufactured and supplied to the powder mill. Since nitro is especially dangerous to produce and extremely hazardous to transport, it is normally manufactured in a separate plant immediately adjacent to the powder mill. Nitro plant output is regulated to meet the needs of the powder mill to avoid storage problems.

Essentially, heated (115°-120°F) glycerine is combined with highly concentrated nitric and sulphuric acids and agitated. Eventually, the nitroglycerine and spent acids are separated, the latter being recovered for reconcentration and further use.

The nitroglycerine is washed and purified to remove acid traces, then stabilized and neutralized, and stored under water. All that remains is to incorporate it in powder — before it blows up. The manufacturing process is certainly far from as simple as the foregoing might indicate. It is not, however, the purpose of this volume to be a text on nitroglycerine manufacture, and the foregoing is sufficient for an understanding of the manufacture of double base powder.

From the dehy press, blocks of nitrocellulose go directly to a "block breaker" house where they are torn apart. Blocks are placed in a copper barrel fitted with myriad teeth on its inner surfaces. Operators retreat to a barricade when the machine is started. The barrel rotates and the block tumbles over the teeth, which tear in into small pieces. The pieces fall through openings in the barrel into a fibre container below. Nitrocellulose is mean, and fires are not uncommon during this process.

When sufficient shredded nitrocelluluose is accumulated, it is transferred to a weighing station and weighed out into proper premix charges. The charges are then moved to the premix area. There the additional dry ingredients are waiting in precise, prepared amounts to fit one premix charge.

Next comes the nitroglycerine. It is delivered only on call from a special storage area located a safe distance from other operations. Special, carefully leveled, elevated wood walkways are built to facilitate nitro movement. The heavy planks are fastened with copper and non-sparking nails and hardware. Nitro is delivered to premix in an "Angel Buggy." a rubber tired wood hand-cart fitted with four small compartments. Each compartment carries 25 lbs. of nitroglycerine mixed with approximately 18 lbs. of acetone, which as mentioned early, desensitizes the nitro. The fellow herding this cart several hundred yards over the elevated walk needn't worry about hospital bills (or even funeral expenses) if he makes a mistake. One bump, and his worries are over — permanently. Some foot traffic also moves on the walks, but when confronted with a man and his Angel Buggy, both glaring white, everyone steps off, mud and snow be damned.

Why four compartments in the buggy? First, they reduce sloshing of the nitro — certainly desirable — then, each compartment contains the nitro charge for one premix bowl.

The nitrocellulose and dry ingredients are first placed in the mixer. Because of its shape, it is called a "figure-eight bowl." They are generally oval in shape, constricted in the center. Each half contains a Sigma blade and the two rotate in opposite directions, feeding the powder mix continuously from one bowl to the other and back in a figure-eight pattern.

The charge is mixed dry for a few minutes, then the nitroglycerine-acetone solvent is fed in slowly while the blades are turning. Gradual feed of the nitro mix is necessary to prevent lumping of the complete mix — just as you must add liquid slowly to flour if you want a smooth mixture. Premix continues for a flexible period of five to ten minutes after solvent addition. The operator must judge when incorporation of dry ingredients and solvent has progressed to the proper degree. No matter how carefully the mixing is done, some small lumps may form. They are undesirable and result in waste, but since they will be screened out later they do not constitute a serious problem.

The premixer is unloaded by means of wood shovels into fibre containers which are then trucked to the ingredients house near the final mix area. In winter the mix must be kept warm. Nitroglycerine is not sensitive while frozen, but once frozen and thawed becomes hypersensitive and will blow with no more excuse than an ill-tempered mistress.

Several premix charges are combined to form a final mix charge. Depending on the powder to be produced, additional ingredients may be added at this point. The charge is placed in the mixer and leveled. At this point a specified amount of alcohol is measured into the charge. Since acetone will have evaporated from the mix since its addition at premix, the mixer operator must judge when more is needed and add it as required. Only experience qualifies a man for this.

The mixer is closed and sealed, the cover being fitted with breakaway devise to release pressure in the event of a flash fire. The mixer is then run for about two hours in what is known as "warm mix." Mix temperature builds up to about 115° and the powder will be soft and gummy. To allow uniform graining later, it must now be cooled, so refrigerant is pumped into the mixer jacket. Mixing continues under refrigeration until the temperature drops to the mid-eighties. Mixing must also continue until the consistency of the powder is correct. The operator judges this.

Upon completion of mixing, the powder is removed from the mixer and is ready for blocking — a very dangerous process. The mix is placed in the same type vertical press described earlier, then operators retire behind barricades. Pressure is built up gradually to over 3,000 psi and held there very briefly. This converts the rubbery powder to a solid, compact block ready for graining. Note that there is no pre-blocking or macaroni press operation as with single base powders.

Graining of double base powders is carried on in practically the same manner as for single base types. Dies and presses are the same, but the cylinder contains a constriction to prevent the powder heel from being withdrawn when the ram is raised for recharging. Since the heel remains in the cylinder, it is covered with acetone when the press is not operating. Cutting of the strands into individual kernels proceeds as outlined earlier. Drying is much the same, but the water dry process is eliminated. Coating, graphiting, blending and packaging remain the same.

With so many lengthy operations involved in powder production, the question of total manufacturing time arises. Processes are continuous and interlocking and the output of any one stage is regulated to meet the demands of subsequent ones. During accelerated wartime production, single base small arms powder progressed from linter delivery to final packing in as little as fifteen days. This may seem like a long time, but represents a vast reduction from the time required a few years earlier. The bulk of this time is consumed in purifying and drying the green powder.

Thus far we've concerned ourselves only with the traditional extruded-grain (kernel) powder. This type has been standard since the earliest days of smokeless powder. Many European powder mills produce essentially the same formula powders in flake form. To achieve the typical square or diamond flake, the powder is not grained as outlined here, but rolled into thin sheets. These sheets are then sliced into flakes of desired size and shape by sophisticated automatic cutting machines. Thickness and size of the flakes govern burning speed in the same manner as do kernel and perforation size in grained powders. Manufacturing time for flake powders does not vary singificantly from the above mentioned.

BALL POWDER

Ball powder, developed before WWII by Winchester-Western, is a relative newcomer to the field. It was used in vast quantities during WWII, the Korean War and the current war in Viet Nam. It is a standard powder for loading 7.62mm and 5.56mm. (.223) military ball ammunition, as well as other ammunition types.

In appearance, ball powder greatly resembles fine shot. It is in the form of tiny solid spheres which are not highly uniform in diameter. It will also be encountered in the form of spheres which have been passed between rollers to flatten them to varying degrees. Though for years offered only within the industry, for the last several years a number of excellent Olin (W-W) rifle, pistol and shotshell powders are available to the handloader. All are produced by Winchester-Western which is, incidentally, the sole U.S. manufacturer of this type.

The chief advantage of ball powder is that it can normally be pro-

Modern Handloading

cessed from start to finish in less than forty hours. Quite a difference from the fifteen days of other types. Test lots have been completed in as little as twelve to eighteen hours, but such high-speed production has not yet been made entirely satisfactory. It also has the advantage of metering and flowing very uniformly through loading machines.

Manufacture of ball powder begins with the nitrocellulose in the already-discussed slurry stage where additional water is introduced. Purification (recovery of residual nitric and sulphuric acids) is speeded up and accomplished in less than eight hours. At this point the thin slurry is mixed with emyl acetate and introduced to a larger distillation unit. If a double base powder is desired, nitroglycerine can be added, or a straight single base propellant can be produced just as easily. Normally some nitro is added.

Nitrocellulose in the slurry combines with the emyl acetate (solvent) to form a lacquer. Since it is not soluble in water, and is also lighter than water, it rises to the surface of the water in the still. The mixture is then agitated by paddles at a predetermined rate. This action breaks the lacquer into an emulsion. Stirring speed controls the size of the lacquer bubbles produced. Agitation stops when the mix is in the proper state and a colloid is added to form a film over the lacquer globules and prevent them from rejoining to form a liquid.

Hardening, the globules settle to the bottom of the still where they are impregnated (still under water) with the desired amount of nitroglycerine. The wet powder is then pulled from the still and fed on a continuous belt through infra-red drying ovens. It is dry in an hour. The now dry spherical powder is coated with deterrent and graphite in the usual manner, then screened to eliminate those balls grossly over or undersize. The rejected screenings are easily salvaged by combining them with solvent, to be put through the process again.

After testing and blending the screened powder, it is ready for packing, to be shipped to the loading plants. Quite a difference from the long, drawn-out process of producing grained powders.

As can be seen from the foregoing pages, the manufacture of smokeless powder is no small task. It requires millions of dollars worth of tools, facilities and real estate. Even when done under the most rigidly controlled conditions, accidents and fires are not uncommon. In view of this it is not hard to imagine what would happen if an individual tried to do it in his basement shop with aid of one of those formulas sold through magazine advertisements. Before you ask the question, I'll answer it. Don't under any circumstances, try to make powder.

CHAPTER 7

Bullets

IN ANY cartridge the bullet is the item that does the desired work. It is the bullet that pierces the target, that lays low game or enemy. All other components of the cartridge are mere incidentals; supporting items with nothing more than the duty of launching the bullet on its way. In truth, exactly the same may be said of the gun itself. The only thing the gun contributes is to provide a platform and mechanism which will launch the bullet along a given path. Both cartridge and gun are simply developments of man's first projectile-launching device, his arm, and serve only to increase the accuracy, power, and velocity of the missile. Obviously, if we could physically hurl a projectile as rapidly and as accurately as the gun/cartridge combination can launch it, we wouldn't have any need for today's sophisticated arms and ammunition.

From the beginning of firearms history, lead has been the predominate material of bullets. Long before that, slingers had found that the unusually high density of lead made it the best choice for thrown missles. A lead pellet, being far smaller for its equal weight in stone or iron, could be thrown faster and flatter, and concentrated its impact in a much smaller area. Consequently, round lead balls were preferred for the earliest firearms, though iron, stone, and even wood or other materials were used in varying degrees. Since the earliest arms were hand cannon with large-diameter, smooth-bored barrels and used very inefficient powder, their effect was far more psychological than physical. Common soldiers attributed a great deal more effectiveness to early cannon and match locks than they really possessed, and many a charge was put to flight by the mere flash, smoke, and sound of enemy firearms, even though not a single ball might take effect. Bullets weren't really very important in the beginning.

However, as firearms developed in the fourteenth and fifteenth centuries, lead became the standard material for small arms projectiles. Artillery, on the other hand, standardized on iron, though handhewn stone balls continued to see service, sometimes sewn into leather coverings.

So long as gun barrels remained smooth-bore, the spherical lead ball remained the standard projectile. There was hardly any need for any further bullet development inasmuch as the barrels in use could not stabilize it nor could they promise any predictable degree of accuracy.

Though rifling had been known for many years in several forms, it did not become widely used until the middle-late eighteenth century. The round lead ball still remained the most practical form of bullet, improved by the addition of a patch of cloth or leather which served not only to simplify ramming the tight-fitting balls down the bore, but aided in enabling the rifling to spin and stabilize the bullet. Contrary to popular belief, the patched round ball for rifles did not originate in the American Colonies. It had been known and somewhat indifferently used in Europe for a good many years before it came to prominence in the so-called "Kentucky" and "Pennsylvania" rifles of the frontiersmen. As early as the 1600's, Danish riflemen used patched balls.

The round lead ball remained supreme until the development in France of the "Minie" bullet by Capt. Claude Etienne Minie of the Chasseurs d'Orleans. Representing the ultimate development of the muzzle-loading bullet, the Minie marked the end of a long struggle to develop a projectile that could be loaded as easily in a rifled barrel as the typical undersize ball could in a smooth-bore combination. The latter remained the world's military standard simply because it permitted rapid and easy reloading in battle. And it must also be considered that infantry tactics of the day did not require any great degree of accuracy. Among dozens of attempts, only the Devilgne system had achieved any degree of success prior to the Minie. Devilgne placed a post centrally in the base of the breech plug and the round undersize ball was dropped in the barrel to come to rest on this post. It was then *hammered* with the ramrod, against the post as an anvil, upsetting

Bullets

it radially to fill the rifling grooves. The ball, being of soft lead, was easily upset to accomplish this, but the several ramrod strokes required still slowed loading considerably. The system did not achieve wide use, and fell out of favor almost immediately with the development of the more convenient Minie which was almost universally adopted within a very short time.

Captain Minie's development was deceptively simple — one of those things about which people might well have said, "Why didn't I think of that?" It utilized a cylindral-conoidal bullet containing a deep conical cavity in its base. The bullet was substantially smaller in diameter than the bore diameter of the barrel and could thus be easily rammed down upon the powder charge, even when the bore was considerably fouled from previous firing. On this subject, you might keep in mind that the black powder of a century or more ago produced a great deal more fouling than the modern products offered to today's muzzle-loading fans. Grease on the Minie bullet, usually combined with the remnants of a paper cartridge and the fouling in the bore served to hold the bullet in position against the powder. Upon firing, Captain Minie's genius came into play. Powder gases expended against the base of a conical clay or wood plug fitted into the bullet cavity. The wedging action of this plug expanded the base of the bullet to fill the rifling grooves. Consequently, the bullet was properly spun and stabilized by the rifling and the goal of rapid and convenient loading combined with rifle-barrel accuracy and power had been achieved at one fell swoop. In very short order it was discovered that the plug in the base of the bullet was not necessary and that equal functioning was obtained in its absence. Consequently, in its most common form the Minie bullet contained simply an open conical cavity in its base.

Though used extensively in other parts of the world during its short life, the Minie ball probably achieved its greatest execution in the American Civil War. There, tactics remained the same as they had been with smooth-bore muskets. Opposing formations closed in tightly packed masses to within a few yards of each other before opening fire, and the high degree of accuracy and greater penetration of the new system simply decimated formations. In that war, the *victors* sometimes lost a *third* of their troops in such deadly close-range engagements.

Oddly enough, this most effective of the muzzle-loading bullet developments had a very short life. By the time it had actually gotten into wide-spread use, the self-contained metallic cartridge was just around the corner. By the late 1860's, most nations were at least in the process of re-arming with breech loading rifles. In some instances, the

Modern Handloading

new cartridges retained the hollow-base feature of the Minie to insure that even undersize bullets would always get a good grip on the rifling.

Up through the American Civil War, sporting shooters and frontiersmen continued to use the cloth-patched round ball for all but the most meticulous target shooting. In order to achieve greater uniformity and accuracy than could be delivered by the round ball, long conical bullets were employed, substituting carefully-wrapped paper patches for cloth. Patches were applied in a number of ways, one layer or two, applied wet or dry, lubricated or not, but generally required the use of a false muzzle to prevent damage during loading. Obviously, such a system wasn't practical for use in the field nor for military purposes. However, it produced the finest accuracy to be had with muzzle-loading rifles. Because of the great amount of force required to seat tight-fitting, paper patched bullets, the nose of a soft or pure lead bullet was likely to be deformed. To avoid this, the best riflemen of the day used *compound* swaged bullets. Such bullets were made in two parts, a body or bearing area of the softest lead which could be easily engraved by the rifling, assembled to a hardened lead-alloy nose which would resist deformation during ramming. The two parts of the bullet might be cast or swaged, but were normally assembled by swaging under heavy pressure. The two parts were locked tightly together by a mortise and tenon joint.

During the same time period, some riflemen also attempted to improve upon the round ball with a type of bullet known variously as "picket" ball or "sugar-loaf." This was simply a more or less conical bullet, often with a rounded base, which did not have any parallel-sided portion to engage the rifling and to insure correct alignment in the bore. This type of bullet was used both patched and naked. Even though some very fine shooting was reported on occasion with this type of bullet, the effort and tools required to insure that it be properly aligned in the bore made it of little practical value in the field. Even when properly loaded, the picket ball had such a short bearing surface that it might not well retain its alignment during initial acceleration down the barrel, resulting in its being tipped in the rifling and thus describing a spiral rather than true parabolic trajectory.

With the coming of the metallic cartridge, thus freeing the bullet from the torture of being loaded at the muzzle and rammed the length of the barrel, various attempts were made to improve its performance and accuracy. Grooved, lubricated lead bullets then became the most common, and were divided into two categories — those with the lubricant applied to the surface of the bullet outside of the cartridge case, called "outside lubricated," and those whose lubricated portion was seated inside the case and called "inside lubricated." Naturally, the latter was much to be preferred since it avoided collection of dirt and grit by the soft sticky lubricant and also avoided having the lubricant rub off in normal handling. Where maximum accuracy was desired, especially in long-range target and hunting rifles, the paper-patch bullet continued in use. Unfortunately, the paper patch could not withstand the forces applied when the cartridges were cycled through repeating gun mechanisms. Consequently, such bullets were reserved for use in top-quality, single-shot rifles. The patch was relatively fragile and easily distorted or torn loose from the bullet. Even the act of seating the bullet in the case could damage the patch seriously.

Recognizing that the soft lead base of the bullet was easily damaged, some makers applied a "metal base" which was, in reality, what we now call a gas check. It was simply a cup stamped from thin sheet copper and swaged on the base of the bullet. It did not extend upward along the bullet sides to protect the bearing surface. Even in the larger calibers, this cup extended no more than ⅛" forward of the base.

So long as black powder remained the principal small arms ammunition propellant, uncovered lead remained the most practical bullet material. Black powder simply would not drive bullets fast enough that any other material was required. In addition, there were many who felt that use of a harder bullet metal was either impossible or impractical. One contemporary writer describes it as "mechanical cruelty" to attempt to force a steel or similar hard-jacketed bullet through a rifled barrel. And considering the relatively soft barrel steels of the day, a hard bullet would probably have greatly reduced barrel life.

The advent of practical smokeless powder, working at much higher pressures and temperatures and producing vastly increased velocities, required the concurrent development of bullets that would withstand the strain. This brought about the basic development of the jacketed bullet as we know it today. It consisted of a lead core to give weight and density, surrounded at the base and over the bearing surfaces by a thin cup of a copper/nickel alloy called then "cupro-nickel." Many other materials were tried, of course, including steel, iron, and plain soft copper.

Cupro-nickel remained the standard material for military jacketed bullet in this country through World War I. Abroad, where wartime copper shortages caused serious problems, the use of steel was more or less perfected to the point where it could be used economically in bullet jackets and would produce satisfactory performance. Cupro-nickel's particular disadvantage was that it had a tendency to oc-

Modern Handloading

REH

REH

Basic bullet shapes. *Top*, l-r: round-nose, flat-base; round-nose, heel-type; round nose, boat-tail; pointed (Spitzer) flat-base. *Bottom*, l-r: flat-nose, flat-base; truncated-cone, flat-base; wadcutter (cylindrical); semi-wadcutter.

casionally foul bores rather badly. Particles of jacket material would be torn off and adhere to the inside of the barrel, and subsequent bullets fired would leave more and more metal as the surface became rough and irregular. This might well substantially decrease the diameter of the bore and deform subsequent bullets to the point that accuracy was virtually destroyed. Removal of such fouling presented serious problems since it adhered very tightly to the surface of the bore.

To solve this problem, ammunition makers developed new bullet-jacket alloys, and eventually settled on what we call "gilding metal" today. It is a copper zinc alloy with the characteristic color of bright copper. It is soft and ductile, easily worked, self-lubricating to a degree, and is not prone to fouling if bores are reasonably smooth. The formula varies from 95-5 to 90-10. Eventually trade names were applied to variations of the basic alloy, such as Winchester-Western's "Lubaloy." Today, gilding metal remains the standard material for bullet jackets with the exception that plain copper is used in many instances for low-velocity projectiles.

Abroad, gilding metal also came into common use, but steel was explored in much more detail. Steel bullet jackets were brought to a

Bullets 85

high degree of development, and during WW II, most German small arms ammunition utilized steel jackets. Generally, such jackets were drawn from a very soft steel and then simply plated or given some other protective coating to prevent corrosion. At the same time, methods for applying a thin coating of copper or gilding metal were developed. While electro-plating was successful to a degree, best results were obtained when a billet of steel was sandwiched between sheets of the coating metal and then rolled to the thickness required for jacket-drawing. The tremendous pressure and heat involved bonded the coating metal securely to the steel, where it remained in place throughout all further manufacturing operations. Thus, protection from corrosion was achieved as well as a reduction in bore/bullet friction. Today, many European bullets use this "sandwich" jacket material and achieve performance characteristics fully equal to those of gilding metal. Also during WW II, the U.S.A. encountered copper shortage problems and developed similar methods of coating steel for military bullet jackets. Both copper-plating and sandwich-type construction were used.

Keeping in mind that copper always becomes in short supply during wartime, most nations have highly developed methods for producing steel-jacketed bullets, well-tested and ready to put to use. Many military ammunition specifications contain alternate provisions for the use of steel for this reason. It is interesting to note that virtually all small arms ammunition produced in the Soviet Union for the past many years utilizes steel-jacketed bullets.

It is only reasonable to assume that we may expect ever-increasing use of steel in bullet jackets in the future. As manufacturing methods become more refined, the much higher price of copper will make steel more attractive to manufacturers.

The first expanding jacketed bullets were simple soft-points with a great deal of lead exposed at the nose. Of rather blunt form and driven at relatively low velocities, they expanded well and penetrated deeply, even on the largest North American game. However, as sharp-pointed designs became necessary for velocity retention, and as velocities climbed, this type became less effective. A blunt soft-point bullet that performed well at 2000-2200 fps would (will) break up badly when driven at 2800-3000 fps. Lead exposure was reduced and cores were hardened to achieve better high-velocity performance, but serious problems developed in that a bullet designed to withstand impact at high velocities at short range would not expand well at, say, 300 yards where velocity had reduced substantially. As velocities and flatness of trajectory increased hunters took longer shots and the problem grew.

Initial efforts in this field consisted of balancing core hardness with jacket thickness and strength so that the nose of the bullet was yielding enough to produce reasonably good expansion at reduced velocities; yet, the body of the bullet was strong enough to resist break-up under high-velocity impact. With additional refinements, this basic design is with us yet today — probably more bullets adhere to it than any other. The refinements consist generally of tapered bullet jackets (thin at the mouth, thick over the body); longitudinal serrations or grooves (inside or outside) and notches or scallops around the jacket mouth; thin, soft metal protective caps over the otherwise-exposed lead at the nose, etc. Nearly all major manufacturers produce superb bullets embodying one or more of these features that will hold together reasonably well for those under-100-yard shots and still produce adequate expansion out to 300 yards in all but the lightest game.

Bullet expansion is promoted two ways, and encouraged by several others. Most common is the simple soft point with the jacket cut away at the point to expose a portion of soft core. Second consists of punching a hole into the point of the bullet along its center line, allowing the jacket to come right up to the mouth of the cavity, the hollow point. Generally, the lower the bullet's velocity, the greater core exposure or cavity size required to insure expansion. At very high velocities even the tiniest exposure or hole can produce violent expansion.

In an attempt to secure uniform and concentric expansion, jacket mouths are sometimes scalloped, cut part way through, cut through, or weakened in various ways as shown.

Various forms of protection have been applied to the bullet point, both to streamline it for better velocity retention, and to prevent handling, feeding, recoil, and firing damage to exposed lead. Probably the best known design is the Winchester-Western "Silvertip." This consists of a thin, soft-alloy cap inserted in the mouth of the jacket to cover the core. Another is the Remington "Bronze Point" which consists of a separate, sharp conical point seated in the core. Upon impact, it acts as a wedge to promote expansion.

Both the above methods have been employed in modified form by various foreign manufacturers with apparently equal success.

However, in the period between the two big wars, European makers attempted to achieve better control over bullet expansion by various means. One was the DWM "Strong-Jacket" which utilized a very thick and heavy section near the jacket base. In its rear third, the jacket was so thick as to reduce core diameter by over one-half. The intention, and reasonably successful, was to insure that the jacket could not roll back completely to the base and that roughly the rear half of the

Bullets

REH

Expanding bullet types, l-r: pointed soft-point; hollow soft-point; hollow-point jacketed.

bullet would always remain intact to insure penetration, even though the forward portion of the bullet might be torn off entirely. RWS took an entirely different approach with its "H-Mantel" which utilized a two-part core and a folded jacket. The rear portion of the core was of quite hard lead, the front portion soft. At a point just below the top of the hard core, the jacket was turned in upon itself and folded under pressure to form in effect a partial partition. It was intended that the rear portion of the bullet from the partition rearward would always retain essentially its original shape and insure penetration regardless of what happened to the other part. This, too, was reasonably successful. In the U.S.A., numerous designs to achieve the same ends were proposed — most popular of which was probably the Remington-Peters "inner-belted" whose jacket contained a thickened portion near the middle. This thick "belt" was accomplished by first drawing the jacket in the usual fashion, then swaging an additional ring of jacket material on the outside, so that when the bullet was brought to proper exterior dimensions, a thick ring was formed. To a degree, this accomplished the same end as the RWS H-Mantel. Many other designs were attempted, but none achieved any truly great acceptance other than the Winchester "Silvertip" which utilized a very soft and thin protective cap over the lead point projecting from the jacket. Charles Newton designed a "wire-point" bullet which used a thin but

Modern Handloading

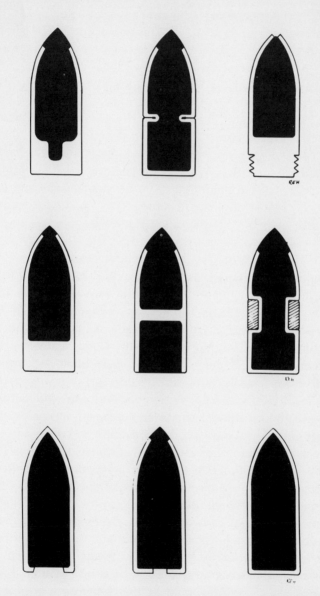

Expanding bullet types. *Top,* l-r: solid base zipedo, pointed, soft-point; H-Mantel (jacket) pointed soft-point. *Middle,* l-r: solid-base, pointed, soft-point; Partition, pointed soft-point; belted pointed soft point. *Bottom,* l-r: full-jacket, open base; soft-point, open base, full-jacket, closed both ends (Remington Power-LokT only)

strong piece of steel wire inserted in the center of the point to reinforce it against close-range expansion.

However, it was not until the 1950's that a truly multiple-range bullet design was developed to the point of commercial success and acceptance. The brainchild of John Nosler, it is called the "Nosler Partition Bullet" and in original form contained a completely solid partition located midway in the bullet jacket. This was accomplished by actually machining the jacket from solid copper rod, a most uneconomical method of manufacture. The forward portion of the jacket was relatively thin and filled with a soft lead core; the rear quite thick and with a hard core. In this form, the front portion would expand at any reasonable velocity, while the rear portion then assumed the characteristics of a full-jacketed wadcutter and remained intact under even the most severe impacts to insure deep penetration. In its present form, this bullet contains a small-diameter hole in the center of the partition. This hole is so small that it has no significant effect upon the bullet's performance. It is there simply as a matter of production simplicity and economy. Nosler Partition Bullets are currently made from lengths of copper tubing which are impact extruded to form the partition. As a practical manufacturing matter, the partition cannot be made completely solid without significantly increasing production costs. Thus, the hole is left there. I cannot personally determine any difference in performance, in this respect, between the two different types. In this writer's opinion, numerous advertising claims to the contrary not withstanding, the Nosler Partition Bullet represents the epitome of development of the modern hunting bullet. It provides all-range performance at a level of consistency and reliability that cannot be obtained with conventional design. Granted, in spite of advanced production techniques and shortcuts, the partition bullet still costs nearly twice as much to make as the conventional type. Consequently, it is not likely to ever displace what I choose to call "open-cup" designs. Shooters who feel that their type of hunting and game justifies the increased cost of Nosler bullets quite willingly pay it.

John Nosler has also developed another controlled-expansion design trademarked "Zipedo." It is a further development of numerous earlier attempts to make a solid-base bullet. "Solid" in this context meaning a jacket base of unusually thick section tapering abruptly to a standard-thickness jacket wall. In addition to the very thick base, the Zipedo is of only bore diameter over the solid portion of the jacket. This portion has rings rolled into it which raise fins to nominal groove diameter. Consequently, barrel friction and engraving pressure is substantially reduced. Then, the rear portion of the ogive and the very

Modern Handloading

front portion of the bearing surface is of conventional diameter, i.e., barrel groove diameter. Here, again, the very thick and heavy base of the jacket resists deformation under high-velocity impact and insures that a substantial portion of the bullet will remain to penetrate deeply, even though the forward portion be completely destroyed.

Other developments have occurred since WW II, intended to increase the versatility of hunting bullets. One often talked about involves bonding the lead core to the jacket material. Where in conventional bullets core and jacket are simply held together by shape and close contact generated by high swaging pressure, true "bonded-core" design involves actual *soldering* of the core to the jacket. The obvious advantage to this is that as expansion begins, the jacket clings very tightly to the core and resists core deformation. Where bonding is not present, the jacket splits or tears as expansion begins and thus no longer offers any great amount of support to the core. It is not uncommon for core and jacket to become separated under high-velocity impact, whereas bonding prevents this occurrence.

It seems possible that in the near future our most efficient game bullets might well combine core-bonding and the Nosler Partition or Zipedo jacket.

The conventional open-cup bullet design will certainly be with us a good long time yet. In spite of being nearly a century old, it is probably the most economical and practical solution to the hunting bullet problem we are likely to encounter for some time. It has the particular advantage that by careful balancing of jacket thickness and strength, core hardness, and core exposure at the nose, it can be adapted to almost any velocity range and/or shooting need. Its performance covers the range of bullets that produce explosive expansion at handgun velocities as low as 1000 fps, up to bullets that will produce deep penetration into heavy game at velocities approaching 4000 fps. In my opinion, that's pretty good.

During the past few years efficient jacketed expanding handgun bullets have been developed. Lee Jurras of Super Vel Cartridge Corp. is the main mover in this field. His developments and successful marketing literally forced the major ammunition makers to follow in his wake. Prior to the early mid-60's, the few expanding-type bullets offered in factory-loaded handgun cartridges were rather a farce. They didn't really expand under normal conditions simply because of poor design and too-thick jackets.

Jurras developed thin-jacket revolver bullets with pure lead cores and ample core exposure which *do* expand well. It is common for .38 Special SV bullets to expand to .65 caliber or more. He further produced hollow point bullets for autoloading pistols — achieving reliable feeding by truncated-cone profile and jacket running to cavity mouth — that expand equally well. I've recovered many .45 ACP SV bullets expanded to .85 or .90 caliber. Jurras' design was necessarily coupled with light bullet weight and high velocity — 20 per cent to 30 per cent less weight and 20 per cent to 40 per cent greater velocity.

Over the past couple years the major ammunition producers have more or less copied Jurras' designs and now offer comparable ammunition.

Manufacture of all modern jacketed bullets proceeds in essentially the same manner. With the exception of the two Nosler designs, bullet jackets begin usually as a strip of high-quality copper alloy made specifically for this purpose. The thickness of the strip will generally be equal to the thickness desired in the base of a closed-base design, or whatever thickness will produce the required jacket thickness at the point of a closed-point, full-jacketed type.

The first operation is "cupping," and for pistol-type, short-jacketed bullets, this one operation may completely finish the jacket. It is accomplished on high-speed automatic presses which punch out discs of the metal and then force them through a second die which forms them into simple, straight-sided, open cups of a specified length. In the case of a thin, short jacket for pistol bullets, the jacket may also be trimmed to length during this same operation by "pinch-trimming" where any unevenness or excess metal is pinched off by a shoulder on the punch passing through the die.

However, most jackets require additional work. The cups are washed and lubricated, sometimes annealed if a hard material is being used, and then run through one or more additional drawing operations which reduce wall thickness and produce the proper internal profile — which is usually a straight or slightly curved taper from mouth to base. In some instances, as many as three or four operations are required to accomplish this, especially when internal serrations or other expansion-controlling characteristics must be formed. After each operation, the cup must be washed, sometimes "pickled" in a weak acid solution, and lubricated.

Throughout all these operations, the cup is gradually reduced in diameter and in finished form, ready to accept the core, must be slightly less than finished-bullet diameter. The jacket is then trimmed to length, usually in an automatic lathe setup. However, if a scalloped mouth is called for, trimming must be done in a pinching die arrangement or variation thereof. If the bullet is to be of boat-tail configuration, the base of the jacket is given this shape in another punch and die operation before it is ready for insertion of the core.

Cores are pre-formed from lead alloy of the required hardness. They are formed in automatic presses and dies in much the same fashion that handloaders swage their own bullets and cores. Generally, the cores are brought to finished weight and of a shape and diameter that will fall readily into the open jacket of their own weight. This isn't usually the exact shape that the core will reach when the bullet is completely assembled.

Cores and jackets are then fed into an automatic bullet assembly press which first inserts the core in the jacket, then swages the two to final form and dimension in precise dies under great pressure. Depending upon the bullet, this may be accomplished in a single die, or in several progressive stages. If a reinforcing cover is called for over the point, such as the Silvertip, it must also be assembled at this time and requires additional steps.

During the assembling operation, the slightly undersize core and jacket are both "bumped-up" under great pressure to final diameter. However, in some instances, the finished bullet is forced through a tungsten-carbide die to burnish it and bring it to final diameter, ironing out any minor surface irregularities that might exist. In the event of a cannelured design, the cannelure will be rolled in place by another automatic machine, and the bullet may then be forced through a final sizing die to remove any irregularities produced by the canneluring operation.

In any event, after final assembly and sizing, bullets are washed and sometimes polished by tumbling or water-polishing to give them that pristine look you see when you take them out of the box.

Naturally, special features require additional manufacturing operations which increase the complexity and cost of the entire process. For example, use of a two-part core requires running the bullet through two assembling machines — the first to properly seat and form the rear core, the second to seat the front core and bring the entire assembly to its final shape and dimensions. The same applies to partition-type bullets. If the core is to be bonded to the jacket, then at some point during assembly this bonding must take place. It must be accomplished after the core and jacket are brought into firm contact under heavy pressure, but before the final shape is produced.

In recent years, another method of bullet manufacture has been developed. Instead of depending upon precisely controlled sheet or strip jacket material, this method utilizes short sections of rod. Raw jacket material of equal quality is cheaper in rod form than in sheets or strips and is less subject to handling and shipment damage.

Typical cast lead lubricated bullets. Right bullet is filled with gas check, left has been improperly resized.

Initially, rod stock of the proper composition and characteristics is cut very accurately into short lengths of the proper size to produce the jacket desired with a minimum of waste. These "slugs" are fed into an automatic press where they are held confined in a die while a shaped punch is driven down into one end of the plug. The punch penetrates the slug and forces the metal under pressure to extrude upward around the punch. The initial operation forms a cup, much like that formed from strips as already described. Depending upon the bullet type and shape and jacket material, the initial extrusion operation may prepare the cup for the rest of the procedures already described, or one or more additional extrusions may be required. However, once the basic cup is properly formed by extrusion from rod, the balance of the jacket-forming and bullet assembly operations remain essentially the same as described above.

The foregoing applies, of course, to sporting bullets and to military, lead-core, *ball* bullets. Special-purpose military types such as incendiary, armor-piercing, and tracer involve essentially the same jacket-forming operation, but become tremendously complicated in core manufacture and final assembly.

After all is said and done, anyone with a punch press and a few dies can make bullets. However, accuracy is more dependent upon bullet quality and consistency than any other single factor. Consequently, it is essential that every refinement possible be made to both material and methods to insure that bullets are *uniformly* of the best quality.

Modern Handloading

In the past few years, Remington-Peters has developed a new bullet-manufacturing process, the details of which are not generally discussed, but which apparently produce a very high degree of uniformity with at least some savings in cost. Apparently, this process consists of first forming the core to final shape and dimension, then building up the jacket by an electroplating process to proper thickness and profile. Following this, the bullet is polished, sized, and altered at the point to secure the desired expansion characteristics. Many tests have proven these "Power-Lokt" bullets to be superbly accurate. At the time of this writing, they are available only in .22 caliber and 6mm caliber, in either full-jacket type for target work and hollow-point for hunting.

Don't get the impression that lead bullets have disappeared. Nearly all rimfire ammunition and most revolver cartridges are loaded with plain lubricated lead bullets whose design and manufacture haven't changed in nearly a century. Our most accurate rimfire ammunition and the world's best match revolver ammunition uses unjacketed lead bullets.

With one or two exceptions such as the .357 Magnum factory lead bullet loads produce well under 1500 fps, most under 1000 fps. A very soft alloy, hardened only slightly by a minute portion of antimony, works well in that range and is easy and cheap to work. Lubricant is vital to accuracy and each company has its own closely-guarded "secret formulas."

Such bullets are made from lead wire which is cut into short lengths which are then fed into automatic presses where they are fully shaped in dies under heavy pressure. Cannelures (lubricant grooves) are then rolled in on another machine, and the finished bullets are lubricated.

When driven fast enough, lead bullets will expand to some degree if of soft alloy. Until just recently, no factory loaded lead bullets designed specifically to produce expansion other than .22 RF hollow-points. Now, though, Winchester-Western offers a lead 158-grain, semi-wadcutter containing a deep hollow point to promote expansion. It is intended primarily for police work and does expand quite well at relatively close ranges when fired from all but the shortest-barreled revolvers. It is offered in .38 Special caliber only.

Gas check lead bullets are little used by the factories. Their manufacture is as already described except that the gas check is joined to the body in the dies. Gas checks are primarily the tool of the handloader and are covered in the chapter on cast bullets.

Factories don't cast bullets today. Most of them haven't done so since well before the turn of the century when efficient swaging machines were developed.

CHAPTER 8

Chamber Pressure

PRESSURE IS good. Without it guns and ammunition wouldn't work at all. Pressure is what drives the bullet out of the cartridge case, down through the barrel, and gives it the velocity and momentum necessary for it to reach the target. It must be made clear that chamber pressure—gas pressure—is an essential part of the gun/ammunition system. All too often people think in terms of "pressure is bad." Well, only *too much* pressure is bad.

Consider the firearm a single-cylinder, internal-combustion engine. The chamber and barrel form the cylinder, the bullet forms the piston, the primer is the spark plug, the propellant powder is the fuel. Until the instant of firing, all these components are static and invalid; nothing moves, no force is exerted, no work is done. At the instant of firing, the flash from the primer (spark plug) ignites the powder (fuel) which by combustion is changed from a solid to a gas which expands in all directions. The bullet (piston) being the lightest of the components and being relatively unrestrained (except by bore and rifling friction) is forced out of the cartridge case (cylinder) and through the barrel by the expanding gases.

From that description, it should become evident that pressure is an essential part of the system.

How *much* pressure is the question. At what point can we say that *excess* pressure exists? Using modern cartridge cases, each general type of firearm has its own general pressure limitations. This isn't to say that the gun is unsafe with 1 per cent or 2 per cent more pressure than the arbitrary cut-off point, but that the safety factor begins to drop rapidly past what we call "working limits." We're speaking, of course, of guns chambered for *centerfire* ammunition only. Modern rifles with front-locking bolts—be they lever, pump, semi-auto, or bolt persuation—are generally designed and manufactured to handle an extensive diet of pressures in the 50,000 to 55,000 psi range. Modern single-shot rifles such as the Ruger and Colt/Sharps have the same working range. Self-loading handguns of the locked-breech variety are generally intended for chamber pressures in the 30,000 to 35,000 psi range. This excludes the notable example of the Colt Government Model .45 ACP pistol which is generally restricted to pressures under 20,000 psi, but follows the aforementioned limits in the smaller 9mm and .38 Super calibers. It also excludes certain older commercial and military locked-breech pistols and all of the unlocked-breech designs except the Astra M400/600 series which is adapted to the standard limits mentioned. Confusing, isn't it? Modern solid-frame revolvers in the magnum class are generally considered capable of handling pressures in the 40,000 psi range, while those not in the magnum class are generally indicated by the manufacturers to be limited to around 20,000 psi. Hinged-frame revolvers are generally restricted to even lower pressures. Factory-loaded ammunition for them never exceeds 15,000 psi, and generally runs substantially lower than that. Even the massive Webley design in .455 caliber was not intended to accommodate pressures over 15,000 psi—regardless of the fact that many handloaders have used much heavier loads in it without damage. The small unlocked, blow-back design pocket automatic pistols are intended for essentially the same pressure ranges as the break-open revolvers.

In short, chamber pressure is essential and good, but only up to the limits imposed by gun design and materials.

An almost unending list of variables and unknowns cause pressure variations. A partial list of the most significant factors would include, but not be limited to the following: Bore dimensions and twist; throat or leade (origin of rifling) profile; case/chamber length relationship; grip of case upon bullet; diameter, weight, bearing surface length, hardness, etc.; case and chamber volume; primer variations; and even variations in firing-pin impact energy.

Consequently, when any of the above variations are introduced into a load already established as producing acceptable levels of chamber pressure, pressures may change either upward or downward. If downward, no harm is done other than a slight reduction in velocity and the attending drop-off in bullet performance. However, any *drastic* increase in chamber pressure can be dangerous if it is of sufficient degree or if it is sufficient to cause an existing mechanical weakness in case or gun to fail. Many handloaders, likely the majority, frequently interchange cases, bullets, primers, etc., and guns, with little or no regard for the effect this action might have upon the pressures being produced. Generally, they recognize that some change might take place, but they feel that it will not be sufficient in magnitude to wipe out the substantial safety factor they know exists. Unfortunately, the amount of pressure change that can take place as the result of component substitution can be sufficient to go well beyond reasonable safe limits. A classic example of this is given in the NRA *Illustrated Reloading Handbook* which cites in detail a series of tests undertaken with the .30/06 caliber cartridge. A standard load developing acceptable pressures and velocity was assembled with 150-grain bullets of a particular make and type. This load was then varied only in the bullet—an additional nine different bullets being substituted while all other factors and components remained the same. All ten loads were then tested for pressure and velocity under identical conditions and in the same combination of test barrel and gun, and instrumentation. The pressure tests results showed a range of 10,000 psi between the minimum and maximum average pressures produced by the different bullets. The lowest pressures of the lot were produced by the U.S. Military M-2 ball bullet. If one were to whip up a 50,000 psi load with that bullet, then switch to the bullet that produced maximum pressure, over 60,000 psi would be produced.

The point that must be made clear here is that had the original load been developed with the bullet that produced the least pressure, and been built up to the maximum working pressures at the start, then random substitution of the other bullets would have produced pressures ranging from 65,000 to 70,000 psi. Such pressures are in no way acceptable in either factory or handloaded ammunition. It must also be considered that the industry standard proof load pressure in this same caliber is only 72,000 psi—and the proof load is intended only as a one-shot test to determine if any mechanical defects exist in the gun. Incidentally, handloaders whose custom is to work up what they call "maximum" loads based upon external evidence of excessive pressures can easily cause actual pressures to go far beyond proof pressures by random interchange of bullets. These pressures can easily wreck the gun and maim or kill the shooter.

Modern Handloading

Random substitution of other components will have a similar but lesser effect and will certainly produce pressure variations, and, these effects could be cumlative. Let's say that you have made a bullet substitution representing the extreme range shown in the NRA tests—then that you have compounded this by substituting case and primer whose differences also tend to increase pressures; then that you fire the substituted load in a rifle with a bore somewhat tighter than before. The three ammunition and one rifle variables, all tending to increase pressure, will combine to produce a very substantial increase which can in no way be predicted. It might well range upwards to 20,000 psi. The effects of that added to a load already in the 50,000 psi range should be obvious. On the other hand, it often occurs that the combination of variables and/or substitutions are self-cancelling—therefore producing no increase in chamber pressure whatever. Again, there is no way in which this can be predicted. It is for these very cogent reasons that the procedures outlined in the chapter devoted to load development should always be followed when substitution of components becomes necessary.

The question then arises as to the validity of chamber pressure values published in various handloading manual data sections. Confusion is compounded when those manuals do not identify the particular make and type of bullet, primer, case, or rifle used in the tests. Generally speaking, the compilers and publishers of such manuals deliberately err on the conservative side. In other words, what is listed as a maximum load with an unidentified group of case, primer and bullet has been deliberately cut back so that random substitution and exchange of components will not generally produce dangerous pressures. Loading data published by the National Rifle Association, whether in the *American Rifleman* magazine or in loading manuals, always positively identify all components and the dimensions of the test barrel used, as well as atmospheric and temperature conditions. A few other publishers of loading data have begun to do this, but many have not. With detailed load data, such as that of the NRA, at hand, one may make intelligent substitutions without fear of creating an actually dangerous condition. This is not necessarily true of the more or less anonymous data contained in some other publications.

The obvious solution to this apparently perplexing problem is to simply reduce the powder charge by an appropriate amount whenever any significant component substitutions are made. Some authorities recommend a 10 per cent reduction in powder charge for some calibers, others may be handled quite nicely with 5 per cent. This reasoning is simple. For example, take a .300 Winchester Magnum load using nearly 80 grains of a very slow-burning powder, as opposed to a 9mm Parabellum load using slightly less than 5 grains of very fast burning Bullseye. Because of the burning characteristics of the powders and the pressure ranges within which the two loads are intended to operate, a 5 per cent reduction is entirely adequate in the .300 Winchester Magnum, while a 10 per cent reduction is more appropriate in the 9mm load.

I would apply the above charge reductions in the case of bullet substitution or when switching to a second gun which might contain an undersize bore or other critical variables. If, on the other hand, one was simply substituting between the two different makes of primers originally intended for the same type of use, a smaller charge reduction would be safe enough.

Let's retrogress a bit and consider individually the pressure-effecting factors already mentioned.

Bullet

An increase in bullet diameter, whether actual and brought about by a physically larger bullet, or effective and brought about by a physically smaller bore will increase pressures. The amount of increase cannot be predicted accurately. Obviously, a small increase in diameter will produce less pressure change than a large increase.

A classic example of how serious an effect can be produced by an oversize bullet was recently delivered to my office in the form of an FN M1950 .30/06 Military rifle in which had been inadvertently fired a standard factory-loaded 8 x 57mm cartridge. The nominal diameter of the bullet that did the damage would have been .322", while the standard .30 caliber diameter is .308". Consequently, we may assume (without being able to recover and measure the 8mm bullet) that the offending bullet was .014" oversize. Generally speaking, a standard .30/06 case loaded with a bullet this large could not be chambered without considerable difficulty—which would have alerted the shooter that some trouble was in the offing. However, in this instance, the shorter 8mm cartridge was easily chambered and fired. It should be noted also that the 8mm cartridge contained a powder charge substantially less than standard for the .30/06. In any event, upon firing, the guide lip on the bolt face was broken off, the case head was melted away and brass brazed to the bolt face, the extractor was bent, the stock was shattered, the magazine floor plate was blown out, and brass and powder particles were embedded in the shooter's forehead and cheek. The gun is, for all practical purposes, destroyed. It could

Modern Handloading

be salvaged for parts, or could be rebuilt at a cost about equal to its current good-condition market value. Granted, this is an extreme example — representing an error that would not normally be made in assembling handloads. Nevertheless, it could occur inadvertently if the handloader did not pay particular attention to the diameter of bullets being used. And it does show clearly what *can* happen.

In addition to bullet diameter, the use of a harder and/or thicker jacket, a harder core, greater weight, or a longer bearing surface will also raise pressures by an appreciable amount. Likewise, reductions in these values will bring about a lowering of chamber pressure.

Probably the most likely areas in which really excessive bullet variations will be encountered are in the occasional attempt to use standard .308" .30 caliber bullets in 7.5mm and 7.35mm barrels because of limited availability of bullets of the correct diameter. For example, the 7.35mm Italian service rifle requires bullets of .298" diameter, and when such bullets are not readily available, some poorly informed handloaders have been known to use .308" bullets. The same situation occurs to a lesser degree with pre-1911 Swiss service rifles which require bullets of .304" diameter. Other possibilities are endless, but less likely to occur inadvertently.

Case Grip on Bullet

Probably the classic example of this occurred many years ago, back in the 1920's, when tin-plated bullets were loaded in match .30/06 ammunition for the U.S. Government. Initially, this ammunition was

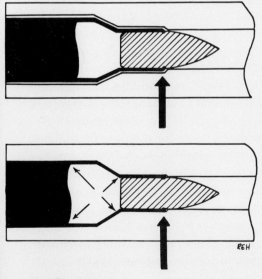

Top: When case mouth is jammed tightly into shoulder at chamber mouth, case is jammed into bullet as shown and pressures rise. *Below:* When there is insufficient clearance between case neck and chamber, bullet is held tightly and pressure rises.

extremely accurate and produced the intended pressures and velocities. Over a period of years, though, it was found that the tin more or less "soldered" itself to the brass of the case neck. This increased bullet pull from the standard value to hundreds of pounds and to the point where it could not even be adequately measured on existing testing equipment. These cartridges, because the bullet was not relatively free to begin moving out of the case when the powder began to burn, generated tremendous pressures and wrecked several guns. The same effect, to a lesser degree, will be produced when overly-thick case neck walls are combined with an unusually small chamber neck diameter or with oversize bullets. This combination will cause the chamber to hold the case neck snugly, even tightly, against the bullet, preventing the case expansion that normally takes place as the powder begins to burn and frees the bullet to be started down the bore.

For these reasons, the tight chamber necks once in vogue among custom rifle makers — and even now popular among bench-rest shooters — require the use of cases whose necks have been carefully reamed or tuned to the thickness that allows the bullet to be started easily if high pressures are to be avoided.

Case Length

Most chambers terminate with a slight shoulder just ahead of the position of the case mouth when a cartridge is fully chambered. If the case is too long, final chambering jams its mouth into this shoulder and in effect crimps it tightly on the bullet. This impedes the bullet's initial movement, causing buildup of excessive pressure in just the same manner as described above.

Case Volume

A thick-walled case has less volume and therefore less space for the powder charge than a case with thin walls. Though during firing it is the chamber that controls the external profile and dimensions of the case, a thicker case will still provide less room for the powder to burn and work than one with thin walls. Extensive tests conducted by various individuals have shown clearly that the volumes of different makes and lots of cases in the same caliber can easily run to as much as 5 per cent or even occasionally more. Reduce the space in which a specific powder charge has to work by that much and you have increased the loading density and the chamber pressure that will be produced. Generally speaking, the amount of increase produced by

Modern Handloading

the volume variations encountered in commercial cases will not produce excessive pressures, even with full-charge loads, unless other factors interfere. However, when one is working with cases reformed from some other caliber, much greater volume variations will be encountered. For example, it is fairly common practice to reform 7.62mm NATO military cases to .22/250 caliber. A case so formed may well have a volume as much as 10 per cent less than its commercially-produced counterpart. Consequently, a full-charge load developed in the commercial case can produce as much as 10,000 psi rise in chamber pressure when fired in the altered military case. All this is brought about simply by the fact that the military case is designed with walls and head substantially thicker than the .22/250; and that the swaging down of this case to smaller dimensions increases wall thickness even further. The result is a case that may appear identical externally, but has considerably less volume. This factor must always be considered when using military cases or any cases that have been extensively reformed or reworked.

Primer

A given charge of powder will produce greater pressure when rapidly ignited throughout by a large, hot primer flash than when sparked by a relatively weak flame. A full-charge load developed with one particular make and model of primer will increase in pressure if a "hotter" primer is substituted. Consequently, if one were to substitute rifle primers — with more priming compound content — for pistol primers in a heavy .357 Magnum load, chamber pressure would be increased. Conversely, substituting a less-hot or "cooler" primer will produce a reduction in pressure. Again, generally speaking, the effect of primer substitution upon pressure is relatively slight, but measurable. This is covered in greater detail in the chapter devoted to primers.

Firing-Pin Energy

A weak firing-pin blow produces slow and erratic (relatively) ignition of the priming compound and this in turn results in less efficient and less complete ignition of the powder charge. In this manner poor primer ignition brought about by a weak firing-pin blow causes poor powder charge ignition and lower chamber pressure. At the other end of the scale, a very heavy firing-pin blow has just the opposite effect. The difference is relatively small, but it does exist, and is particularly encountered in revolvers where alterations have been to the mainsprings and firing-pins; and also in self-loading pistols with very light weight inertia-type firing-pins.

Chamber Volume and Dimensions

As mentioned above, the actual volume within which the powder must burn is dependent upon case volume; and case volume is dependent upon the external dimensions of the case at the instant of firing and those values are controlled by the chamber. A tight chamber has the effect of reducing case volume and consequently increasing loading density and chamber pressure. All other factors being equal, a tight chamber will produce greater pressure than a sloppy, loose one.

Throat or Leade

In modern practice, the bullet is permitted a modest amount of free travel within which it picks up velocity and momentum before engaging the rifling. This initial unrestricted movement allows powder gases to begin expanding early and thus eases somewhat the pressure peak.

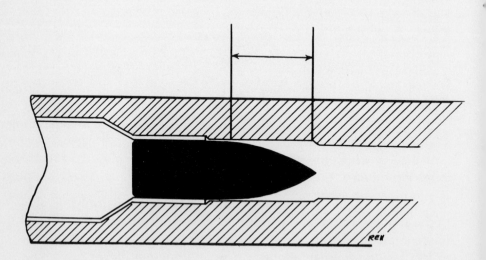

Distance indicated by arrow is amount of free travel or lead before bullet engages rifling. If too short, pressure goes up — if excessively long, barrel is said to be freebored.

This amount of free travel is generally established by the rifle manufacturer based upon the longest and bluntest bullet which will be loaded in a given caliber. Consequently, the amount of free travel does not remain constant for other loads and other bullets. Shorter or more sharply pointed bullets will travel farther before engaging the rifling. All other factors being equal, the greater the distance the bullet travels before engaging the rifling, the lower pressures will be. Many wildcatters and some foreign gun manufacturers deliberately "free-bore" barrels for a substantial distance for the sole purpose of holding peak pressures down and allowing maximum velocity to be obtained. The term "free-bore" is usually applied whenever excessive bullet travel is involved before engaging the rifling. However, inasmuch as there is no *standard* distance the bullet of any given caliber must travel, this is a rather ambiguous term.

In addition to the leade profile furnished by the manufacturer, one must consider that the throat is advanced by extensive shooting with modern high-intensity cartridges. The origin of the rifling is simply eroded away to the point where it substantially increases the distance the bullets must travel before engagement. Consequently, a given load will produce a particular pressure when fired in a new barrel, but after, say, one or two thousand rounds have been fired in that same barrel, less pressure will be produced. It is the practice of some meticulous shooters to compensate for erosion by simply seating the bullet farther out of the case as erosion progresses in order to maintain a standard amount of free travel. However, to retain exactly the same performance, the powder charge must also be increased to maintain a standard loading density.

All of which is fine to know, of course, but how does the casual handloader learn whether a particular load and set of conditions is producing safe or excessive pressures? Short of a pressure-measuring laboratory setup — which is not generally available — there is no method by which the individual can determine the *amount* of pressure being produced. There are, however, visible symptoms of pressure that is in excess of what the particular gun/load/component combination can handle.

Primer

It has often been stated that the condition of the fired primer can be "read" to determine pressure. This is true only when an exact comparison can be made from the same make, model, and lot of primers. The simple comparison of a fired factory-load primer with one of your handloads is useless and may even be completely misleading. The only practical way in which a primer comparison would indicate pressure levels is when two loads are fired with identical components — except for powder charge — in the same gun at the same time. Then, there enters the necessity for knowing positively that at least one of those loads produces safe pressures. Personally, I prefer to do this by taking a box of factory-loaded ammunition of the caliber desired, then pulling the bullets and dumping the powder from ten of them and reloading with my own bullets and powder charge combination. The two ten-round samples, thus produced are then fired and the primers are examined under low magnification.

The greater the pressure developed, the more will the fired primer be impressed with the tool marks and surface irregularities of the bolt face; the more the outer edge of the primer cup will be flattened to a relatively sharp corner; and the more the center of the primer cup will extrude into the firing-pin hole in the bolt face, even to the extent of driving the firing-pin back and blowing out and allowing gas to escape back into the interior of the bolt. In extreme instances, the edges of the primer cup will flow outward, filling the radius around the mouth of the primer pocket.

So, if my handload fired primers show more of the above pressure evidence than the factory loads, I know pressure is greater. I don't know how much; just that it's greater.

In addition, there is the symptom of primer leakage to be considered. When any degree of gas leakage exists around or through the primer, excessive pressures are being generated by that load. The only exception to this is a damaged or defective primer or primer pocket which of itself allows gas to escape. Thus, at least five and normally ten rounds should be fired for any given test.

Even so, distorted, extruded, and flattened primers are not *always* a positive indication of excessive pressures. Granules of powder, especially the spherical ball type, may find their way through the flash hole into the primer pocket under the primer. Under some conditions, they can increase pressure inside the primer — but not in the chamber — sufficiently to cause primer distortion that indicates excess pressure. This can be avoided when loading ammunition specifically for primer-oriented pressure tests by a thin disc of paper or foil in the bottom of the primer pocket. Also, a condition of excess headspace will usually produce badly flattened primers even though pressures are quite normal, or even moderate. This occurs, as described elsewhere in this book, when primer ignition drives the case forward, allowing the

Modern Handloading

primer to back out of the case head, followed by the case being driven rearward as the powder charge burns, thus deforming the primer. A classic example of this is shown in the accompanying photograph wherein 9mm Parabellum cartridges producing standard pressures of under 33,000 psi were fired in a pistol chambered for the 9mm Bergmann-Bayard cartridge. This combination, though often recommended by people who should know better, creates a condition of over 1/10" excess headspace. The long firing pin drives the cartridge forward and ignites the primer and allows the primer to back almost completely out of the case. Then, the case is driven back against the breech face, smashing the primer flat and sometimes even producing primer leaks and serious case head distortion. Note in the photographs that in some instances fragments of the case have actually been blown clear. The fellow who claims he can identify excessive pressures by simple visual examination of case and primer would conclude that the ceses shown had been fired at pressures well above 75,000 psi, where in reality roughly 32,500 psi were generated.

Yet, it is possible to learn *something* from examination of fired primers. But this can only be done when a valid basis for comparison of the same primer and case can be established as outlined above with factory loads and primed cases.

Case

Firing any cartridge causes the case to be deformed by gas pressure. When that deformation passes certain limits, then pressures are becoming excessive *for that particular batch of cases in that particular gun*. All metals are plastic and begin to flow under certain amounts of pressure. Within the gun/ammunition combination, the brass cartridge case is by far the weakest and softest and is therefore the first to fail when pressures do become excessive. Case failure is not sudden and abrupt at a particular pressure level. As pressure increases, the brass slowly begins to flow and as pressure goes on up, flow becomes more evident, finally reaching the point where the least-supported portion of the case ruptures and gas begins to escape; at the same time the gas melts and erodes away the edges of the failure, increasing the flow of gas. So, if we can spot the point at which the brass first begins to flow, we know that we are entering the danger zone and that further increases in pressure are certain to wreck the case and probably the gun.

This point where the case begins to flow cannot be identified with a particular pressure level. The pressure level will depend to a large degree upon gun design, case design, hardness of that particular case or lot of cases, and the manner in which the bolt and chamber support the head of the case. Soft and/or poorly-supported brass will begin to flow and eventually rupture at much lower pressures than will a hard case tightly enclosed and supported.

Probably the most commonly-used evidence of excessive pressure is "head expansion." This refers to an increase in diameter in the *solid* portion of the case head directly in front of the rim (in a rimmed case) or extractor groove (rimless case) or of the forward portion of the belt (belted case). This measurement must be taken behind the so-called "pressure ridge" formed where the case wall ceases contacting the chamber wall. The accompanying drawings show this quite clearly. The pressure ridge itself is an indicator of pressure, and the farther rearward on the case it appears, the greater the chamber pressure. However, it is extremely difficult to measure accurately and to interpret correctly, so is a less valid indicator than head expansion.

Generally speaking, head expansion of .001" or more indicates that *for that particular case* the point of excess chamber pressure has been reached. To be of any value, this measurement must be taken with a top-quality micrometer reading in tenths of thousandths and must be taken at several points around the case head perimeter. Often the rim of the case (even in rimless designs) and the encroaching pressure ridge will prevent an accurate reading being taken with standard round-anvil micrometers. Under those circumstances, rectangular or wedge-shaped anvils must be used to permit reaching down in between the pressure ridge and the rim to measure the head at the proper point.

Thus, to determine whether excessive pressures are being produced, a new, unfired case should first be measured at the head, and then loaded and fired, and afterward measured again at the head. If more than .001" increase in diameter occurs, you are getting into the excessive pressure zone; and danger mounts as diameter increases beyond that limit. The accompanying drawings show where each type of case must be measured. Fire at least five rounds and measure all to reduce error as much as possible.

What makes this possible? It is really quite simple. When chamber pressure is sufficiently high, the solid case head is compressed longitudinally, causing the brass to flow radially and thus increase its diameter, just as your wife flattens a ball of dough for a crust with the heel of her hand. If the pressure becomes great enough, the case head will be shortened so much that the thin walls will advance beyond the mouth of the chamber and be blown out.

Modern Handloading

Other evidence of excessive pressures can also be found on the case head. As pressure builds up, brass will be forced to flow into the surface irregularities and tool marks on the bolt face and into any depressions such as extractor cuts, ejector slot, ejector hole (in the case of plunger-type ejectors), feed ramps, etc. When this happens, the effect can be both seen and felt. It can be felt in the increased force required to lift the bolt handle because of the greater friction brought about by the intimate contact between case head and bolt face. Bolt lift becomes particularly hard when brass is actually extruded into the various cuts already mentioned. Once the case has been extracted, bright burnished marks will be seen on the cartridge head and there may be slivers of brass sheared off where it extruded into bolt face recesses. Some burnishing of the case head is always present because normally the bolt rotates during extraction and unlocking while the case remains fixed in the chamber. However, an excessive amount of burnishing and any extrusion whatever into the bolt face recesses indicate that you have at least started into the area of excessive pressures. Slight smearing or flattening of the headstamp markings may also be encountered, even before the symptoms just described. This also indicates encroachment into the excessive pressure zone.

It should be evident that at least to a degree, all of these pressure signs will depend upon the hardness of the brass. A soft case will expand radially and will be pressed into the bolt face irregularities by a lower pressure than a hard case. Regardless of the actual pressures involved, when this type and amount of case head distortion develops, pressures are entering the excessive zone *for that particular case*. All other factors being equal, hard cases will safely withstand greater pressure than soft cases.

Expansion of primer pockets also indicates excessive pressure. As indicated elsewhere in this volume, even normal loads will eventually expand primer pockets. However, excessive pressures will often cause the primer pocket to open up enough in a single firing that a new primer will not be held securely. This can be felt, sometimes even seen, when repriming the case. In extreme instances, the pocket will be expanded so much that the fired primer will either be loose, or will drop out of the case during extraction and ejection. Such pocket expansion may or may not be accompanied by gas leakage around the primer — which is made evident by sooty streaks on the case head or, in extreme instances, molten brass sprayed on the bolt face and a partially melted primer cup. Any time the primer pockets are expanded so that a fresh primer enters with very little force, the pressures involved in the previous firing were in the excessive zone.

There is one other area in which the exterior of the case may show evidence of excessive pressure. Some older rifles and most self-loading pistols have portions of the chamber cut away to provide room for extractors, ejectors, or feed ramps. Generally, such cuts leave portions of the case *ahead of* the solid web poorly supported. Consequently, before pressures become sufficiently heavy to cause case head distortion, the case walls will bulge outward into these poorly supported areas. The bulges will usually clearly outline the cuts which they match and they become particularly prominent in the various guns chambered for the .45 ACP pistol cartridge. Any pronounced bulges ahead of the solid case head indicate the practical pressure limits for that particular combination of case and gun have been exceeded. The one exception to this rule is a very slight and shallow bulge of the case into the feed ramp portion of self-loading pistols. A slight bulge there is normal, as is the pressure ridge mentioned earlier, and does not indicate excessive pressure until it becomes pronounced and clearly outlined. However, even though the case does not fail initially at the bulge, repeated full-length resizing will work-harden and otherwise weaken the brass at that point so that eventually it may fail even with the standard-pressure load. Thus, rupture of a case at that point after many reloadings does not necessarily indicate excessive pressures.

Generally speaking, cases fired in revolvers are not subjected to pressures sufficiently high to produce the symptoms already described. With modern guns and brass, the only clear indicator of excessive pressures is a slight difficulty in beginning extraction, that is, a fairly hefty tap is required on the extractor rod to start the cases from the chambers. Primers may also extrude into the firing-pin hole in the recoil shield and thus indicate excessive pressures; although, this is more often an indication of an oversize firing-pin hole than of excessive pressures. Old, balloon-head, thin-walled revolver cases will bulge into the extractor cuts in the inner side of the chamber when excessive pressures are generated.

Aside from evidence visible on the cartridge case, some excess pressure indicators may be found in the gun itself. Certainly, any distortion of the chamber or barrel or locking surfaces indicates excess pressure, whether or not any such evidence is visible on the case or primer. Likewise any abnormal force required to unlock the breech and extract the fired case is indicative of excessive pressures. Again, soft brass abnormally accentuates such symptoms. Also, any roughness in the chamber might also cause hard bolt lift and give an erroneous impression of excess pressures, but this can be verified simply by

Modern Handloading

looking at the case which will be impressed with any pits or tool marks of sufficient magnitude to have caused the difficult bolt lift.

It should be pointed out here that many other factors control the pressures produced by a given powder charge. They are concerned primarily with the powder itself, and not with the other components or the gun. All of these are discussed in detail in the chapter devoted to powders.

CHAPTER 9

Headspace and Handloading

"MY RIFLE'S got headspace. Reckon I ought to get it fixed." We hear remarks of that nature quite often. Unfortunately, more often than not the speaker doesn't really understand the term "headspace." Of course, his rifle's got headspace — every metallic-cartridge arm has headspace. Put into its most elemental form, the word "headspace" refers simply to that space into which the cartridge fits in the barrel and breech of any firearm.

Perhaps some of the confusion arises from the use of the word "head" in this descriptive term. There is legitimate reason for the word having developed as "headspace", for at the time that the basic methods of gauging and measuring both cartridges and guns were being developed, only the classic *rimmed* type of cartridge case was involved. Consequently, the term headspace evolved simply because only the head (rim) of the cartridge case and the area into which it seats into the gun were involved. Since then, though, four additional types of cartridge cases have been developed, all of which require a different type of headspace measurement. And, as a matter of fact, among the world's current cartridge list, the other types far, far outnumber the venerable rimmed case for which the term was coined. If we were to use instead the term "cartridge space," or "case space," perhaps so much confusion would not have arisen.

Before delving further in this matter, let's look into just what headspace means to the handloader. First we must realize that the handloader has complete control over the headspace condition. This is due to the fact that headspace is not purely a condition of the gun, but rather a condition of the relationship between the cartridge case and the gun in which it is chambered. Thus, when the handloader varies dimensions of cases by resizing, trimming, etc., he is actually altering the headspace condition of the gun/cartridge combination. Because of this, a careless or uninformed handloader may *create* dangerous headspace condition even though the rifle meets factory or arsenal headspace specifications perfectly. It is possible to induce case failure through an *excessive* headspace condition by too much resizing of the case; it is equally possible to cause a failure to chamber by creating a less-than-minimum headspace condition by *not* resizing fired cases sufficiently or by using improperly adjusted or badly worn dies. Many's the time I have had presented to me fired cases showing classic conditions of excessive headspace and been forced to listen to the owner's tirade against the gun manufacturer, and then found the fault to be entirely with the handloading practices of that individual, while the gun was not at all at fault.

So, it may be seen that headspace is actually one of the most important factors to be considered by the handloader. Failure to give it proper consideration during handloading can in extreme instances result in a completely wrecked rifle, not to mention a probably injured shooter; or in ammunition that cannot be chambered and fired. All of this in a perfectly-made gun.

Simply defined, headspace is the distance between that portion of the chamber which supports the cartridge case against the initial firing-pin blow (that portion which arrests forward movement of the case within the chamber) and the face of the breech bolt or breech block.

However, this cannot be a single, precise measurement. It is economically impractical for either ammunition or guns to be made to exactly the same dimension continuously. Tolerances (allowances above and below a stated dimension) must be allowed in both cartridge and gun manufacture. Practical realities force manufacturers to establish both a minimum and a maximum chamber size and to accept any chamber which falls between those two sets of dimensions; likewise for cartridges of any given caliber. This means that the "maximum" cartridge must be no larger than the "minimum" chamber for the same caliber. There can be no overlap in this area, or a shooter might very well pick up a box of *maximum* cartridges which would *not* chamber in a *minimum* rifle. Statistically, the odds against such a happening are very slim, but even the remotest probability of it happening must be avoided, for the results could be fatal for a soldier or police officer, even for a hunter.

Likewise, to digress just a moment, resizing dies inject another dimensional factor, and must be made so that the maximum die will reduce a fired cartridge case (regardless of the chamber in which it was fired) to such dimensions that it will chamber in a minimum rifle.

So, we find headspace being expressed as a minimum and a maximum, as for the .30/06, 1.940" minimum-1.946" maximum. That is for the *rifle* chamber, while the same dimension on the cartridge case will be 1.934" minimum-1.940" maximum.

Minimum rifle headspace and maximum cartridge headspace are standardized among both gun and ammunition manufacturers and specifications for the same are published in this country by the Sporting Arms and Ammunition Manufacturers Institute. These specifications appear in the form of "Minimum Chamber/Maximum Cartridge" drawings available to the industry from SAAMI. Maximum chamber headspace and minimum cartridge headspace are not in truth standardized throughout the industry. Different manufacturers establish them at different levels, depending upon their own personal views of quality control, tool life, and other factors they consider pertinent. Even so, these values differ only very slightly throughout the domestic arms industry. Certainly not enough to cause any problems in mixing makes of guns and ammunition.

Most concern is over "excessive headspace." This condition exists in the gun when the dimension mentioned above is greater than the maximum allowed. It exists in the ammunition when the distance between the rear face of the case head and the front face of the supporting surface is *less* than the prescribed minimum. In spite of the fact that excess headspace may exist because of conditions in *either* the gun or the ammunition, most shooters seem to prefer to blame the gun. In my own experience I have found this to be just the opposite. In all fairness, it must be admitted that under-minimum cases are encountered far more often than over-long chambers. Considering that many thousands of rounds of ammunition are manufactured for each gun turned out, this would normally be expected. Statistically it is far more likely that symptoms of excess headspace are the fault of the ammunition rather than the gun.

Gun headspace is checked by hardened male steel gauges of the plug type. They are inserted in the chamber and then the action is closed upon them. The action must *not* close on the "no-go" (max) gauge and must close on the "go" (min) gauge. Such gauges are available from Wilson and Forster-Appelt, as well as other makers. Naturally, all the principles of proper gauge usage must be applied or erroneous readings will be obtained. Because of the tremendously powerful camming action of most modern rifle actions, it is entirely possible to *force* a maximum gauge into a correct chamber with relatively little effort. A very light touch is required to use such gauges correctly.

Conversely, headspace on cartridges or cartridge cases is checked with a female-type gauge into which the case is inserted. Such gauges are a replica of the chamber in that particular caliber and have minimum and maximum length steps machined into the face. A case whose head falls below the lowest (min) step is too short and will create a condition of excess headspace in a correct chamber. A case whose head protrudes above the upper step is too long, creating a condition of under-minimum headspace and may well refuse to seat in a correct chamber. So long as the case head falls between the two steps, it is of the proper length for use in a correct chamber.

Different types of the cases requiring different methods of headspace measurement have been mentioned. The line drawings here show quite clearly the difference in the various types.

Case gauge showing case with excess headspace in place.

First we have the classic rimmed case. Headspace is measured in a rifle from the rear of the barrel breech to the face of the bolt, and is, consequently, a very short measurement. Headspace on the case is simply the rim thickness from front to rear.

With rimless cases, rifle headspace is measured from a point on the shoulder in the chamber to the bolt face. Case headspace is a measurement from the same point on the shoulder to the case head.

Belted cases *appear* quite different from rim cases, but in reality

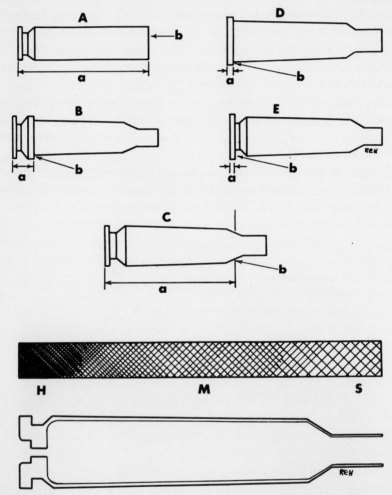

Headspace Measurement: (A) straight rimless case, headspaces on mouth; (B) belted case, headspaces on front of belt; (C) Rimmed case, headspace on front of rim; (D) Semi-rim case, headspaces on front of rim; (E) Rimless case, headspaces on shoulder. Hardness gradient of case. Shaded strip represents varying hardness. Darkest area hardest, of finest grain.

are not. They may be treated simply as a rimmed case with a thicker-than-usual rim. Consequently, headspace is measured from the front of the belt and from the bottom of the counterbore in the chamber against which the belt seats.

Then, we have an abortion called the semi-rimmed case. It utilizes the deep extractor groove turned into the head, but its "rim" protrudes very slightly beyond the diameter of the case proper. Theoretically, this very slight rim protrusion is intended to facilitate headspacing. In reality, I have found that the semi-rim seldom provides any real support for the cartridge case. A classic example is a Colt .38 Super Automatic caliber in which almost invariably the thin rim will slip right past the supporting surface it is supposed to engage. Consequently, it is necessary for additional support to be provided. In the case of a bottleneck caliber, the shoulder performs this function, and in a straight case, the mouth does the job. In handloading, then, one must more or less disregard the headspacing function of the semi-rim and utilize either the shoulder or the case mouth to obtain the proper cartridge/chamber relationship.

Last, but by no means least, we have the classic straight cartridge case epitomized by the .45 ACP pistol and .30 M1 Carbine cartridges. Typically, the case is of true rimless configuration and either of cylindrical form or tapered only very slightly from head to mouth. The mouth is square in the profile, not turned over by any crimp, and seats against a shoulder at the forward end of the chamber. Conse-quently, gun headspace is measured from that shoulder within the chamber to the bolt or breech face — and in the case is measured directly as overall length from mouth to head.

While tolerances do differ among calibers and types of cases, the spread from minimum to maximum is usually around .006" in both case and gun. This means that in extreme instances, we may have as much as .012" clearance or "slop" yet still be in standard limits.

As can be clearly seen, the handloader has complete control over the headspace of a bottle-neck rimless case through resizing die adjustment. If the die is simply screwed into the press until it contacts the shell-holder at the top of the ramstroke, it may well set the shoulder back sufficiently to create a condition of excess headspace in a specific rifle. This condition is avoided by smoking the shoulder (with a candle or oil lamp) of a fired case from the rifle in question, then screwing the die into the press *only* until it barely marks the soot film on the shoulder. Then, that fired, resized case is chambered in the rifle — and if it *does not* chamber freely, the die is turned successively further into the press until the shoulder is set back only sufficiently that the bolt will barely close freely on the case. Further details on this will be found in the section dealing with resizing. There also will be found details on altering cases to compensate for an excessive-headspace condition in the gun.

The handloader has also a degree of control over headspace in straight cases. There is no way to compensate for excessive gun headspace, but over-long cases may be trimmed to prescribed length or a particular lot of cases may be trimmed to uniform length. Routine case trimming procedures are all that are required.

The other three types of cases, rimmed, semi-rimmed, and belted suffer from a basic problem as they are currently manufactured. At least, it can be a serious deficiency insofar as the handloader is concerned.

Based on the assumption that the headspacing surface (belt, rim, semi-rim), relationships are taking care of the safety angle, manufacturers have allowed a great difference in location of shoulders between case and chamber. I have measured belted magnum and rimmed cases in which the case shoulder fell fully 1/16" rearward of the chamber shoulder when the cartridge was properly seated against its headspacing surface. When a cartridge is fired with that much shoulder clearance, gas pressure expands the case to *fill* the chamber. The brass for this expansion must come from somewhere, and the forward and outward expansion involved results in the case wall being stretched and greatly reduced in thickness back where it joins the solid web of the case head. In extreme cases, this results in an incipient or partial separation, and in all instances results in a substantial weakening at that point. Even in the event of a complete separation from this cause, usually no significant hazard to the gun or shooter exists, for the rear portion of the case adequately obturates the chamber and confines powder gases to the chamber and barrel. Unless, of course, the gun is defective or of poor design.

Apparently, ammunition and gun manufacturers' reasoning behind this condition is that no safety hazard exists, and that they *really* make ammunition for only one-time use. From that viewpoint, this excessive amount of shoulder clearance is acceptable — but it is *not* acceptable to the handloader. Initial firing greatly weakens the case, just ahead of the web, and this weakened condition may cause it to separate completely after only one or two subsequent reloadings. The economy of handloading virtually disappears under those conditions.

If you think you might get around this problem simply by purchasing unloaded brass and then reloading it to suit your own needs, think again. Cases sold commercially, either primed or unprimed for the

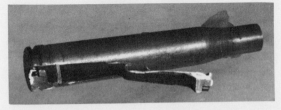

Top: Typical incipient separations. Left example produced by excess shoulder clearance with correct belted headspace; right by simple excess headspace. *Bottom:* Case stretched and thinned to breaking point by repeated resizing and firing; repeated fore and aft movement of shoulder. Headspace of rifle was correct.

handloader, are made to the same dimensional specifications as those used at the factory for loaded ammunition. Consequently, nothing is gained by loading those cases as they come from the factory. However all is not lost, for in the chapter on case forming, detailed information is given on modifying new cases so that damage resulting from the first firing will be minimized.

And, of course, it goes without saying that proper die adjustment for all resizings is essential for rimmed, belted, and semi-rimmed cases just as outlined previously for rimless cases.

The older rifles chambered for rimmed cartridges, especially the lever-action type, are often encountered with grossly excess headspace — sufficiently so that cases are nearly ruined in the first firing. If such guns are to be fired to any great extent, they certainly should be repaired. Many methods exist — setting back the barrel, installing new locking blocks, chrome-plating locking surfaces, etc. All are effective, either singly or in combination, and are not difficult to accomplish, but they do usually run up a pretty good gunsmithing bill. For only occasional firing, it is entirely possible and practical to correct the gun's excessive headspace condition by altering the cartridge cases. Let's take a Winchester M1886 in caliber.40/82 as an example. A well-used specimen may easily show excess headspace by .020" or more. Simply lay new, *unprimed* cases on a fairly heavy, smooth steel plate and lightly tap the circumference of the rim with a small hammer — rotating the case on the plate all the while — so that the forward edge of the rim is peened (caused to flow) forward. Continue alternately peening and trying the case in the chamber until the action closes *hard* upon the case. When this occurs, the peening will have increased the thickness of the rim forward sufficiently to compensate for the gun's excess headspace. Cases so treated may then be used safely and without being damaged for further reloading by initial firing. In some instances, the heavy firing-pin and hammer blow of those big old guns will drive the case forward with enough force to reduce the peened rim thickness after only a few firings. When this occurs, simply re-peen the rim until the action closes hard upon the case again. For other methods of increasing rim thickness — which may be used to compensate for excessive gun headspace — see the chapter on case forming.

The shooting gentry in general appears to have the opinion that any gun with excess headspace is more or less unsafe. As we pointed out, since excess headspace is a condition of *both* gun and cartridge, this is not necessarily so insofar as handloading is concerned. With only one exception — that of the straight rimless case — certain handloading procedures may be utilized to compensate completely for the condition that exists within the gun. This involves only moving the supporting surface of the case sufficiently forward that it contacts the supporting surface of the chamber while allowing no more than .006" clearance between the case head and the face of the breech bolt or breech block. This is accomplished either by increasing rim thickness as above, or by moving the shoulder forward in bottle-neck calibers. In *all* bottle-neck calibers, regardless of head type, moving the case shoulder forward to contact the chamber shoulder as just described will effectively eliminate virtually any degree of excess headspace.

Some years ago, this writer conducted an experiment with an M-98 Mauser rifle in .30/06 caliber to determine just what degree of excess headspace could be compensated for by proper handloading techniques. Rifle headspace was successively increased in .005" increments until the actual measurement was 1.990 — and that is .044" over the 1.946" maximum; .040" over the military *field* maximum. At every stage of excess headspace, it was possible to eliminate the symptoms thereof and the resulting case damage by moving the case shoulder forward to contact the chamber shoulder.

All modern guns confine the case head well enough so that normally-encountered amounts of excess headspace are not hazardous to gun or shooter. Excess headspace becomes dangerous only when combined with gun defects which allow the case head to protrude from the chamber so far as to expose the unsupported case walls, which will then blow out into extractor or ejector cuts or other open areas. When that happens, high-pressure/high velocity gas escapes to wreck the gun and, possibly, injure the shooter.

All of the foregoing has not been intended to encourage the continued use of guns possessing excess headspace. Certainly, they should be repaired, if for no other reason than that they may at some time in the future be fired with standard factory ammunition by someone who does not know that the undesirable condition exists. It is, however, perfectly safe and practical to continue such guns in service with properly handloaded ammunition if one chooses to do so. It should be pointed out that if more than a relatively few rounds will be fired in such a gun, the added inconvenience and cost of loading special ammunition will soon far exceed the cost of proper repairs. It is extremely important, though, that you not sell or otherwise transfer to someone else a gun with excess headspace without making him fully aware that the condition exists and that it should be repaired unless he utilizes the same corrective loading procedures that you have employed.

CHAPTER 10

Case Preparation and Resizing

AS ALREADY mentioned, the brass case (and other types, too, though not quite as well) contracts or "springs back" after firing — but not quite does it return to its original pre-firing dimensions. It shrinks enough to permit easy extraction, but no more. It still comes out of the gun larger than it went in. Generally unless excess pressures were developed in firing, a fired case will re-enter the chamber in which it was fired. It will not, however, freely enter another chamber that is smaller. Consequently, only if the case were fired in an absolute minimum-size chamber could it be expected to enter all other chambers of the same caliber without resizing. Incidentally, quite a few chambers are slightly out-of-round — and such chambers may not accept their own fired cases unless precisely oriented.

Resizing consists simply of forcing the fired case into a die shaped like the chamber but slightly smaller so that the case is reduced in all pertinent dimensions back to approximately minimum chamber size. Then, it can be depended upon to chamber freely in almost any gun you might encounter. This operation is known simply as full-length resizing.

Another form of resizing length often used when the fired cases will be used again in the same gun is called neck-sizing. Just as the name implies, it consists of forcing *only* the case neck into the die

SAECO "stubby" neck-sizing die
which covers *only* the case neck.

which reduces it in diameter. Some short dies, only long enough to
accept the neck of the case, are used for this purpose, but generally
typical full-length dies are used with chamber-like cavities to guide
the body of the case while the neck is being reduced.

Neck-sizing may be conducted to either the full length of the neck
or only a small portion of the mouth. The shallow bit is adequate for
target or varmint ammunition to be single-loaded, but will not hold the
bullet securely enough for magazine feeding or rough handling. However, all other conditions being ideal, a minimal amount of neck-
sizing will improve accuracy because the fired-formed case neck
assures best alignment of the bullet in the bore.

Alternatively we have what is often called neck-sizing but is done in
a full-length resizing die. Actually, this is not *true* neck-sizing for it
involves a partial resizing of the case body as well. It simply isn't
possible to force the case far enough into a full-length die to resize the
neck without also producing some resizing of the body. This destroys
the close case/chamber fit thought to be an advantage of neck-sizing.

Partial resizing has another disadvantage not often recognized. It
can actually make the case difficult to chamber. This is easily understood when we consider how the sizing die really works and how the
case must react to it. As the case is forced into the die, the walls are
squeezed inward. This has the same effect as if you took a handful of
clay and squeezed it. Just as the clay would elongate and squirt out
the sides of your closed hand, the shoulder portion of the case moves
forward. In full-length resizing, as the case is forced completely into
the die, the die shoulder pushes the case shoulder back to its proper
position. However, when the case is only partially sized, the shoulder

does not contact the die and remains in the forward position. This results in excessive head-to-cone length, and when the case is chambered, its shoulder jams tightly against the chamber shoulder and prevents the bolt from being closed and locked. It may seize and lock the whole works up until a mallet or boot heel can be applied to the bolt handle.

Since resizing involves heavy metal-to-metal contact, lubrication is essential. Try ramming a dry brass case into a dry sizing die, and you're in trouble. With extra effort, you can probably force the case at least most of the way in the die — but there it will stay. The rim won't withstand the force necessary to pull it back out. Adequate lubrication prevents this, and also greatly reduces the effort required for resizing. Just any old oil that happens to be handy, or grease, for that matter, won't do the job. A lubricant with high film strength is necessary because of the high pressures involved. This is the reason simple household oils won't work. Their film is broken by the pressure of die/case contact, and all the lubricant is forced out from between the two. The result is much the same as if there was no lubricant present.

At one time a fellow had to scratch around a little bit for a good resizing lubricant and drugstore lanolin was more or less standard. Nowadays, though, nearly all makers of loading dies and tools offer excellent lubricants made specifically for this purpose. Considering the quantities used, this stuff is actually quite economical. However, lack of it need not present a problem — there are a good many entirely suitable lubricants readily available. One of the old-time favorites can be had at most of the larger pharmacies and drugstores and works very nearly as well as the special types. It is simple anhydrous lanolin in a stiff paste form. Plain medical-type green soap (liquid form) also works quite well and is readily obtainable at drugstores and pharmacies. If nothing else is available, and it's on a weekend when all the drugstores are closed, pick up a can of STP Oil Additive at your favorite service station. It makes an excellent sizing lubricant, and, ounce for ounce, is really cheaper than some of the others just mentioned. Other crankcase oil additives work about as well. In a pinch, any high film strength grease or oil can be used.

The usual tendency is to use far too much lubricant. A very thin film over the case body only, from head to rear of shoulder, is adequate. Even if this film does not cover the entire surface of the case body, enough of it will be transferred and spread around inside the die that occasional bare spots will not cause any difficulty. If an excessive amount of lubricant is placed on the case, it will accumulate quickly on the inside of the die and will be forced to flow forward and will

Typical case lubrication setup. Roll cases over a cloth stamp pad impregnated with lubricant.

accumulate at the die shoulder. Inasmuch as the lubricant is not compressible to any significant degree, it will puddle up there and cause dents in the case shoulder. These dents are shown here clearly and are usually referred to as "oil dents." In moderate form, they cause no harm except to the case's appearance. Firing irons them right back out again. However, if enough lubricant has accumulated, the bulges will be so deep that the brass splits, spoiling the case.

We've mentioned that lubricant need not be applied to the *outside* of the case neck. Actually, after only a few properly lubricated cases have been run through the die, a very small amount of lubricant will have been carried up into the neck and will provide all the lubrication needed there. More will cause oil dents.

Lubricating the inside of the case neck is another matter. A typical fired *bottle-neck case* will contain a slight amount of gritty powder fouling inside the neck. At its as-fired diameter, this neck will pass freely over the expander plug in the die. Once reduced, though, it is then expanded considerably as it is pulled over the plug while being withdrawn from the die. Expander plugs are usually fairly long in relation to their diameter, and while fairly smooth, still create a great deal of friction with the case neck. Given a slightly rough plug with long bearing surface and sharp edges (as many of them are) and gritty residue in the neck, enough drag can be exerted to actually pull the neck off. Lubrication will reduce this problem considerably, but is not necessarily the best solution.

A classic method of lubricating the inside of the case neck is to merely press the mouth down against a stamp pad moistened with lubricant, causing a very tiny bead of lube to well up inside the mouth. This is then distributed over the inside of the neck as it passes over the expander. If more than a very tiny amount of lubricant is thus used, it may cause contamination or dampening of the powder charge so care must be exercised. Dry lubricants are better — admitting of less probability of error — and both powdered graphite and finely-divided mica have been used for many years. A snug-fitting brush dusted with dry lube is simply passed through the case neck, depositing there sufficient lubricant to do the job.

Other approaches can be taken to the problem of inside neck lubrication. Simplest is the use of tungsten carbide expander balls with a very short bearing surface. These are available in several makes of dies, including Bonanza, in which dies the position of the ball can be varied to place its passage through the case neck at the point where maximum leverage is being exerted as the case is withdrawn. The carbide greatly reduces friction and case withdrawal effort is also reduced. Thus no lubricant is necessary.

Conventional steel expanders can be greatly improved in this respect by altering them as shown in the drawings. All sharp edges must be removed and a gentle taper ground or stoned so that only about 1/16" of bearing surface remains. A high polish on the expander will also help.

If either altered or carbide expanders are combined with cleaning of the case necks, lubrication is no longer necessary. If only a moderate number of cases are to be cleaned, then simply a few passes through the neck of each with a tight-fitting brass bristle bore brush will do the job nicely. Alternatively, chuck a bristle (not wire) bore brush in a drill press or electric hand drill clamped in a vise and quickly pass the necks once over the spinning brush. But, when large numbers of cases are involved, tumbling is the answer. While many tumbling mediums will do a good job, I've had excellent results in the case necks proper with red rouge-impregnated ground nut hulls distributed by J & G Rifle Ranch for use in its tumblers. When cases are tumbled sufficiently long to clean and polish them externally, the necks are automatically cleaned to the point that no lubrication is needed with properly shaped and finished expander plugs.

Before resizing, cases must be cleaned and free from grit. Grit or dirt of any sort will be carried into the die and imbedded either in the case or in the die. In either instance, it results in longitudinal scratches in the die cavity which will scratch every subsequent case resized therein. The result doesn't necessarily weaken cases, but it makes them look like hell.

Cases need not be smooth and brightly polished before resizing, just clean. So, unless appearance is important to you, simply washing them in detergent will do the job. A quick and simple method consists of boiling the cases in water and detergent for a few moments, then dumping them into a large strainer or collander to drain, then rinsing them under the hottest running tap water. Drain, dump them out on a large, heavily-napped towel, and rumble them around the towel a bit to get rid of as much excess as possible. They will then dry of their own heat in fairly short order. However, if it becomes necessary — like for an urgent loading project — to dry the cases in a hurry, they may be placed in the center of an oven set at no more than 250°. I dump them out on a cookie sheet and set them in an oven for thirty minutes or so at 200° to evaporate residual moisture out of primer pockets and flash holes. Always check for water there before repriming.

One little trick to eliminate separate lubrication of cases is to wash them in *soap* (not detergent) and hot water, and then let them dry *without* rinsing off the soap film. Once the cases are thoroughly dry, the thin, soapy film does an acceptable job of lubricating the case during its passage through the die. Medical green soap works well for this. And, the dry soap film inside the case will not contaminate powder or primer.

In my own opinion, by far the most practical and efficient method of cleaning cases is to tumble them. Ground nut hulls, hardwood sawdust, and a number of other commercial tumbling and polishing mediums will do a good job of cleaning cases. Cut steel shot may also be used in conjunction with water or a soap solution, but this makes for a complicated process insofar as the average handloader is concerned and also requires a water-tight tumbling rig. That's something few of us care to spend the money for. For several years now I've obtained unusually good results with the rouge-impregnated ground nut hulls mentioned earlier. They do not clean the case quite as rapidly as harsher abrasives, but they do not produce any measurable dimensional change, and the final finish is even brighter than new. The most practical way of tumbling seems to me to be loading the tumbler and switching it on before retiring for the evening, and then switching it off to take out the cases the next morning. Overnight tumbling won't harm the cases, and produces a beautiful finish.

Generally, cases are tumbled before resizing, simply because it's the most practical way to put them in condition for the job. However, time permitting, a short period of tumbling after resizing will serve two very useful purposes. First, it will remove the residue of whatever lubrication was employed, and it will also clean out the primer pocket. This doesn't do away with the need for some form of cleaning before resizing. In my little shop, I tumble twice — over-night before resizing, and a couple hours after resizing to clean off the lube and clean out the primer pockets. The first tumbling prepares the cases for resizing, as is necessary, while the second does an equally fine job of further preparing them for priming and loading. It cleans out the primer pockets, thus insuring uniform primer seating and subsequent ignition. It also leaves cases perfectly dry for clean and convenient handling, and at the same time eliminates any possibility of either primers or powder being contaminated by residual resizing lubricant. I've seen many a primer "killed" by lubricant.

Of course, there are numerous other methods of cleaning cases, most of then involving acids. One of the most common consists of dipping cases a few seconds into a 2 per cent solution of sulphuric acid (H_2SO_4) to which a trace of chromic acid has been added for brightness. Cases should be contained in a stainless steel basket and agitated while being dipped, thoroughly rinsed, then neutralized in water containing baking soda, and finally dried.

Other acid cleaners are offered commercially. All I've tried do a good (too good, sometimes) job. Some are self-cancelling; that is, they are compounded to stop working in a very short time and thus not eat away enough brass to ruin the case. Others will completely destroy a case submerged long enough. Since they are all also highly toxic and more or less dangerous to handle, I prefer to *not* use acid cleaners. Tumbling is far safer and more convenient, not to mention cheaper.

If you don't have a tumbler and don't want to use acids, yet want bright cases, buffing or brushing will do. Just don't overdo it or cases will be weakened. A few handloaders I've known use a soft, thin-wire brush spun by an electric drill. They roll individual cases lightly against the brush by hand. The same can be done on a soft buffing wheel. Some people slip cases over a spinning mandrel and use a wad of fine steel wool to do the cleaning and polishing.

You can doubtless think up a half-dozen more methods, but keep in mind that any which remove brass will weaken the case. Tumbling in a gentle medium is still best in the end.

Assuming that cases are clean, they can — and must be — inspected. Splits, cracks, excessive stretch marks, evidence of excessive pressure,

Modern Handloading

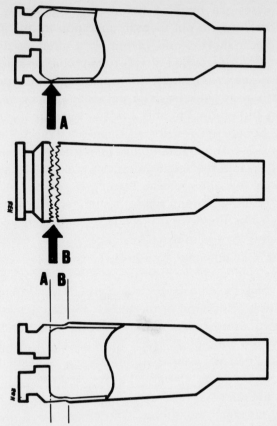

A shows thinning of case wall produced by repeated firing when shoulder is allowed to be blown forward each time. Also characteristic of cases fired under conditions of excess headspace.

B shows complete head separation that eventually develops from A above.

Third drawing shows transition between solid head and flexible case. walls after firing. A is original head diameter which firing does not change (except under excessive pressures) and B is point where walls are forced against chamber to form the "pressure ridge" often mistaken for head expansion. Resizing reduces B, but has no effect on A.

primer leaks, and severe dents or deformation are cause for scrapping the case. Small dents and minor neck deformation will be ironed out by the resizing die so need not cause scrapping.

When visual inspection discloses considerable expansion ahead of the base or signs of substantial stretching, there may be internal damage. This will be in the form of a thinned ring ahead of the base as described in detail in the chapter on headspace. It can only be discovered by using a bent wire probe to locate the ring inside the case. If the ring exists, set the case aside or reserve if for light loads. Don't use it for full charge loads.

Case Preparation and Resizing   **129**

Cases fired in automatic and semi-automatic weapons — especially military types — often have severe neck dents caused by the case striking the receiver as it is ejected. Such cases often will not enter a resizing die, yet can be salvaged if otherwise in good condition. Simply tap a tapered punch into the mounth to round it out. If the neck doesn't split as a result of that or of subsequent resizing, it may be used. In severe instances, annealing will stave off quick failure later where the brass was bent.

Auto weapons, especially the gas-operated types, often deform case rims. The M1 Garand rifle and the Belgian FN FAL are noted for this. A segment of the rim is bent rearward by the extractor and prevents the case from entering a shell holder. Such cases can be dropped over a solidly-supported rod, then the rim segment tapped back into position with a light hammer.

This damage, incidentally, is caused by excessive extraction power combined (usually) with a dirty chamber. Extraction load becomes more than the rim can stand, so it bends. Rotating-bolt designs, with their superior initial extraction, damage cases less in this respect then do others.

Occasionally fired cases will be found bearing distinct longitudinal, dark, powder fouling stripes, sometimes with the brass expanded as into grooves. These marks are produced by *fluted* (grooved) chambers found in Tokarev, CETME, and G-3 assault rifles. The flutes allow gas to flow between case and chamber walls to ease extraction. If cases are just a wee bit softer than ideal, they will clearly show brass bulged into the chamber flutes. It has been frequently said that such cases are not suitable for reloading. I've found that when properly cleaned and lubricated they resize easily and will withstand several reloadings. Such cases aren't likely to be encountered except in 7.62 x 54R Russian and 7.62mm NATO calibers.

One other form of fluted chamber *does* ruin cases so they can't be reloaded. The HK-4 .380 ACP pistol chamber contains three deep, rectangular grooves. Cases bulge badly into these grooves and often split in the process. Don't bother trying to salvage such cases. In other calibers the HK-4 uses a smooth chamber and does not harm the cases.

In the final analysis, your handloads are functionally only as good and as safe as the cases in which they are assembled. At best a bad case can cause an inaccurate shot or a minor gun malfunction; at worst it can cause serious injury. With that in mind, it becomes obvious that cases must be carefully inspected and prepared for resizing.

Modern Handloading

CHAPTER 11

Priming
the Case

THE PRIMER is the key to the whole operation — the "sparkplug" so to speak. While admittedly none of the components is of much use without the others, the primer's function is the first in the actions of the cartridge, and it is also the first *active* function to take place. As with the chicken and egg, we could have all sorts of arguments about which comes first and about which is the most important — nevertheless, we must start somewhere with *action*. And, the primer's action is the first in the chain of events that place a bullet on target hundreds of yards away.

The primer *per se* is discussed in considerable detail in another chapter. Here, we will concern ourselves primarily with the assembly of the primer to the case and the manner in which incorrect assembly or incorrect selection and fit may affect performance of the cartridge.

Generally speaking, U.S. handloaders (and increasingly, in other countries) are concerned only with the Boxer-type or anvil-type primer which contains its own anvil held friction-tight in the mouth of the cup. Berdan-type primers with their separate case-contained anvil are a different matter entirely and will be discussed farther along.

First, the only thing that holds a primer in the case for proper ignition is friction between the primer cup and the walls of the primer pocket in the case head. Consequently, an "interference fit" is required. That is, the primer cup is slightly larger than the hole into which it is pressed.

The assembled primer is a relatively fragile item. The cup is of thin, soft brass, the anvil of quite similar material, while the thin, flat pellet of priming mix is hard and brittle. If the primer fits too tightly in the pocket, then the priming pellet may well be compressed or distorted and fractured or split into fragments. In extreme cases, it may actually crumble. Consequently, there are limits to the amount of interference fit that can be tolerated. We must have enough to hold the primer securely in place, yet not so much that the priming pellet is damaged.

Simple cracking or fracturing of the pellet will not ordinarily produce a misfire, hangfire, or *detectable* difference in accuracy — except, possibly, in the very best of bench-rest rifles shot by championship marksman. Even so, a single crack will affect to some degree the burning rate and consequent jet of flame and incandescent particles that ignite the propellant charge. Though small, this effect should certainly be avoided.

Few handloaders have measuring instruments of sufficient sophistication to *accurately* check the inside diameter of primer pockets. It isn't really necessary anyway, for the best guide to whether primers and pockets are of the correct relationship is the amount of pressure required to seat the primer properly. This is a form of subjective *feel* that one can acquire only with experience. Generally speaking, this experience is best acquired by a bit of practice in priming, say, 100 *new* cases (which have never been primed before) with primers made for them by the same manufacturer. One quickly gets to recognize the feel of this with his own particular loading tool. It should be pointed out that the feel will vary from one make or model of tool to another and is most prominent in the small tools designed only for priming. Some handloading presses, particularly those utilizing compound linkage, offer such a great mechanical advantage and so much friction that it is almost impossible to *accurately* feel a primer enter the pocket.

Some handloaders seat primers by slamming the press handle down hard, making no attempt to feel in into its seat. This not only often distorts the cup and fractures the pellet, it *can* ignite the primer. Believe me, it's a hell of a shock to have a primer detonate as you slam down the press handle. The case and shell holder confine the blast so there is little probability of injury, but ammunition so loaded isn't likely to be much of an asset on the range or in the field.

Primer pocket gauges are available, but one is better guided by the way the primers feel during seating than by actual dimensions. The

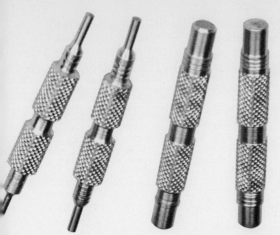

Primer pocket (R) and flash hole gauges.

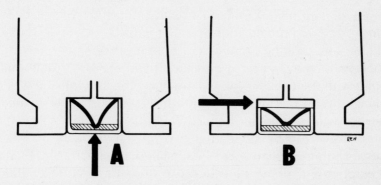

"A" shows primer properly seated to contact bottom of pocket for precise, uniform ignition. "B" shows primer *not* resting on bottom of pocket, therefore free to be driven forward by firing pin blow, cushioning impact and impairing ignition.

accompanying drawing shows the general dimensional relationship of primers and pockets as manufactured domestically.

Assuming that the diametrical fit of primer to pocket is correct, the job is still far from finished. Failure to seat the primer deep enough may produce at worst a hangfire, and at best, slightly erratic ignition. The reason for this is really quite simple. The priming mix pellet inside the cup must be crushed against the point of the anvil in order to be detonated. Unless the anvil is resting on the bottom of the primer pocket — against the solid metal of the case web — it may be driven forward by the force of the firing-pin, consequently reducing the sharpness of the firing-pin blow, which in turn will produce less than perfect ignition of the priming pellet. Ignition of the priming pellet is brought about when the firing-pin actually crushes by friction the priming mix against the anvil. Obviously, if the anvil is not strongly supported, the required amount of crushing may not be brought about, even though the firing-pin indents the cup properly. The farther the anvil is free to move under these conditions, the more erratic the ignition of the priming pellet and, consequently, the propellant charge will be. If the anvil is free to move a sufficiently great distance, a complete misfire will be produced. Usually, such a misfired primer will ignite on the second or third firing-pin blow, for by that time the anvil has been driven down against the bottom of the pocket and is resting on or very near the bottom of the primer pocket. The only thing enabling it to resist the firing-pin blow is the modest amount of friction created by its legs friction-tight inside the mouth of the cup. That isn't enough for proper ignition.

Because of the foregoing, combined with manufacturing tolerances in overall length of the assembled primer, depth of primer pocket, thickness of case rim, all combined with cumulative tolerances in the loading press or priming tool, it is not possible to depend upon a fixed adjustment of the priming tool to seat primers uniformly against the bottom of the pocket. In order to obtain maximum uniformity of ignition, with the anvil resting uniformly on the bottom of the pocket, it is necessary to *feel* the primer into place. Additional resistance will be felt as the legs of the anvil touch the bottom of the pocket, and the priming stroke can be stopped at that point. With a bit of practice, remarkably uniform seating can be obtained in this manner.

Alternatively, if primers are seated too tight against the bottom of the primer pocket, the anvil is forced deeply into the cup — perhaps the cup is even collapsed somewhat — and the priming pellet is fractured or crushed between the anvil and the cup, the latter being supported solidly by the priming punch. Light compression — without fracturing — of the pellet between anvil and cup actually increases sensitivity. However, if the pellet be fractured or crumbled, ignition will be erratic at best, and at worst complete misfires or hangfires will be produced.

One must make certain that the primer is not seated too hard. Generally speaking, if seating pressure is heavy enough to deform or mark the primer cup in any way, you run the risk of damaging the primer pellet. Again, if maximum accuracy and uniformity is your goal, develop the proper feel for this job.

Of course, someone is bound to ask, "How in the hell can the manufacturers do it uniformly? We know they don't have time to seat primers by hand." A good question, really. The factories use high-speed automatic priming machines which regulate the depth to which the primer is seated from the face of the case head. All this means is that primers are seated to a uniform distance below the case head. And, of course, I just said that you shouldn't do that in handloading.

If you were always loading fresh, new, unfired cases, and were making the primers to match those cases, you could get by quite well by seating primers thusly. When working with new cases, primers fit quite tight in the pocket and are squeezed in a bit at the bottom, thus gripping the anvil tightly. In this condition, the anvil does offer sufficient support for acceptably uniform ignition even if not against the pocket floor. However, for two reasons, each subsequent time you re-prime that same case, the primer pocket is slightly larger in diameter and thus a looser fit on the primer. Look first inside a freshly decapped once-fired case and note the scratches and scrapings caused by the sharp-edged mouth of the primer cup. Most likely you'll see a thin ridge of brass scraped up into a ring at the point where the primer cup mouth came to rest. This is repeated every time you re-prime that same case—even through the brass cup is soft, it does have sharp edges, and they will gouge the brass of the case. Second, *some* increase in primer pocket size is produced by the pressures developed within the primer cup during firing. Even full-charge Magnum pistol loads develop pressures high enough to enlarge primer pockets over several firings. In some high-intensity rifle cartridges, especially with hand-loads that go over the advised limit, a primer pocket may be enlarged to the point of uselessness in only three or four firings.

Quite likely you can't measure the initial amount of pocket diameter increase with the best inside micrometer at your disposal. However, you probably can measure the increase after as little as four or five full-charge firings of a load in the 50,000 psi bracket.

If your tools won't allow you to measure this increase, then try this simple test: Take a single new case and components and tools to the range; load and fire this one case with a full-charge load at least eight or ten times; after each decapping, pay very particular attention to the amount of effort required to seat a fresh primer correctly. Invariably, you will find that after the second or third firing, it becomes progressively easier to seat a fresh primer. Eventually, you will reach the point where hardly any pressure at all—you can even do it with your thumb—is required to seat a primer.

Well before the pocket becomes that loose, a potentially dangerous condition is reached. When the primer is not quite snug in the pocket, gas will begin to leak out around the cup. A slight smudge of soot or powder fouling on the case head is the danger signal. When gas begins to leak around a primer (remember, it may be at over 50,000 psi) it acts on both the case and the cup just like a thin trickle of water starting over an earth dam. That high-speed gas jet can in a mere milisecond melt or erode away a passage that becomes progressively larger and allows a substantial burst of gas to rush violently back into the bolt through the firing-pin hole and on back into the shooter's face. In some of the more modern guns with completely shrouded bolts and firing-pins or closed receivers, no harm will be done—but where an open passage exists, that gas can come directly on back into your shooting eye. (More than a few careless handloaders and shooters have lost the sight of an eye in just that fashion. They were careless in two ways—first in permitting the gas to escape at all; and second, not wearing shock-resistant, shatterproof shooting glasses. For over twenty years now, this scribe has not fired a single shot without first donning proper protective glasses. Several times during that period, those same glasses have been marked up by jets of gas and molten brass which would certainly have seriously impaired—if not destroyed —my shooting eye. Don't ever fall for that old cliche, "It won't happen to me." If you shoot long enough, eventually it *will* happen to you for one reason or another. Never, repeat, never shoot without protective glasses.)

Because of this growth in pocket diameter, the anvil must seat on the bottom of the pocket in order to resist properly the firing-pin blow.

Thus far, we haven't even mentioned the different sizes and types of Boxer primers generally available. Inasmuch as the chapter on primers covers the subject in detail, we'll consider here only the difference between rifle and pistol types. Small rifle primers will fit in the pockets of handgun caliber cases made for small pistol primers— and vice-versa. The same applies to the large size. Under certain conditions, a switch can be made safely. There are two basic differences between rifle and pistol primers. Because of the lower chamber pressures involved and the generally lighter firing-pin blows possible, pistol primers are assembled in thinner and/or softer cups. Auto-loading pistols in particular deliver relatively light firing-pin blows which simply cannot indent the thicker rifle primer cups sufficiently to produce uniform ignition. Some of the large-framed revolvers such as the old Colt New Service and the M-frame Smith and Wessons have a sufficiently heavy hammer and long travel to indent even rifle primers rather well. All the same, most smaller-frame revolvers and most auto-loading pistols will not produce consistent ignition with either large or small rifle primers. Their firing-pins just don't hit hard enough —especially when fired double-action, during which the hammer travels through a far shorter arc than when thumb-cocked. In short,

Modern Handloading

poor ignition or misfires result from rifle primers in handgun ammunition.

Using pistol primers in rifle cartridges can lead to much more serious consequences. The thinner cup simply does not have the strength to withstand the pressures generated in modern high-intensity loads The result of using a large pistol primer in, say, a full-charge 7mm Remington Magnum load is quite likely to be a "blown" primer, dumping hot gas back through the bolt to the shooter's face. That can be disastrous.

However, where chamber pressures are down in the range of handgun cartridges, pistol primers may be safely used in rifle loads. This applies primarily to black-powder loads in obsolete calibers and to lead-bullet loads producing velocities less than 2400 fps.

The other difference is in the amount of priming mixture contained in the primer. Large quantities of slow-burning powder as found in the larger rifle calibers require considerable primer energy and heat for proper ignition. This means a relatively large amount of priming mix. The extreme of this is represented by so-called "magnum" primers intended originally for the largest belted cases and their unusually large powder charges. Pistol primers on the other hand, need ignite only a very small charge of fast-burning powder. Much less primer heat and energy is required, so they contain less mix than rifle primers.

Magnum primers just mentioned exist in both rifle and pistol types. Generally speaking, they produce a flame of greater intensity and longer duration and this improves ignition of the harder-to-ignite and slow-burning powders used in high-intensity cartridges. An example of this is found in comparing the .38 Special and .357 Magnum. The .38 operates at less than 15,000 psi with a small charge of easily-ignited, fast-burning powder. Standard pistol primers ignite this charge very well. The .357 operates at over 30,000 psi with about three times as much of a much slower-burning powder. The .38 Special and most other handgun cartridges are handled very well by the standard primer; but the mass of the large charge of the .357 absorbs much more of the heat and energy of the primer flash during ignition. It's like the difference between lighting a piece of paper and a stick of wood. The magnum primer furnishes the additional heat and energy to properly ignite the magnum powder charge.

The same relationship exists in standard and magnum rifle primers, a good example being the .30/06 or .308 and the .300 Weatherby Magnum. In fact, the first magnum-type primers were produced by Federal for the hot Weatherby calibers.

Incidentally, a *standard* rifle primer is *not* a suitable substitute for a *magnum* pistol primer. Both are designed for different sets of conditions.

Not all primers of magnum characteristics are identified as such. CCI tacks the Magnum label on its primers, but others don't. Catalogs do identify them by type, so there is no need for confusion. Our table here lists all the major makes and models fairly well.

The Berdan primer found in much European ammunition depends for its proper functioning upon a fixed anvil that is an integral part of the bottom of the primer pocket. The primer is simply an open-mouth cup which is pressed into the primer pocket until flush or only very slightly (.002-.003") below the face of the case head. When so seated, any gun with normal function will crush the priming pellet against the fixed anvil and assure proper ignition. Since the Berdan primer is located from the head, it is quite practical to set primers face-up on a smooth steel or glass plate and press or gently drive the case over them until the case head contacts the plate. Since one seldom will have occasion to load much ammunition with this type of primer, the plate method eliminates the need for special tools.

Berdan primers do not come in the same sizes as the Boxer type. The most common size measures approximately .217" in diameter and is found in most modern bottle neck rifle cases. With care, this size may be seated with the standard large-size flat-face primer seating punch. The smaller size, usually found only in pistol and small-caliber rifle cartridges, measures about .180" in diameter and may also be seated by careful use of the small, flat-face punch. Two other sizes will be encountered fairly often, the large British size, measuring .250" in diameter, and the European 6.45mm size which measures .254". Attempting to seat these two sizes with even the largest standard punch will generally result only in damage to primers. Fortunately, most makers of loading tools and priming tools can supply on special order the proper size punches to fit both of the large primers.

Regardless of whether you are using a hand tool, a loading press, or a bench-type priming tool, always use a primer punch whose contour and diameter match that of the primer. In years gone by, primers were about equally divided among flat-face and round types. Over the past decade, those makers which previously used the rounded face have almost all switched to the flat-face. However, when rounded types were in vogue, loading tool manufacturers supplied punches contoured to fit both types, and many tools are still fitted with the old round-face primer punch. When the round-face punch is used to seat flat-face primers, it exerts pressure only around the perimeter of the primer cup, leaving the center of the face unsup-

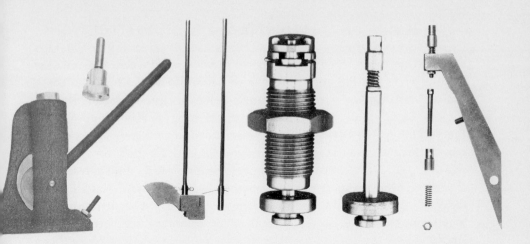

From left: Lachmiller bench priming tool; typical gravity-feed auto priming attachment by RCBS; auxiliary priming unit which screws into press like a die — primer seating punch fits in shell holder; universal-type priming arm with extra parts for both size primers.

ported. This results in the anvil moving up into the cup as it bottoms into the pocket and often raising a dimple in the face of the primer. Since the priming pellet is quite brittle, this usually results in its being cracked or crumbled — which would not have happened if the primer face were solidly supported so that the pellet did not flex under the thrust of the anvil. Use of a flat-face punch to seat round-face primers produces just the opposite result. The punch contacts only the center of the primer face and, depending upon the amount of force required to seat the primer properly, will flatten the cup which has the same effect of breaking up the priming pellet. If you're using only fresh, modern primers and purchased your loading or priming tool in the past few years, doubtless your punch and primers are properly matched. Nevertheless, if you obtained primers of unknown ancestry, or swap for some old, used loading equipment, make certain you check both primers and punch to be sure that they match.

Occasionally in the cheaper loading tools improperly finished primer seating punches will be encountered. Often we see handloads which have a slight depression occupying about 80 per cent of the cup face. This results when the perimeter of the end of the punch is beveled so much that the face contacting the primer is smaller in diameter than the flat area in the center of the cup. Unless the face of the punch is large enough to extend out over the radius of the cup, it will depress the center of the cup and damage the priming pellet. The primer cup

Priming the Case 139

is best suited to resist the loads imposed on it by seating out at the edges over the radius, and if the punch doesn't contact that area, the primer is going to be damaged. Consequently, punches with a pronounced bevel or chamfer around the edge of the working face are to be avoided.

Another defect frequently found in cheap punches is a sharp teat in the center of the working face. These punches are usually made on screw machines or turret lathes and this sharp point is simply the result of failing to clean up the cut after the punch has been cut off from the stock from which it was turned. A very slight dimple there probably does no harm — but the prominent sharp point I have encountered too often certainly weakens the primer cup by partially penetrating it, setting things up for a primer leak. If large enough, this teat can also cause damage to the priming pellet.

On many occasions we've been asked to help handloaders who simply couldn't seem to get a decent job of primer seating done. They read all the books and tried everything under the sun, yet primers were still cocked or canted in the case and were often deformed badly by the pressure required to seat them. In most instances, it was discovered that either sloppy shell-holder fit and alignment or improper priming punch alignment was responsible. The typical shell-holder grips the case only through about ⅔ of the rim circumference and if the slot into which the rim fits is substantially larger than the rim, the case may slither around from side to side or tip as the primer first begins to enter the pocket. When this happens, the primer is often squeezed off-center or one edge of the cup catches on the mouth of the pocket causing it to be cocked and deformed. Before primers can be seated properly, it is essential that the case be held snugly, and that case, shell-holder, primer, and primer punch all be concentric. Often case mis-alignment is found to be due to dirt and powder granules packed into the groove in which the case rim fits. Over a period of time, a few granules of powder will be spilled during various loading projects and will fall into the shell-holder without being noticed. Eventually they become packed in from repeated insertion of cases and build up to the point where the case cannot be pushed all the way into the holder. The primer pocket is then off-center insofar as the primer and punch are concerned, and primers are damaged in seating.

The same result can be produced by dirt in the ram cuts into which the shell-holder seats. It will prevent the shell-holder from seating fully, causing it to fail to line up with the primer punch.

At any time that you encounter unusual resistance in seating primers, quit right there and check all of the points mentioned until you find the problem. Damaged or deformed primer cups not only result in erratic ignition and misfires for the reasons already discussed, but they are also much more likely to leak and allow gas to blow back into the shooter's face. This is particularly true when a primer is cocked or seated at an angle — one side of the cup is often crumbled up or split when this happens and offers virtually no resistance to gas when the cartridge is fired.

PRIMER CHART

	Alcan	CCI	Federal	Herter	Norma	Rem.-Pet.	Win.-West.
Large Rifle	Max-Fire	200	210	120	LR	9½	8½-120
		250M*	215M*	500		9½M*	
Small Rifle	Max-Fire	400	200	6½	SR	6½*	6½-116
		450M*				7½	
Large Pistol	Max-Fire	300	150	111	LP	2½	7-111
		350M*		400			7M-111F
Small Pistol	Max-Fire	500	100	1½	SP	1½	1½-108
		550M*				5½	1½M-108
Shotshell Caps	PC209[k]	209B		H209FWW[c]			
	PC57[k]	PC57					
Shotshell[a]	WW209F	109	209	H209W			209
Shotshell[b]	G57F	157		H57PR		57	
	220					97[d]	
			410			69[f]	
Berdan Rifle[f]	175B (1.75")[l]						
	210B (.210")						
	217B (.217")						
	250B (.250")						
	1794 (.254")						
Berdan Shotshell	645B (.254")						
Percussion Caps	G10F					10 (.162")	
	G11F					11 (.167")	
	G12F					12 (.172")	
Winged Musket							
Caps	G4F[g]						

NOTE: Large rifle and large pistol primers measure .210"; small rifle and small pistol measure .175".
(a) For Winchester-Western, Monarch, J. C. Higgins, Revelation and Canuck cases.
(b) For Remington-Peters paper cases.
(c) Long battery cup type for Win chester-Western, Federal or herter plastic shells.

(d) Norma makes a .216" Berdan primer.
(e) Battery cup; used in 12-ga. plastic trap and Skeet loads.
(f) Copper plated; used in 410 and 28 ga. shells.
(g) Fits old Springfields, muskets. etc., of yester-year.

Priming the Case

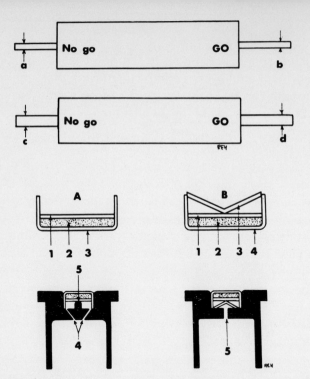

Top: Primer flash hole gauge. "a" is minimum hole diameter, "b" is maximum. "a" must enter flash hole, "b" must not. "c" and "d'" are comparable for primer pockets.

Bottom: A is Berdan type primer and pocket. 1 is foil disc, 2 is priming compound, 3 is cup, 4 is multiple flash holes, 6 is anvil integral with case. B is Boxer type, 1 is foil disc, 2 is priming compound, 3 is anvil, 4 is cup, 5 is single central flash hole.

Few handloaders get very deeply into the game without using military fired cases. Regardless of origin, nearly all military rifle and machine gun cartridges include some form of crimp applied over the primer after it is seated. In U.S. ammunition this is usually a circular crimp encircling the primer, produced by a hollow punch. Much foreign ammunition has instead a 3- or 4-point stab crimp as shown. Some pistol and submachine gun cases are also crimped, but a great deal is not. The crimp is intended to keep primers from backing out when fired in guns with long headspace.

Both types impede decapping to some degree because the mouth of the primer pocket is turned slightly over the edge of the primer. The crimp must be forced aside by the primer as it is expelled. Since the cup is thin and soft, a thin or pointed decapping pin may pierce the primer rather rather than force it out. This can be avoided by use of so-called "heavy-duty" decapping pins which are flat-tipped and of as large a diameter as the flash hole will accommodate. Such pins will remove 99 per cent of the crimped-in Boxer primers you'll encounter.

Modern Handloading

All of which is simply leading up to the point that those same cases cannot be reprimed until primer crimp is removed. Brass intruding into the pocket mouth can be cut away, or it can be forced back where it came from. Reamers are made for the first method, shaped and dimensioned so the original mouth radius will be reproduced and so that the pocket will not be enlarged. Lyman makes an excellent hand tool of this type. It may also be cut away with a pointed, sharp knife blade, but care must be taken to avoid removing too much metal. Experiment to learn how *little* of the crimp can be removed and still allow distortion-free repriming with normal pressure.

Most authorities recommend swaging rather than cutting out the crimp by forcing a shaped punch into the pocket. I question that this is necessarily any better than properly-done reaming or cutting, but it is faster and more convenient. One quick thrust into the pocket and the crimp is forced back to duplicate the original pocket mouth.

Pocket-swaging setups which hold the case by its rim are not good. The force required to seat the punch often deforms the rim. Best results are gotten with the type containing a rod over which the case is solidly supported. RCBS makes this type for use in a loading press. A simpler version consists of a base and a punch which is just driven into the pocket with a hammer. It's cheap and easy to use, and the results are fine if the punch is carfully handled.

Once the crimp is removed, such cases are just as serviceable as the uncrimped variety. Swaging does work-harden the case head slightly, but not enough to cause any problems.

Left: Best type of primer pocket swage, which supports inside of case head on a solid rod central in the die. Button swage at bottom fits into shell holder and reforms pocket. *Right:* Ring crimp around primer in U. S. military .45 ACP ammunition; Commerical case is smoothly radiused at pocket mouth.

High primer such as this can cause premature firing in autoloaders.

High primers — protuding beyond the case head — not only cause poor ignition, but can be dangerous. On repeating or semi-automatic actions (either rifle or pistol), the protruding primer may be struck hard enough by the bolt to be ignited as the cartridge is chambered. Ignition under those circumstances occurs *before* the breech is locked and will usually wreck the gun and will often injure the shooter badly. Semi-automatic arms with heavy operating springs and consequently high bolt velocity are particularly susceptible to such premature firing if primers are not seated deeply enough. Keep primers down.

High primers will also jam revolver cylinders and prevent locking in some other manually operated actions which lack the camming power to force the primer into the case.

Anyway you look at it, priming is a very important part of hand-loading. There is probably more that can be screwed up in priming than in any of the other operations.

CHAPTER 12

Seating Bullets

SEATING THE bullet is the final act in the creation of a handloaded cartridge. It is sort of like corking a bottle of fine wine. By placing the bullet securely and properly in the slender case neck, you have completed the assembly; sealed the parts into a whole. Aside from little things like inspection and packaging, the job is complete when the bullet is thrust home to the proper depth.

It all seems so simple. All you have to do is to jam that bullet into that hole, more or less to a specified distance, nothing to it at all. Sure, you can assemble ammunition that way, and it will go "bang" and the bullet will go sailing off into the blue. And that's just where it will go, too — into the blue, just as likely as into the target.

Seating a bullet doesn't consist simply of plugging that hole in the mouth of the case. Certainly, the case is the package and the bullet is the lid — but the manner in which that lid is affixed to that package has a great deal to do with the accuracy that may be obtained from the completed cartridge.

It is necessary to understand just what function the case performs *for the bullet*. First, by the depth to which the bullet is seated in the case, the bullet's relationship to the leade (origin of the rifling) is established. The amount of free travel the bullet is allowed after it begins moving and before it strikes the beginning of the rifling affects both accuracy and chamber pressure. The effect on pressure is discussed in the chapter devoted to that subject, but the effect on accuracy is a more nebulous one. We cannot make a general statement that allowing the bullet X amount of free travel will produce maximum accuracy. The amount of free travel that will place the shots within the smallest group on the target will vary according to the type of bullet, chamber pressure, caliber, even the individual gun, and many other factors. Heavy-barreled, single-shot rifles utilizing lubricated lead bullets quite often produce their best accuracy with the bearing surface of the bullet virtually touching the origin of the rifling. Jacketed bullets in high-intensity cartridges often perform better if they are given a relatively good "running start" before being forced into the rifling. Generally speaking (and I said that we couldn't do that), most bullets will produce acceptable accuracy when loaded to factory-standard overall length for that particular caliber and bullet weight. However, in handloading, we soon learn that variation from the standard will often — but not always — produce measurable improvements in accuracy. Consequently, in the beginning, it is wise to seat bullets to standard depth, then as you become more proficient and sufficiently skillful with the gun, to edge the bullet closer to the rifling until increased accuracy is or is not produced. Eventually, you will find a seating depth for a specific bullet in a particular rifle that will — all other factors being equal — produce just a hair more accuracy.

The case neck also serves to hold the bullet concentric with the bore — that is, the center line of the bullet precisely on the centerline of the bore. Only when the bullet is so aligned can maximum accuracy be obtained. If the bullet is at a bit of an angle to the bore or is off-center to the bore, then less than top accuracy will be produced. Neither factory-loaded ammunition or cases resized full-length can completely meet this condition. Primarily, they can't meet it because the entire case is slightly smaller than the chamber and, therefore, the cartridge lies upon the lower edge of the chamber by virtue of its weight. This puts both the case neck and the bullet slightly below and at a slight angle to the center line of the bullet. So long as the case is concentric throughout its length and it is a relatively close fit in the chamber, the amount of misalignment produced is usually not perceptible. It is significant that top bench-rest shooters eliminate this misalignment by using only cases that have been fire-formed in that particular chamber and by "orienting" (placing them with the same point upward every time) in the chamber and by neck-sizing only a very small band at the mouth of the case to hold the bullet. They further refine the process by turning or reaming the neck of the case so that its wall thickness is uniform throughout its circumference. Thus, with the case virtually a push-fit in the chamber, it is oriented for every shot, and has uniform neck wall thickness, then the bullet will be aligned as accurately as possible with the bore. That the better bench-resters often shoot 10-shot groups measuring less than ¼" center to center at 200 yards under ideal conditions is ample evidence that these refinements in bullet seating pay off.

However, ammunition so highly refined is virtually useless in a hunting gun or a conventional target rifle, though it can sometimes be used for varmint shooting when one is not in a rush to take a second shot and can coddle both gun and ammunition a good bit.

Hunting and general target-shooting ammunition has other requirements it must meet — several of which are dependent upon the manner in which the bullet is seated in the case. First of all, in a magazine-fed arm, the overall length of the completed cartridge must be short enough to feed freely through the magazine box, yet long enough that it receives proper guidance into the chamber by the feed ramp or bullet guide. Secondly, the case neck must hold the bullet with sufficient tightness that the bullet will not be pushed back into the case by impact with the feed ramp during rapid reloading; and the joint between the two must be strong enough that feeding impacts will not bend the bullets and case neck out of line with the case body. (Interestingly enough, there were in use at one time several military automatic weapons which actually consistently bent the cartridge case during feeding. The outstanding example of this was the Danish Madsen machine gun which bent the cartridge through several degrees while chambering it, but yet by the time the cartridge was driven fully home into the chamber, it had been bent back straight again, the only evidence of the bending being dents in the brass.)

Particularly in self-loading and full-automatic weapons, the bullet must be tightly gripped by the neck through at least one full caliber length of the bullet bearing surface. This means, for example, that 3/10" of a .30 caliber bullet must be gripped by the neck — and that means that the case neck must be at least 3/10" long, as must the bearing surface of the bullet. When less grip than this is involved, cartridges are prone to become bent during the high-speed impacts applied during automatic feeding. Military ammunition likely to be used in self-loading or automatic weapons normally has the case mouth crimped deeply into a cannelure in the bullet to add further strength to the union for this purpose.

In addition, the case neck must hold the bullet tightly enough so that successive recoil impacts will not force bullets deeper into the necks of the cases in the magazine. Considering that the .458 Winchester Magnum develops a recoil force of approximately 800 G's (that is 800 times the acceleration of gravity), virtually all factory-loaded sporting ammunition also has the case mouth crimped into the bullet.

Visualize for a moment what happens to cartridges in a typical box magazine when the gun is fired: First, the cartridges tend to remain where they are as the gun recoils rearward; causing the front of the magazine box to strike the points of the bullets a severe blow, which drives the cartridges to the rear, causing them to strike the rear of the magazine box a severe blow as the rifle's recoil movement halts; after which the cartridges will bounce forward to again strike the front of the magazine box; and, finally, perhaps bounce back and forth between the front and rear of the box several times before coming to rest. Considering that one or two cartridges may remain in the bottom of a magazine during a dozen or more shots and subsequent recoil impulses, it is easy to see that a strong assembly of bullet to case is absolutely essential for hunting use.

The foregoing applies primarily to bottle-neck and rimmed cases which may be crimped into the bullet without interfering with case function. Cases which headspace on their mouths cannot. As in the .45 ACP and 9mm Parabellum, crimping such cases to any significant degree on the bullet destroys the headspacing surface. Factories solve this problem by very tight fit of bullet in case and sometimes by placing a cannelure in the case positioned so the base of the bullet rests upon it. Some foreign makers use a "stab-crimp" consisting of several indentations of the case into the bullet.

The handloader may use any or all of these methods, though the stab crimp is injurious to cases in that the stab points soon become holes. Since these problems apply almost exclusively to handgun cartridges they are covered in detail in the handgun chapter.

Often, depending upon the hardness and thickness of the case neck walls, mere friction of a tightly-fitted bullet is not sufficient to keep the bullet from being driven deeper into the case. For that reason, most larger-caliber bullets supplied by both independent and major manufacturers for handloading are provided with a deep, narrow cannelure, located so that the case mouth may be crimped heavily into it.

In box-magazine rifles, considerable improvement can be gained by fitting a device known as a "bullet point protector". This consists of a U-shaped piece of thin sheet metal which simply slips into the front of a magazine box. The outer ends of the U are bent into shoulders which restrain cartridges from moving forward. The bullet point protector shoulders are located exactly in front of the cartridge case

Modern Handloading

shoulders and thus the impact of recoil is taken by the case rather than the bullet. Identical results are obtained by some custom gun-makers who fit metal shoulders permanently to the inside of the magazine box wall at the proper point. Factory-made rifles are not so fitted simply because production realities require the same magazine box to be used for a number of different calibers with different shoulder locations.

Tubular-magazine rifles introduce a slightly different problem insofar as bullet seating is concerned, but one readily solved by heavy crimping of the case upon the bullet. With several cartridges in the magazine in line astern, recoil causes the rifle to move rearward while the cartridge column tends to remain in place and thereby heavily compress the magazine follower spring. Then, as rifle recoil movement slows down or halts, the magazine spring asserts itself and drives the entire column of cartridges violently to the rear. This type of reaction greatly multiplies the impact suffered by the individual cartridges. And makes it especially important that a heavy, secure crimp be used.

Heavy caliber revolvers fired with full-charge loads present an opposite problem. In them, the cartridges are forced to recoil rearward with the gun by virtue of their rim contact with the rear face of the cylinder. This, combined with the fact that revolver recoil velocity is higher than in many rifles results in a tendency for the bullets to remain in place while the cases are drawn off of them to the rear. Prevention of this requires not only a very heavy crimp, but also in calibers such as the .44 Magnum, a very tight assembly of bullet to case. Some authorities recommend having the as-resized inside diameter of necks as much as *.003-.005" less* than bullet diameter prior to seating. This results in a considerable "interference fit" and a very tight grip of case upon bullet. Even then, a heavy crimp is still required to retain the bullets securely in place in the case. When for any reason recoil causes revolver cartridge bullets to "walk out" of the case, bullets will protrude from the face of the cylinder and jam against the barrel breech, preventing cylinder rotation. Consequently, it is essential that bullets be held sufficiently firmly in the case to withstand at least five and preferably more successive recoil impacts. This is especially important for ammunition to be used for police or defense purposes.

A heavy crimp is also helpful in magnum-type revolver cartridges loaded with large charges of slow-burning powder for another reason. The smooth throat of revolver chambers offers little resistance to bullet movement, so the added resistance of a tight crimp aids ignition and initial combustion of the powder charge. It restricts bullet movement just enough to allow more rapid build-up of initial chamber pressure which increases combustion efficiency.

All this discussion of crimps and the needs therefore cannot be ended without pointing out the best way to crimp a case mouth upon a bullet. Most reloading dies incorporate bullet seating and crimping into a single operation. This is practical only when very light crimping is involved and when cases are of uniform length. And when cannelured bullets are being used.

Combining seating with crimping results in the mouth of the case being turned inward by the crimping shoulder of the die before the bullet is seated to full depth, the bullet continuing on into the case as the mouth is being turned in. This results in the case mouth being jammed into the bearing surface of the bullet before reaching the cannelure in many instances. Where a lead bullet is involved, this means that often the case mouth shaves a ring of lead from the bullet and pushes it on ahead. This ring of lead can be thick enough to prevent chambering, not only in revolvers, but in auto-loading pistols and in some rifles. It can also result in a portion of the case mouth "digging-in" and cutting deeply into the bullet.

Where jacketed bullets are involved, the case mouth usually will not cut into the jacket, though this has occurred in thin-jacketed types, but will nevertheless cause some deformation of both case mouth and bullet which can reduce accuracy.

By far the better solution, though it requires additional work and time, is to seat all bullets to the correct depth first, with the die backed out of the press so that the crimping shoulder does not touch the case mouth. Then, re-adjust the die to produce the proper crimp, and back off the bullet seating screw so that it does not touch the bullet. All cartridges can then be run through the die and the cases crimped as a separate operation. This produces the best crimp with least bullet deformation.

As briefly indicated above, a uniform crimp can be had with a single die adjustment only when case length is uniform. If you adjust the die to produce a proper crimp on a case picked at random from the lot with which you are working, then it will produce excessive crimp on a case longer than that particular one, and less than the desired amount (maybe even none at all) on shorter cases. This can only be avoided by trimming cases into separate batches of very nearly the same length, then adjusting the dies separately for each batch.

Actually, two types of crimp exist. Most often used by handloaders is the roll type and it is typical of that found on factory-loaded revolver cartridges. The segmental or stab-like crimp found on factory-loaded rifle ammunition cannot be duplicated with conventional

Modern Handloading

handloading equipment. The roll-type crimp is best for use on all rifle cartridges and heavy or full-charge revolver ammunition. However, auto pistol cartridges which headspace on the case mouth — such as the .45 ACP — will have the headspacing surface destroyed by even a moderate roll crimp. Therefore, by far the best results will be obtained on this type of cartridge if a straight taper crimp is used. It is produced simply by forcing the case with bullet fully seated into a slightly tapered hole which compresses the mouth of the case slightly into the bullet. No bullet cannelure is necessary.

In spite of the above long-winded dissertation on the need for crimping bullet and case together under many circumstances, the bench-rest and varmint shooters have adequately proven that top-quality rifles normally produce their best accuracy with smooth, uncannelured bullets and with no crimp whatsoever applied. So, at least in rifles, where recoil effect and feeding do not make it necessary to crimp the case on the bullet, one should avoid the practice in the interest of maximum accuracy.

In preparing cases, be they rifle or pistol, for seating lead or gas-check bullets, the mouth of the case needs first be chamfered lightly on the inside. This is done with a simple countersink-type reamer or with one of the several tools made for the purpose, but can be done equally in all but the smallest calibers with a sharply pointed and a reasonably sharp knifeblade. Excessive chamfering simply weakens the mouth of the case and encourages early failure from cracking. I find it simplest to hold the case, mouth up, in my left hand, and hold the knifeblade in my right hand with only about ⅜" protruding from between thumb and forefinger. The point of the blade is inserted into the case mouth and the case is rolled through one full revolution while the knifeblade is held stationary, peeling away a thin, uniform shaving of brass completely around the inside of the mouth. Only a slight sliver of brass need be removed, just enough to break the sharp corner of new cases. Do not attempt to remove the residue of a previous crimp in this manner — to do so will result in far too much brass being removed. Excessive crimp residue is best removed by flaring the case mouth slightly either before or after chamfering.

Chamfering alone is generally considered to be sufficient to enable jacketed bullets to be seated properly. However, I have not found this to be entirely true. Sharp-cornered bullet bases will still give problems, even though the mouth be properly chamfered. Also, if the bullet is not perfectly aligned, it may crumple or split the case neck if only chamfered. These problems can be alleviated by flaring the case mouth slightly, either by running it up over a tapered expander plug in the die, or by simply dropping a tapered punch of proper diameter into the case mouth and by giving it a very light tap. In fact, in many instances, merely the weight of a heavy machinist's punch dropped freely into the case mouth will produce enough flaring so that bullets may be started into the case by hand. Alternatively, use the Lyman two-diameter expander plug which is made primarily for lead bullets. Properly adjusted, the larger-diameter portion of this plug enters the case mouth only 1/16", expanding that portion to a diameter large enough to freely admit the bullet, while expanding the balance of the neck just enough to give proper neck tension on the bullet. When cases are thus expanded to two diameters, it is extremely quick and simple to manually start the bullets in a block of fifty cases and then run them through the seating die without any trouble whatsoever. I find this by far the most practical method of preparing the cases for seating either lead or jacketed bullets.

The shape of the bullet seating stem or screw must match the profile of the bullet being used for best results. Just a day or so before this was written, a young lad brought me in a handful of his pet 9mm Luger loads, assembled with jacketed hollow-point bullets. He was complaining bitterly that they didn't seem to expand like the factory loads utilizing identical bullets at the same velocity. A brief glance showed the trouble quite clearly. He had used a typical round-nose bullet seating punch, and the bullets were quite a tight fit (as they should be) in the cases. Since those bullets were designed with a pure lead core and a very thin jacket for maximum expansion at practical handgun velocities, and were of truncated-cone profile, the round-nose bullet seating screw simply squeezed the hollow-point almost closed when sufficient pressure to force the bullets into the cases was applied. With the cavities almost closed and the jacket folded together over the cavity, it is not at all unreasonable that the bullets did not expand as intended.

This type of problem is encountered frequently. A set of reloading dies is purchased, often without even specifying the type of bullet which will be reloaded. The maker then simply supplies the more or less standard shape punch. The proud owner then proceeds to load all manner of different shaped bullets without any regard to whether the seating punch fits them correctly. Almost without exception, except for the one bullet for which the punch was made, bullets will be deformed to some degree. The more force required to seat the bullet, the more deformation. I have seen wadcutter bullets seated with pointed-bullet stems, resulting in the bullet being jammed so tightly into the stem that it was actually withdrawn from the case as the latter was removed from the die. At the other end of the scale, I've seen wadcutter stems used to seat pointed, or very soft-pointed bullets which were, as a consequence, smashed almost out of recognition. When any significant quantity of any particular bullet is to be loaded, it is by far the best to order the matching seating punch. However, if only a few bullets are to be used, then a typical round-nose or semi-wadcutter punch can be temporarily modified to do the job. The modification consists of simply warming the punch and then filling its nose cavity with melted sealing wax — then pressing the desired bullet into the still-soft wax and then leaving it there until hard. If the bullet is carefully aligned with the axis of the punch, the wax-lined cavity will do an admirable job. And, it can be melted out with the heat of a cigarette lighter when no longer needed, leaving the punch in its original configuration. Alternatively, the punch cavity can be filled with a quick-setting epoxy compound and given the same treatment to make it fit the bullet. Afterwards, application of heat will break down the epoxy and it may be picked out, restoring the punch.

Today's crop of jacketed pointed, soft-point, and hollow-point rifle bullets is less susceptible to deformation from improper-fitting punches than are the older types with that generous exposure of soft lead. Worst of today's crop are the short or half-jacketed varieties of hand-gun bullets with large amounts of dead-soft, pure lead core. They can almost be squeezed out of shape between thumb and finger — and a typical attempt to seat such bullets with an incorrect punch will result in their shape being completely destroyed.

Seating die adjustment is really quite simple. There is no mystery involved at all. Simply place a resized case in the shell holder and run the press ram to the upper limit of its stroke; then screw the bullet seating die down into the press and over the case until the crimping shoulder (if present) is felt to touch the case mouth; back the die out ¼ turn and lock it in place. Next, back the seating screw out part way, then start a bullet in the case and run it into the die — cautiously so as to barely start the bullet — then screw the seating plug down in successively small amounts until the full tool handle stroke seats the bullet to the desired depth. Lock the seating screw there, then proceed to seat your bullets.

If crimping is desired, first make certain that the bullet is seated as above so that the mouth of the case falls just barely below the upper edge of the cannelure in the bullet. Then, back the seating screw out of the die about four or five turns and lock it in place. Then, screw the die successively farther into the press in ⅛-turn increments, trying a case with fully seated bullet at each stage, until the desired degree of crimp is produced. Once that is accomplished, lock the die in place and proceed merrily on your way. Keep in mind, though, what was said earlier about uniform case length being essential to uniform crimp. Some handloaders of my acquaintance prefer not to regulate degree of crimp by die adjustment, but to forget about uniformity of case length and merly crimp by *feel,* with the die screwed an extra turn into the press so as to be able to crimp even the shortest case of any particular batch. If a fellow is willing to spend a few hours at practice, he can become quite proficient at regulating crimp in this manner. Personally, I prefer trimmed cases and relying on die adjustment for uniformity.

Often sufficient flaring of the case mouth to permit easy starting of bullets will leave an out-turned, sharp lip after the bullet is seated. Crimping will, of course, remove this. However, if crimping isn't desired, it can still be removed easily. Simply adjust the crimping die to roll the flare in toward the bullet without actually producing any measurable degree of crimp.

Actually, proper seating of bullets is neither difficult nor time-consuming. It is only when one starts reaching for maximum accuracy that many of the foregoing items must be given careful consideration. For normal hunting and typical high-power target work, the simple full-length resizing and one-stage bullet seating without crimping will deliver all the accuracy 99.999 per cent of today's shooters can use or appreciate.

CHAPTER 13

Inspection and Testing of Handloads

THE JOB is never done when you've dumped the powder charge and seated the bullet. All too often, the typical handloader feels that as soon as he has rammed the bullet home he's ready to dash off to the range or the field and do some serious shooting. It isn't really all that simple, and those unaccountable misses or even dangerous reactions can quite frequently be laid at the door of failure to take a long, hard look at the completed ammunition. This doesn't mean just glancing casually at the cartridge to make sure that the bullet isn't seated cross-wise or that the case isn't smeared with grease.

Proper inspection actually means determining that everything about the cartridge is right. Not all of this can be done after it is completely loaded. As we've mentioned elsewhere in this volume, one of the most important things is to be certain that you have used not only the proper amount of powder, but also the proper make and number. You would probably be surprised at the number of people, often in a hurry, who reach blindly up on the shelf, dump powder from a can into the measure, and proceed blithely on with the loading — and never dis-cover until it's too late that the wrong powder was used. Now, if you've dumped in H4831 instead of IMR3031, no harm is likely to occur — but you'll play hell having that gun shoot where it looks or develop the velocity you expected under those conditions. However, make the

Always double-check to make sure you are using the correct powder. These two look much alike, but mistake Bullseye for Unique in a heavy charge for this .45 and you'll blow the gun apart. Powders *don't* interchange, so you must be right.

same mistake in reverse and inadvertently substitute IMR3031 for a full-charge load of H4831 in one of the high-intensity magnum cartridges, and you have just set the stage for a blown-up gun and, quite possibly, serious injury to the shooter. Consequently, determining that the powder charge is exactly what you intended it to be is the first stage of load inspection, and it must be accomplished before closing up the case with the bullet. It consists first of checking and then double-checking the powder can label against the loading data you are using, then double-checking the powder measure against a good scale to make certain that it's throwing the correct charge weight. But, even that isn't enough — before the bullet is seated, each powder charge must be visually examined to make certain that all bulk is the same in the case. This is best done by taking a filled loading block into which charges have been dropped, then inclining the block toward a light source so that the level of the charge in each case can be clearly seen. Any charges that bulk higher or lower than the norm can be instantly spotted. They should be dumped and re-thrown. Then, if they still bulk incorrectly in the case, the volume of that particular case may differ considerably from the others, or there may be some foreign material occupying part of the space.

Once the bullet is seated, then it is time to give the complete cartridge a thorough inspection. Let's begin at the head of the case.

Modern Handloading

Primer must be seated flush or below the case head, not protruding like this.

Is the primer seated flush with or slightly below the case head? If it protrudes at all, the stage is set for at least poor ignition; and in extreme instances, especially in semi-automatic guns, an inadvertent firing as the breech closes. Is the primer marred, crushed, or deformed? If so, ignition is bound to be less than standard. If otherwise acceptable, cartridges with the aforementioned primer defects may be fired for plinking purposes — or simply to salvage the cases — if carefully and gently chambered in a bolt-action rifle.

Check the case head, rim, and extraction groove. Is the rim straight or has it been bent either by the gun which last fired it or by resizing? Are there burrs of sufficient size to interfere with extraction thrown up into the extraction groove by either the shell-holder or the last gun in which the case was fired? If the case was badly bulged by previous firing in an oversized chamber, the resizing die may have wiped a thin ring of brass back over or into the extraction groove. This can cause extraction trouble, but it also indicates more than a little weakening of the case in that area.

Move on up the case, watching for dents, splits, or other visible damage to the case body. Minor dents won't cause any trouble; firing will iron them out. But, cracks or splits should be set aside. Use them for plinking, where a case separation isn't likely to cost you a trophy, then discard the cases to avoid inadvertently using them again.

Examine the shoulder closely for oil dents or folds. Neither is unsafe for one more firing, but the case may be of no further use afterwards. An overlong case or incorrectly adjusted seat-crimp die will sometimes cause the shoulder to be punched inward or folded upon itself. This ruins the case for further use and probably will make it impossible to chamber that particular round. Take a close look at the case neck. There should be no splits or tears and if a crimp is applied, it should be uniform around the full perimeter and should not have lead or jacket material piled up in front of it. Naturally, bullets should be held in the

Check all loaded rounds to insure they will enter the chamber freely. The Colt .45 auto allows removal of the barrel to simplify this check, but most guns do not, so cartridges must be run through the action.

neck as nearly concentric with the case as possible. A simple check for this is to merely roll the loaded cartridge across a flat surface (such as a piece of plate glass or steel plate) and watch the bullet point — if it swivels in a circle more than 1/16″ in diameter, that round is not likely to shoot very close to center of impact. Look also for damage to the bullet, perhaps from the use of an improper seating stem. Last, but not least, make certain the bullet is seated to the correct depth, and that *all* bullets in the lot are seated to uniform depth. Perhaps for "depth" there, we should read "overall length." Cartridges loaded too long will not feed through conventional box magazines, and in some instances those loaded too short will not be properly guided into the chamber. And, of course, while doing all this you have made certain that any sizing or bullet lubricant has been wiped off with a cloth lightly moistened with lighter fluid or some other volatile solvent. On second thought, don't use lighter fluid; stick with non-flammable solvents, and you will have eliminated one fire hazard.

In spite of all the foregoing inspections, it still isn't possible to be absolutely certain that each and every round will feed and chamber correctly when the adrenalin is racing through your system and you are frantically trying to get a fast second shot off to salvage a poorly hit or missed trophy. (And, if I might digress just a moment here, be assured that the steps to be described next also apply to factory-loaded ammunition. I recall vividly missing a fine white-tail buck early one morning, then discovering that the next factory-loaded round in the magazine simply wouldn't enter the chamber, no matter how hard I struggled with the bolt handle.) So, the only way you can be absolutely certain that each cartridge will feed correctly when the time comes is to run them all through both the magazine and chamber before you leave for the hunt or range. Back in the days when we all shot mostly Springfields or Mausers with the original military safety and some of the earlier sporting rifles with similar safeties, it was possible

Modern Handloading

to fill the magazine and run the cartridges through the chamber without hazard. One simply placed the safety in the middle or "safe but not locked" position and proceeded with the job. Those old three-position safeties made sure an accidental firing didn't take place. Today, though, that very valuable feature doesn't appear on many new rifles. Consequently, the only completely safe manner in which you can jack live ammunition through the chamber is to first remove the firing mechanism. Unless your particular rifle permits that, don't under any circumstances attempt to chamber-check your handloads inside the house or in a built-up area. Take it to the range or other safe place where the accidental discharge you're sure won't happen can't cause any trouble if it *does*. Run *every round* you'll be taking on a hunt through *all* of the guns in which they'll be used.

When it comes to checking handloads through pistols and some rifles, there is no way they can be made completely safe and yet it is impractical to remove the firing mechanism as in bolt-action rifles. Again, when that happens, and you really need to be certain every round will feed, you have no choice but to go to the range or other safe place for the job. Believe me, I know; somewhere there's a .45 caliber bullet hole in the wall which proves it can happen, no matter how careful I was. It shouldn't need to be said that handloads to be used in handguns for defensive or service purposes should always be run through the gun to make absolutely certain no problem will be encountered when the chips are down.

But, even that isn't all there is to inspection. At every step along the way, from the first look at a fired case, you should be inspecting. Other chapters devoted to the various loading operations contain numerous inspection points. If all are attended to, you'll begin assembly with perfect, thoroughly checked cases, primers, and bullets. And, if you are to have safe, reliable, and accurate handloaded ammunition, continuous inspection must be the byword.

So much for inspection, but what about testing? There is no other way to determine what ammunition will do in a particular gun except to fire it in *that* gun. If that seems like unnecessary attention to a perfectly obvious point, rest assured that it is not. Hardly a season goes by that I do not hear of someone — sometimes even an advanced handloader — who has at the last minute whipped up a new load and dashed off madly after game without ever making certain it will perform properly in his gun. One fellow I know put several rounds over the back of a trophy whitetail buck this past season simply because he had changed bullet and powder charge ("but only a little bit, so it can't matter much") and hadn't rechecked performance in his rifle. The year before, it was a fellow who switched loads in his gas-operated semi-automatic rifle and didn't "waste the time" to make certain the new fodder would function the action. It didn't, and he lost a wounded elk as a result. Then, there was the police officer who ran off a bunch of handloads for his 9mm submachine gun and didn't make the simple tests to insure they would operate the gun reliably. Needless to say, when the occasion came to use that gun and ammunition in serious social intercourse, he had a fine single-shot combination. His little oversight could have easily gotten him killed.

In any event, I hope that the necessity for some degree of testing has been brought out by the foregoing anecdotes.

If you chamber a round and pull the trigger and it goes "bang," you've performed a test of sorts. You've proved that the cartridge will fire in that particular rifle. Don't laugh, I've seen the time when just such a simple test would have saved a lot of trouble, particularly in rifles with the Remington-type spring-clip extractor combined with the rimless case whose shoulder had been set back overmuch in resizing. That little extractor simply shoved that too-short case so deep into the chamber that a minimum-protrusion firing-pin couldn't set the primer off.

So, the first test is simply to determine that the ammunition will fire in the gun. The second, especially in the case of semi-automatics or lever or pump-guns with limited extraction power, is to insure that the fired case will extract easily and that the action will feed the next round from the magazine. All this can be done by simply firing into a convenient creek bank at a range of ten feet or so.

The second and no less important test is that for accuracy. By this I do not mean that the ammunition/gun combination must shoot ten-shot groups within one minute of angle — not by any means. The combination must be shot sufficiently to determine that it produces the degree of accuracy required for *your* particular purpose. If your game is 300-yard prairie dogs, then even 1 MOA isn't good enough — on the other hand, if you are after Kodiak at twenty-five feet in an alder thicket, anything that will hit a gallon bucket at fifty yards will do the job nicely. Needless to say, you'll require a proper sight setting, "zero," if you will, before any reasonable accuracy testing can begin. In a way, this can be a form of testing. Your gun may be completely and properly zeroed for a previous load, yet a good bit off for a similar but new handload. Many a handgunner and police officer is greatly surprised to find that even a heavy six-gun shoots substantially lower with high-velocity, light-bullet handloads than with slow-poke factory

Modern Handloading

fodder. And, the difference is often enough to be highly embarrassing, even fatal. No one can predict exactly what a new load will do insofar as shifting center of impact is concerned. Nothing less than thorough targeting will determine this.

With handguns, I find twenty-five yards range adequate for preliminary accuracy testing; with rifles it's 100 yards. At those relatively short ranges it's possible to get a pretty fair idea of what the load and gun will do. If it's too bad, you have saved the time that would have been spent in setting up for the longer ranges. If the results are acceptable, you can then move on out and do additional shooting. It should go without saying that you must have knowledge that the gun in question is capable of a certain level of accuracy before you can learn anything about the handload being tested. If a rifle has proven it can shoot within, say, one or two MOA in the past, then you'll know four- to six-inch groups are the fault of the ammunition when you're doing your job. On the other hand, if a gun can do no better than four to six inches, you'll never know if the ammunition is any better than that.

The more you can do to eliminate your personal error, the better you are able to test ammunition. This means simply shooting from a solid rest of some sort — preferably a comfortable bench rest, though you may prefer shooting from a prone position over a bedroll or similar rest — with wind, light, and temperature conditions as near ideal as possible. A cloudy, windy, day with temperatures down near zero is a hell of a time to learn anything about ammunition. Neither you nor the gun can do its best. If you would learn the most about your ammunition's performance, take the time to read up on the better techniques of bench-rest and other target shooting. Pay particular attention to getting each shot off under uniform conditions of wind and light, and save wear and tear on both your eyes and legs by using a spotting scope and a fairly high-magnification scope on the rifle. Above all, don't settle for just a couple of reasonably good five-shot groups. You really aren't learning anything about the ammunition's performance with so small a test. Ten or a dozen such groups will tell you many times more about the load's *average* behavior. With that many groups assembled, then you can calculate a reasonable average and know what to expect in the future. If only a couple of groups are fired, they may represent *both* the top and bottom limits of the ammunition's capabilities, or they might represent *either* the top or bottom and thus give you a completely erroneous impression. Substantially more about this aspect of testing will be found in the chapter devoted to load development, so we'll not dwell upon it here.

There is one other form of testing that can be of considerable value, especially when you are hoping to duplicate the performance of an existing load — which may be a handload, or a factory standard. It may be that you want to achieve the same flatness of trajectory, but you don't have a chronograph available so as to be able to duplicate velocity. No real problem — simply shoot 100- and 200-yard groups with the load you want to duplicate, with the gun properly zeroed, then re-zero with the handload to be tested, and shoot the same series of groups with the same sight picture. Assuming the shooting is done with the rifle zeroed at 100 yards, the amount of drop shown at 200 yards will indicate quite clearly how closely the trajectories are matched.

There isn't any really good way to test high-velocity rifle bullet performance these days. If you've plenty of time and money, you can build up large blocks of Duxseal (large enough to be certain you can hit them somewhere centrally) and shoot into them at the various ranges. The bullets can then be recovered and examined. Duxseal does not duplicate the effect of animal tissue exactly, but it does provide a sound basis for comparison. Consequently, if you have a bullet that you know performs well on game, both it and any others you may wish to compare may be fired into Duxseal and the recovered bullets will give a clear indication of their relative effectiveness. With Duxseal at sixty-five cents per pound retail, and twenty-five or thirty pounds being necessary for a block to be hit reliably at 100 to 200 yards with most rifles, this can get a wee bit expensive. Also, in order for results to be reliable and consistent, the Duxseal blocks should be at a temperature reasonably close to 70° at the time of the shooting. This rather restricts the times when such tests can be conducted unless there is a heated building near the range where the blocks can be conditioned and then taken out briefly for shooting. All in all, a hell of a lot of work. Still, a great deal can be learned in this fashion, and if you are serious enough about digging out such information, it is not an entirely worthless project.

On the other hand, handgun bullets of all types can be readily tested for penetration and expansion in Duxseal. Most of my own testing has been done with this material indoors at a range of ten feet. With no more space than this required, the average basement or garage is quite adequate (providing the neighbors don't say too much about the noise) and the temperature of the testing medium is easy to control. Since the handgun is essentially — in my opinion — a close-range weapon, I feel the ten-foot range is quite adequate for bullet performance testing. However, if you must shoot an elk at fifty yards with a .44 Magnum, and want to learn just which bullet is likely to do the best job, then you will simply have to build up large blocks of the material and shoot the tests out of doors at that range.

Even some of the lightly-regarded, low-power handgun cartridges have surprising penetration in Duxseal with non-expanding bullets. This includes all of the full-jacketed auto-loader and revolver cartridges, and most of the plain lead bullets driven at so-called standard velocities. Even the latter types do not often expand to any measurable degree and retain their point shape well. Consequently, they will often penetrate eight inches or more of Duxseal. This means simply that you must either use relatively thick blocks of the material, or place some sort of backstop behind the blocks. Generally, when shooting indoors, I keep a twenty-five-pound block behind the much smaller block (about ten pounds) into which I'm shooting. Thus, when a bullet does penetrate the test block completely, it is caught and held with very little penetration in the large block. On the other hand, bullets designed expressly for close-in defensive work generally expand violently and penetrate only two or three inches at most — that is, they penetrate no more than that *if* they expand as their designers claim. Some do not, and this results in penetration almost equal to that of the non-expanding bullets. For example, the Norma 230-gr. .45 ACP bullet with exposed lead nose and fairly deep point cavity *looks* as if it would expand well — yet it expands very little in Duxseal and penetrates roughly six inches. At the other end of the scale we have the Super Vel 190-gr. jacketed hollow-point bullet which expands violently to .85-.90" diameter and penetrates only two or three inches. Those results, incidentally, were obtained both with factory loads and with handloads intended to duplicate factory velocities and pressures. The point of all this is simply that it is essential that you either *know* how much penetration to expect from the load being tested, or that you provide adequate backstop material of some sort in the event that the bullets do pass through the test block. The importance of this can hardly be overemphasized when the temptation is so great to do this sort of thing indoors during the cold winter months when shooting outside holds little appeal.

Personally, I have found the use of Duxseal to compare handgun bullet performance highly enlightening. Not only is it superb for comparing the performance of factory bullets in handloads (and also in factory loads) but it is ideal for determining what bullet shape and what bullet hardness will produce the maximum shock or wounding effect from your particular gun.

Another test often conducted by handloaders but not really recognized as such is simply that of examining closely the fired case. Discounting evidence of excessive pressures which is covered in detail elsewhere, a fair amount can be learned about both ammunition and gun in this way. A heavily-expanded area just ahead of the solid web indicates an over-sized chamber; annular grooves on any part of the case indicate a poorly-reamed or finished chamber; and measurement of the diameter of the case now will tell you whether the chamber is oversize at any particular point. And, it is possible by comparing the shoulder of a fired case with that of an unfired specimen to determine that the chamber is improperly cut. And, of course, any large amount of excess headspace in the rifle can be determined from examination of the case as described elsewhere. If the neck of the fired case will not permit a proper-diameter bullet to be inserted by hand, either the neck walls have become too thick and require reaming, or the case has grown too long and the mouth is being pinched inward by the forward part of the chamber. Both of those defects will raise pressures and reduce accuracy. And, you might look for long, deep, longitudinal scratches or scrapings on the case body which can indicate that the feed-rails or lips need smoothing and polishing or that the chamber mouth is too sharp and needs similar attention. A cratered or extruded primer when no other indications of high pressure are visible usually indicates excessive firing-pin protrusion and/or loose firing-pin fit in the bolt face. A careful thirty-second examination of a couple of fired cases can easily tell you numerous things about the gun and ammunition that you might never have otherwise learned.

In any event, the handloader who simply loads up a batch of cartridges and then goes out shooting them willy nilly in plinking at target or at game without having conducted reasonably intelligent tests is really wasting most of his time and effort. He may even wind up wasting the entire cost of an expensive hunting or match expedition. Keep this in mind the next time you are loading, and follow through with the testing and you'll be surprised at how much more you know about the performance of your ammunition.

CHAPTER 14

Load Development

AS IT comes from the factory, store-bought ammunition is designed and produced to give *acceptable* performance in every extant gun in the same caliber. Also, because it is mass-produced as are the guns that use it, it must incorporate all manner of tolerance, both in dimensions and performance, so that it may be used in an extremely wide variety of guns.

For these and many other reasons factory ammunition cannot hope to extract the maximum performance out of any one gun. Neither can one single load be expected to perform equally well on everything from jack rabbits to elk. Reloading in and of itself represents no performance improvement over factory loads. Only when you carefully tailor load to gun, to target, and to purpose can you say your product is better than Bridgeport's or Alton's. So speaking, we get into load development.

Load development is a pretty loose term applied to those processes by which one matches the various components and performance of the reloaded cartridge to the gun and to the purpose the ammunition is intended to perform, and to the ranges and conditions under which it will be used.

In short, one must come up with the correct bullet, correct case, right powder (and the right amount of it), the best primer, the best bullet seating depth, the right primer seating pressure, proper amount of case crimp on the bullet, etc., etc. — all within the limitations imposed by the gun and the use to which it will be put.

Even the development goal is not constant. The bench-rester seeks pure accuracy and will sacrifice all other performance factors to that end. The varmint shooter also wants maximum accuracy but must temper that demand to obtain satisfactory degrees of velocity, wind-bucking ability, flatness of trajectory, explosive bullet expansion, and even barrel life. The plinker tempers all other demands with economy. The fellow after elephant or Cape buffalo at short range will sacrifice all other factors for absolute reliability and smashing power and yards of penetration. The competitive handgunner wants maximum accuracy teamed with minimum recoil. And on and on into the night. Far too many people think of load development as being aimed purely at increasing accuracy. That ain't necessarily so. Many a load development project is undertaken for entirely different reasons.

When one considers the vast variety of bullets, powders and primers available today, load development can be an almost never-ending process. There are, however, guide lines one can follow that will simplify matters. Then too, there is a practical limit beyond which continual fussing with minute variations isn't really worthwhile. It's possible to shoot out a barrel before reaching that *ne plus ultra* load.

First of all, there are two entirely different goals in load development, and they require different techniques, equipment, and methods. Pure accuracy of the highest order is the bench rest shooter's main concern. He cares not whether the bullet will kill game, nor does his ammunition have to withstand rough handling or the stresses of recoil and magazine feeding. Neither would he really care if it took a half hour to load each cartridge. The bench rest shooter will put up with virtually any inconvenience to achieve the utmost in accuracy, disregarding everything else firearms and ammunition were originally intended to do.

Load development for the average shooter means a load for hunting — one that delivers maximum accuracy with a bullet that will kill his game properly. Equally important, the loaded cartridges must function unfailingly through his rifle, and they must not develop dangerous pressures under any climatic conditions he may encounter.

Let's take a typical example that isn't aimed at pure accuracy. I had arranged to go on a two-week brown bear hunt in southeastern Alaska. Not only was a good trophy wanted, but a photographer was being taken along to get closeups of *live* bears. The trip would naturally be expensive, and the photog valued his hide rather highly.

With all this riding on perhaps a single shot, it was well worth considerable effort to make certain the ammunition was the best available. The .300 Winchester Magnum caliber was chosen, in a light, full-stocked, 20" barrel carbine built by DuMoulin on an FN action. The new gun arrived about six weeks before the hunt, and was appropriately fitted with scope, Pachmayr Lo-Swing mount, and auxiliary iron sights.

Minimum requirements the grizzly load must meet were established as: 180 gr. bullet at 3,000 fps; four-inch five-shot groups at 200 yards; easy extraction of fired cases; perfect feeding from the magazine; maximum bullet expansion coupled with broadside penetration of the adult brown bear.

The Nosler Partition Bullet (180 gr.) was selected on the basis of its capability for the forward half to expand rapidly out to long range, while the rear portion continues on, relatively undeformed, to achieve maximum penetration. This same bullet had also produced one-minute-of-angle and better accuracy in two other rifles. All load development was to be centered around this bullet. It was decided ahead of time that minor concessions in accuracy and velocity could be justified in order to retain this bullet whose well-established performance was considered essential. Two hundred of the same lot of the bullet were procured, more than an ample supply for both load development and the hunt. After all, it only takes one shot to down a bear.

New, unprimed cases were selected. Those with necks of uneven thickness were set aside for less demanding tasks. Primer flash holes were sized with a No. 80 drill, and burrs on the inside of the case head (produced by the punch that formed the flash holes) were removed with a small countersink. Cases were trimmed to uniform length and chamfered lightly, inside and out, at the mouth. Though cases were new, necks were trued up by resizing and expanding.

Published test results indicated that CCI Magnum (large rifle) primers would produce extremely uniform velocities in this large 75-grain case. So CCI Magnums were chosen.

Having settled on bullet, case and primer, only the powder type and amount remained to be chosen. The initial choice could be made from published data, but would require extensive shooting to confirm it.

The combination of large capacity, sharply bottle-necked case and a heavy bullet requires a very slow burning powder if high velocities are to be obtained at reasonable pressures. Several such powders are available: DuPont 4350; Hodgdon 4831, H-450, and H-570; Hercules

ReLoder 21; and Norma 205. It was decided to try all six.

From existing loading data, starting charges were selected. Each manufacturer had published charge recommendations for the .300 Win. Mag. with 180 gr. bullets. For each powder, a charge just below the middle of the recommended range was selected. Since all of these powders are similar in burning rate, charges ran much the same — 73.5 gr. 4350, 75.0 gr. H-450, 75.0 gr. 4831, etc.

Ten rounds of each starting load were prepared, with all possible attention being given to detail. Rifle, ammo, tools and components were then carried to the range. Since, at that time, I had a fully equipped loading shop installed in a Chevrolet Corvan, it was possible to do all further load development right on the range.

With rifle solidly sandbagged on the bench, it was first zeroed at 100 yards with factory loads. It had not, up to this moment, been fired at all. Once laid on, the factory loads made three-shot groups of 1″ and slightly less. This is unusual, and would convince anyone except a true gun nut that he should hunt with the factory stuff and forget reloading. However, I wanted the added performance of the Nosler bullet, so work proceeded apace.

Firing with the 75/4831 load produced two five-shot groups of 1½″ and 2⅛″ respectively. Respectable, but not good enough. Shooting was done slowly, about one shot per minute, to avoid excessive barrel heating and attendant "walking." While the barrel was cooling from previous firing, ten rounds were loaded with 76 gr. of 4831, a compressed charge in the lot of cases being used. This load produced groups significantly larger than 75/4831. No further charge increase was possible while keeping cartridge overall length short enough to feed through the FN magazine. Consequently, ten more rounds were loaded with 74/4831. This load grouped 1 1/16″ and 1⅝″, best of the lot with 4831. However, at this charge level, the desired 3,000 fps velocity was not being produced. For the time being, 4831 was set aside in the hope that one of the other powders would give better accuracy at the desired velocity.

Next in line in the loading block was 74.5/4350. The first five spread 2½″, auguring no good. A little barrel-cooling time and the second group was no better. Varying the charge a grain either way produced no earth-shaking improvement, so 4350 was likewise set aside.

Hodgdon's H-450 came next, the first five going into 1¾″ — but one of those was called out, the other four just nudging an inch. This with a 75.0 gr. charge. So, tap the foot while the barrel cools a bit, hoping the next group will do as well, without the wild shot. It did, with all five neatly clustered, barely under 1¼″. Dropping the charge

Modern Handloading

a grain does just as well, and increasing it one grain produces a three-group average of 1 1/16″. Not quite minute-of-angle, but the best by far, and only a hair over half the group size I was willing to accept.

But how is it for velocity? Unload the trusty Avtron T-333 chronograph, and when set up, it indicates a three-shot average of 2,961 fps. Well, that isn't quite the 3,000 we wanted, but no bear will be able to feel the 39 fps shortage.

While no pressure gun is handy, it can be determined from careful examination of the cases and primers that pressure is not excessive. Fired primers are still nicely rounded at the edge, and cases show no measurable (only about .0005″) head expansion. Nor are there any bright spots on the rear face of the head. Extraction was very easy, requiring virtually no effort.

Enthusiastic over these results, I almost ignore the H-570 and Norma 205, but decide it's best to shoot them anyway, even though the bullets cost twelve cents each. Neither does as well as 76/H-450, so I'm happy.

Now that we know the 76/H-450 load will damn near make minute-of-angle five-shot groups on the range, what will it do under field conditions? There's only one way to find out. Clean and dry the bore, make certain barrel has cooled to air temperature, then fire one very careful, lonesome shot, and walk away. Go down and check the lake, or, perhaps, do some pistol shooting while the barrel cools again. At least a fifteen-minute wait is required (better thirty) before another shot is fired.

This goes on for five three-shot groups, not all in the same day. Luck is with me, (so are the components and that fine DuMoulin barrel) and the five-group average is just a hair under an inch. This is superb accuracy from a light hunting gun, and luck of the draw plays a great part. The next identical gun shipped by DuMoulin (or anyone else) might come nowhere near such performance.

Each of the test shots was fired from a cold (air temperature) barrel because that duplicates normal first-shot hunting conditions. A rifle may well produce a close five-shot group at point of aim when fired fairly rapidly, but a different group at a different point when each shot is from a cold barrel. Duplicating the latter condition for a final test will uncover any such differences.

So that particular job of load development is done. Of course, after a few hundred rounds, the barrel may change its taste somewhat. If it does go sour later (and hasn't actually been shot out) and the stock isn't at fault, it can *probably* be brought back into life with further load development. As it is, I have a new rifle/load combination that will deliver less than minute-of-angle accuracy from a cold bore. It will put that first shot of the day somewhere within a half-inch of point of aim at 100 yards. Such a combination gives one tremendous confidence when going after mean, ugly game that fights back. It also makes the photographer's wife feel much more secure. After all, he's armed only with a pair of Leicas, and might need help. In any event, the man who goes through the foregoing load development procedures is far more likely to make a first-shot kill than the John who grabs a box of drugstore cartridge and heads forthright for the timber.

When no loading data exists for the cartridge with which you are working, you can still come up with starting charges in one of three ways. Transposition of available data from another caliber is probably the most common. When this is done, use data from a cartridge having the same bullet diameter and weight, and the same or less powder capacity. Because of the many variables involved, reduce charge weights of such transposed data by 5 per cent for starting loads.

As an example of this procedure, the .284 Winchester came out in 1963, and no loading data was to be had when my first test gun was completed (before any factory guns were shipped). Perusal of references indicated the 7mm/06 had virtually the same capacity, and same bullet diameter. Starting loads were taken right from 7mm/06 data, reduced 5 per cent. Subsequently, when laboratory-tested data for the .284 was released, it was virtually identical to that developed in my gun. The same situation existed in 1965 with the .350 Remington Magnum. Initial loads were taken from the same-capacity .35 Whelen data which existed in plenty.

If you are mathematically inclined, there is another approach available. Smokeless powder has known burning rates, energy content, etc. Standard formulae can be used to calculate the amount of any given powder required to produce a specific performance level.

However you may not have a large enough library to find data that fits your needs. Then too, you may not be an accomplished mathematician, so the calculations route is out. All is not lost. Get a *Powley Computer For Handloaders* and read the instructions carefully.

This computer was designed by ballistician Homer Powley to allow the average handloader to calculate basic loads for any rifle caliber. It is essentially a slide rule of simple construction. Procedures for its use are quite simple, repeated here in part verbatim from the Powley *Instruction Manual*:

"GOOD WORKING PRESSURE"

Your computer is suitable only for modern arms in first class condition. When this is the case it has been arranged so that the pressures you obtain as measured by copper crusher gauges will be well within the middle of the 40,000 to 50,000 psi range. This has been arranged with the idea that you will be able to have barrel life of several thousands of rounds and you can anticipate no difficulties from sticking cases, leaking primers, short case life and other similar headaches and intermittent troubles resulting from higher pressures. We are not attempting to work in the vicinity of maximum loads because experience indicates that many of the troubles just enumerated are likely to occur.

Since errors in gun measurements and slide settings are possible, we suggest that you check the computer selections against recommendations in any of the several loading handbooks and manuals available to you. Should you not be able to find the particular load in question, we suggest that you follow the standard handloading procedure of approaching an untried load by starting upwards from 10 per cent less powder charge.

There are instances where you, on your own, may consider it desirable to increase powder charges above the levels indicated by your computer. If you have a gun which has extra strength built into it and you are willing to sacrifice some barrel life, you can find loads in the handbooks and manuals which will be suitable for such situations.

Each gun has its own peculiarities so that with a given set of components you may obtain a different pressure and, therefore, velocity than another similar gun. For this reason, the Expansion Ratio-Velocity Tables on your computer can be regarded as giving you only a close estimate of the velocity to be expected. However, they have been included for your use because when you are making changes in loads, the DIFFERENCES indicated in the velocity table are quite accurate. Further more, they enable you to evaluate the performance of many guns which may not be available to you and you wish to know more about them.

PART II
USING YOUR COMPUTER

Please pay particular attention to all directions and especially the important Definitions appearing on the computer itself. It is very important that you make your own measurements of barrel length and water capacity of your cases. Case capacity will vary with seating depth which is probably not the same as for published loads with which you may be familiar. Case capacity varies with the brand of case being used; 3 to 5 per cent variations have been seen. Your computer then will use your own measurement and allow such variation. Use the average of several readings because case capacity will also vary among the ones you are using.

You will find it very helpful if you write down on a piece of paper the results at each step as you calculate.

TYPICAL EXAMPLE

You have a .308 Winchester. The case capacity is 51.5 gr. with a 150 gr. bullet. The powder charge is then 44.3 gr. from the slide. You also find the Ratio of Charge to Bullet Weight is 0.295. Reading from the Sectional Density Table you find the value of 0.227 for the 150 gr. bullet in .30 caliber. 4064 is the selected powder. You have a 24-in. barrel but this is not the number to use with the computer; you need the distance the bullet travels which is the effective barrel length for your seating depth. From tip of seated bullet to muzzle measures 21 5/16 on your cleaning rod. The bullet is 1-1/16 long to make the effective barrel length 22-⅜. This is close enough to 22.4 in. for the computer to tell you the Expansion Ratio is 9.0. The Expansion Ratio-Velocity Tables with a Ratio value of 0.30 predict a velocity of 2730 f/s from using 44.3 gr. of 4064. Before using this load in your gun, be sure to check in one of the Handloading Manuals to see that this is a load which has been used before. Errors in measurements and slide settings are always possible, so ALWAYS take the precaution of running a check on your figures whenever you can.

SPECIAL CONDITIONS

We have tried to make your computer for any caliber, bullet, gun or case combination you may wish to use. However, it may happen that you will use a caliber, for instance, which is not on the Caliber Slide. If this is the case, you will have to figure you own Expansion Ratio, taking careful note of the Definition. It is not difficult. You will use the average between bore and groove diameter, in inches. The case capacity in grains of water can be converted to cubic inches by remembering that one cubic inch of water contains 253 grs.

If we haven't anticipated a future bullet weight in your caliber, you will have to figure Sectional Density. Square the diameter, in inches, of the bullet and then multiply by 7,000; the bullet weight, in grains, can then be divided by this product to give the Sectional Density.

If your case capacity is greater than 140 grs. your powder charge is figured at 86 per cent of the case capacity. You must then figure your own Ratio of Charge to Bullet Weight by dividing the weight of powder by the weight of the bullet. Also figure Expansion Ratio, then use your computer as usual.

If the Expansion Ratio of your gun comes to a value less than 4.0 your gun will not perform properly with loads selected by your computer. Special purpose guns have been made for military uses in this area but pressures are extremely high, barrel life is very short, muzzle blast is quite severe and accuracy is not good.

POWDER NUMBERS

As you can see, Powder Numbers do not fill the slide and we have used letters of the alphabet to fill in the vacancies. Whereever a letter appears it means that there is no exact powder currently available for the combination you wish to use. This does not mean that we cannot come up with a very satisfactory load, however.

In the first place, any powder appearing to the LEFT of Arrow 2, will produce less pressure and velocity than we want. Any powder on the RIGHT of the arrow will produce HIGHER pressure than we want.

Remember this on those very few occasions when Arrow 2 happens to fall exactly on the dividing line between two powders, as between 4198 and 4227, for instance. When this does happen which way you go does not matter from the pressure standpoint because the change in pressure will not cause trouble. Therefore, the general rule when this happens is to use the powder on the LEFT of Arrow 2. If your Expansion Ratio is less than 5.5 then use the powder on the RIGHT of Arrow 2. For a 180 gr. bullet in .308 Winchester Arrow 2 is between 4320 and 4064. The Expansion Ratio is 9.1 so use the indicated amount of 4320.

What do you do when the first powder selection is a letter and not a number? We will take up each case in reverse order:

"G" POWDER IS THE FIRST SELECTION: This is "off scale" for your computer, so consult the Handloading Handbooks.

"F" POWDER IS THE FIRST SELECTION: Take the Grains of Powder indicated by Arrow 1 and ADD to it 5 per cent of this weight. Reset Arrow 1 to this increased amount. Use this weight of 3031. You will next find a new Ratio of Charge to Bullet Weight on the basis of this new weight of 3031. Use the new ratio for finding velocity. See under D POWDER for a similar numerical example.

"E" POWDER IS THE FIRST SELECTION: Take Grains of Powder indicated by Arrow 1 and ADD to it 4 per cent of this weight. Reset Arrow 1 to this increased amount. Use this weight of 4064. Next find a new Ratio of Charge to Bullet Weight on the basis of this new weight of 4064. The new ratio is used for finding velocity. Example: You want to use a 110 gr. bullet in a .308 Winchester. The case capacity is 53 gr. so Arrow 1 indicates the powder charge to be 45.5 gr. Arrow 2 gives E powder. 45.5 x .04 is 1.8 gr. New powder weight is 45.5 plus 1.8 or 47.3 grs. of 4064 which you will use. Set Arrow 1 on 47.3 to find the new Ratio of Charge to Bullet Weight is .430. Using this Ratio you find the velocity at an Expansion Ratio of 8.9 is about 3190.

"D" POWDER IS THE FIRST SELECTION: This will be handled nearly the same as F above. Take Grains of Powder indicated by Arrow 1 and ADD to it 5 per cent of this weight. Reset Arrow 1 to this increased amount. Use this weight of 4350. Next find a new Ratio of Charge to Bullet Weight on the basis of this new weight of 4350. Use the new Ratio for finding velocity. Example: You are going to use a 180 gr. bullet in .30-06. The case capacity is 61.5 gr. so Arrow 1 indicates the powder charge to be 52.9 gr. Arrow 2 shows D powder. 52.9 x .05 is 2.7 grs. 52.9 plus 2.7 is 55.6 gr. of 4350 which you will use. Set Arrow 1 on 55.6 and find the new Ratio of Charge to Bullet Weight is .308. A Ratio of .31 with an Expansion Ratio of 7.5 gives a velocity of 2680 f/s.

"C" POWDER IS THE FIRST SELECTION: Here is where we have to use the powder on the right of Arrow 2. Take Grains of Powder indicated by Arrow 1 and SUBTRACT from it 5 per cent of the weight. Reset Arrow 1 to this smaller amount. Use this weight 4831. Next find a new Ratio of Charge to Bullet Weight on the basis of this new weight of 4831.

The new Ratio is used to find the velocity. Example: You have a 100 gr. bullet for the .264 Winchester. The case capacity is 82 grs. so Arrow 1 shows 70.5 grs. 5 per cent of 70.5 is 3.5 grs. 3.5 grs. SUBTRACTED from 70.5 leaves 67.0 grs. of 4831 to be used. Resetting Arrow 1 on 67.0 shows the new Ratio of Charge to Bullet Weight is .670. From the Expansion Ratio-Velocity Tables you find this corresponds to a velocity of 3440 f/s at an Expansion Ratio of 5.0. You have to read between the lines in the table because the velocity at .65 is 3400 and at .70 it is 3500.

"B" POWDER IS THE FIRST SELECTION: There is no choice here but to go to the left of Arrow 2, regardless of the Expansion Ratio. Take Grains of Powder indicated by Arrow 1 and ADD to it 5 per cent of this weight. Reset Arrow 1 to this increased amount. Use this weight of 5010. Next find a new Ratio of Charge to Bullet Weight on the basis of this new weight of 5010. Next find a new Ratio of Charge to Bullet Weight on the basis of this new weight of 5010. Use the new Ratio for determining velocity. Example: You want to use a 140 grs. bullet in the .264 Winchester. The case capacity is 79 grs. so Arrow 1 shows 68 grs. of powder and Arrow 2 is at B powder. 5 per cent of 68 is 3.4 grs. to be added to 68.0 so you will use 71.4 grs. of 5010. Resetting Arrow 1 at 71.4 now shows a new Ratio of Charge to Bullet Weight of .510. With an Expansion Ratio of 5.1 and the other ratio of .51 the velocity is indicated to be about 3080. Here again you have to read between the lines in the table.

"A" POWDER IS THE FIRST SELECTION: This is "off scale" for your computer and there is no data available. In this region DuPont made a powder with number 7013 but we have no data.

5010 powder is quite similar to another which has been made with the number 7005. Evidently these were made during the past war primarily for .50 caliber machine guns. There are few rifle cartridges where this would be selected but we can give one example which has been successful. This was a 6.5mm 139 grs. bullet in a case necked down from the .300 Weatherby with an Expansion Ratio of 5.0 due to a barrel of overall length of 30-in. The case capacity is 95 grs. and the indicated powder charge is 81.7 grs. of 5010. Velocity is shown on your computer as about 3220 f/s. This gun was designed by Lt. Col. Paul Wright of Silver City, New Mexico.

There are applications for your computer which come on occasionally and are regarded as "rough". Yet your computer will handle these just as easily as though it were a problem pertaining to cartridges which have been standard for fifty or more years. Robert Hutton had a .378 Weatherby necked down to .30 caliber and wanted to check a 77 grs. bullet. The overall barrel was only 26-in. so the Expansion Ratio was only 4.5. The Sectional Density of this bullet is low, only .106. Your computer would have selected "D" powder at a weight of 107 grs. Ordinarily with such a low Expansion Ratio we would want to decrease this charge and use 4320 powder on the right of Arrow 2. However, Robert Hutton did use 4350 on the left. Therefore, let us increase 107 grs. by 5 per cent which makes a charge of 112 grs. This gives a Ratio of Powder Charge to Bullet Weight of 1.45, more powder than bullet weight. Your computer gives a velocity of 4480 f/s for this combination. Robert Hutton actually fired 115 grs. of 4350 and the chronographed velocity was 4615; not a bad check for such unusually extreme conditions.

CHANGING POWDER CHARGES

By now it has probably occurred to you that, for most cases, your computer will develop relatively "mild loads." For various reasons, the question always comes up as to what happens when you change the amount of powder. If you change the weight by a given percentage, inside the same case, the velocity changes by the SAME GIVEN PERCENTAGE. However, at the same time, the pressure changes by TWICE THE GIVEN PERCENTAGE. That is to say, if you reduce the powder weight by 10 per cent the velocity goes down by 10 per cent and the pressure goes down by 20 per cent. If you raise the powder charge by 3½ per cent the velocity goes up 3½ per cent but the pressure goes up by 7 per cent. And so on.

For the most part, your computer still gives you a little leeway in going up on powder charges and more leeway in going down. Suppose that, for some mechanical reason, your cases are sticking so you want to cut down the pressure by 10 per cent. Then cut the powder charge by 5 per cent and you can expect the velocity will go down only 5 per cent. All of these figures apply to small changes in loading only. Furthermore, if you are otherwise working at maximum loads, do not use these figures.

As an example of loading changes, Lt. Col. Wright was willing to sacrifice some barrel life in order to match the velocity of another load which gave a velocity of 3400 on an Avtron Chronograph for Robert Hutton. The computer indicated 81.7 grs. of 5010, as you just read. The load was increased to 85 grs. an increase of 3.3 or 4.0 per cent. The computer velocity indication was 3220 so a 4 per cent increase, as above, brings this up to 3360 f/s. This is close enough to the requirement, unless we want to split hairs. The pressure, on the basis of 3400 f/s velocity, is indicated to be 51,500 psi. The new load showed the same drop over 500 yds. as the load to be matched.

BARREL LENGTH AND VELOCITY

You can use your computer to solve a problem which very frequently comes up. You want to know what the velocity change would be if you had the barrel shortened. You have already found the Expansion Ratio for your gun as well as the Ratio of Powder Charge to Bullet Weight. For a good working pressure these are the only two factors involved. You will not change the powder charge, so all you have to do is figure a new Expansion Ratio with the proposed shorter Barrel Length. As you look over the Expansion Ratio-Velocity Tables you can see that a gun with low Expansion Ratio and high Ratio of Powder Charge to Bullet Weight will be affected much more by barrel shortening than will a gun with both high Expansion Ratio and low Ratio of Powder Charge to Bullet Weight. This is the reason that you cannot specify a certain percentage change of muzzle velocity per inch of barrel unless you refer to only one particular gun in the first place. You can use the Speer Ballistic Calculator to see what effect a new velocity would have on the trajectory and Remaining Velocities and Striking Energies.

Basic procedures are printed right on the calculator and consist generally of the following steps:

1. Determine weight in grains of water case will hold with bullet seated.
2. Set this weight at "start" index mark.
3. Read weight of powder charge at Arrow 1.
4. Read ratio of charge to bullet weight at bullet weight.
5. Set sectional density at ratio of charge to bullet weight.
6. Read powder number at Arrow 2.

At this point you have a basic, efficient load that generates pressures in the 50,000 psi range, and will produce velocities approximately those of similar loaded cartridges. Minor changes may be made in powder charge, primer and bullet (not in weight or seating depth) in the interest of accuracy.

The calculator is also set up to carry on and determine the velocity the calculated load will produce in any given length of barrel. So continue as follows:

7. Measure barrel length (muzzle to base of bullet — not complete length). Insert cleaning rod until it contacts tip of seated bullet. Mark rod at muzzle, measure this distance and add to length of bullet. This total is true bullet travel in the barrel.
8. Set barrel length at caliber.
9. Read expansion ratio at case capacity (grains of water).
10. On reverse side of calculator, set ratio of charge to bullet weight.
11. Read velocity at expansion ratio.

All of the foregoing may sound complicated, but once case capacity has been determined, no more than a minute is required to come up with any one load. In this device, Homer Powley has provided the first scientific tool for the handloader. With it, even the most inexperienced loader can produce a safe, efficient load without wasting any effort or components. Also, the dyed-in-the-wool wildcatter can readily determine whether his brain child is worthwhile — without firing a single shot. Working backward with the calculator, it is possible to determine what barrel length is required to produce a specified level of performance in any caliber.

Admittedly, the computer is susceptible to human error, and will show only one load for any given set of conditions. Actual mathematical calculations will provide more precise answers, but may in many instances require long sessions with paper and pencil. Use of the computer to select basic loads is far simpler. Refinement of the basic load can then be accomplished within the framework of good loading practices.

This scribe enthusiastically endorses the use of Powley's Computer and strongly recommends every handloader obtain and use one. It is especially valuable to wildcatters and for odd and obsolete calibers for which data has not been published.

Handy though it is, the Powley Computer is no help in load *development*. In merely supplies a single, full-charge load with a given bullet weight. A starting point, actually, from which load development may proceed. With the vast amount of data published on most calibers, it's hard for the neophyte to know where to begin. Rather than take pot luck, try the following.

No purpose is served in load development by attempting the use of outlandish or unsuitable components. The expert may sometimes achieve special effects from odd-ball combinations, but they are more hazard than help to the average individual. For every cartridge there are several powders suited for use with each common bullet weight. To deviate from those powders is to ask for lousy performance or, in some instances, to create hazards.

Consequently, the first step is to choose a bullet weight, and then to select a powder known to be suitable for it. There is no single gathering of data that lists *all* powders for all combinations. The simplest method of choosing a powder is to examine reputable published loading data. In, for example, the Speer handloading manual (and others), data for a half-dozen powders is listed with each bullet weight for each caliber. From this it is possible to select the most efficient powders for any given purpose.

Let's use the venerable .30/06 with 150-grain bullet as an example. We see maximum loads listed with six powders, and we may assume all were loaded to approximately the same pressures. The least velocity is given by a large charge of H-450, indicating that powder is too slow-burning for efficiency. A measure of this efficiency may be obtained by dividing the velocity produced (in fps) by the charge weight in grains. Doing the same for the other loads listed will produce other indexes of efficiency. The powder with the highest index is the most efficient in terms of velocity produced per unit of powder weight. It will be noted that the velocity of the most efficient powder is not necessarily the highest of the lot. In reality two or three powders will turn out to have almost identical indexes of efficiency. You can't go wrong with any of them.

Having selected a powder in this manner, choose the charge weight from the tables that produces the velocity you want. Do *not* begin with the maximum charge given.

Load at least ten rounds with the charge chosen, seating the bullet to standard depth. Shoot these loads in five-shot groups, to check accuracy. Generally, it won't be too bad. To improve it vary powder charge both ways in small increments. One of the charge variations will produce better accuracy than the others. From that point you may vary bullet seating depth — or even bullet make, so long as weight isn't changed — until maximum accuracy is obtained. At that point you may want to try other makes of primers One *may* do a bit better than others.

Once you've gotten this far, it's wise to load up, say, 100 rounds and consume them in carefully-fired five-shot groups to insure that performance is consistent.

If the above doesn't perform to suit you, then switch to the next suitable powder and repeat the process. Usually it won't be necessary to try more than two or three each of powder and bullet in order to produce a satisfactory load.

As indicated, if top velocity is your aim, then one of the less efficient powders may be best. However, for a less efficient powder to do this, a much larger charge is required. That means shorter barrel life, not much shorter, perhaps, but some. More powder, other factors being equal, means more erosion. And it's erosion that shortens tube life.

The most important factor to be kept in mind during load development — other than safety — is that *only one* factor can be varied at a time. Nothing is gained by changing *both* powder and primer simultaneously. No matter whether the results are better or worse, you can't tell which change was responsible. Change one thing at a time and do plenty of shooting in between to be certain of the result.

While not necessarily a part of load development, some additional comments fit in nicely at this point. Every round fired in the .300 Winchester Magnum example given was loaded in a fresh, new case, and for a specific purpose. By firing new cases in that particular gun, at essentially the same pressures my final load would develop, I assured myself of a supply of *tested* cases for the hunt. Perhaps you'll consider that superfluous effort — that new cases would surely be better than fired ones. Not in my opinion, and I'll cite examples. Over a score of years back I loaded some fresh factory .30/40 cases. Once in the field, two of them blew their primers (and it *was* a *safe* load), one of them tying up the action. Later I had the same thing happen with new belted magnum cases. Only three or four years ago, I answered a rush invite to hunt Texas white-tail. Leaving the house at the ungodly hour of 0300, I grabbed a fresh box of factory loads off the shelf. Once on the stand, the gun was loaded. Later, a shot was fired, and a fast follow-up attempted. No dice — the bolt wouldn't close on that second round. With my usual luck, that box had contained the once-in-a-million oversize cartridge, leaving me with a tightly jammed bolt that would neither close nor open until a vigorous boot heel (and much profanity) was applied. Now, had my target in that instance been a perturbed lion or armed and aggressive man, I'd have been in a hell of a fix. New cases are no more perfect than factory ammunition.

Because of those experiences, all the new cases loaded for my guns are fired *before* they participate in an expensive hunt. Likewise, *every* round (new or reloaded) is cycled through magazine *and* chamber before a trip is started. Such precautions can easily make the difference between success and failure, or life and death.

All load development is performed in much the same manner as outlined in the preceding pages. It need not be quite so extensive. If your requirement is only accuracy necessary to bag a white-tail at fifty yards in Michigan brush, there's no need to spend hours to produce one-inch groups. In most instances, you'll be able to use more than one bullet. And it isn't really necessary to shoot ten rounds of each load. If you're a fair-to-middlin' rifleman, you can get by with three rounds at a crack.

Remember always that every rifle (and pistol, too, to a lesser degree) is a rule unto itself. Its reaction to a given load can be predicted with no more accuracy than can that of the shapely young thing sipping her daiquiri over the piano bar. Only after you've tried will you know the results. Every component of the cartridge, and every operation you perform on it, is a variable, and each variable will produce *some* effect within the gun and on target. For this reason, never vary but one thing at a time. If you change primers and powder charge at the same time and get hideous groups, you'll not know which was responsible. Even so simple a thing as .040" change in bullet seating depth must be conducted separately. To do otherwise is to waste your time and supplies.

So the next time you get ready to crank out a box of fodder for your favorite smoke pole, stop and think a bit. Is the load you've been using *really* getting the best out of that barrel? I doubt it. Get with it and tailor the load to the gun and the job. Have at it!

CHAPTER 15

Home-made Bullets

IF MUCH of this book so far has referred primarily to the use of store-bought bullets, it's because they are easiest to talk about. Unlike the days before the 1950's, such a wide variety of factory-made bullets is available today that no one *must* make his own. Where a handloader might once have had a choice of only one or two bullets in a given caliber, today he may choose from at least a dozen sold right over the counter. Back in the 30's and 40's that wasn't so, and many shooters found it *necessary* to make bullets.

Today no one need make his own bullets. Virtually any kind of shooting need can be met by a wide variety of store-bought projectiles. Only in the two areas of economics (lowest-cost shooting) and ultra-precise accuracy demanded by bench-rest shooters is there any true advantage in making one's own bullets. The first requires casting bullets of lead and its alloys and can, under certain conditions, produce bullets that cost you absolutely nothing but time and effort. At the other end of the scale, jacketed bench-rest bullets may cost their makers a dollar or more each when made meticulously at home on tools of great precision. In the one we sacrifice a bit here and there in the sake of economy; in the latter, accuracy is paramount, and damn the cost.

From an accuracy viewpoint, one-inch 100-yard groups are not uncommon with cast bullets. Admittedly, the average is a good bit larger. Bench-rest groups beggar description, often being single holes at 100 yards which can hardly be discerned from a single bullet hole.

Aside from achieving economy and accuracy the factories can't supply, home-brewed bullets can restore many a foreign or obsolete caliber to service. Odd calibers not produced for many years can be made up without too much trouble if you are a real enthusiast.

Let's take cast bullets first; they are easiest, cheapest, and are useful in all manner of ways.

Lubricated lead bullets cast of proper alloy and sized correctly can be driven with good practical accuracy at velocities up to a bit over 2400 fps. Some shooters report velocities as high as 2700 fps, but 2400 fps is a reasonable limit. At the lower end of the scale they can be accurate enough for some purposes driven as low as a mere 400 fps. Generally, though, they are used at around 700-800 fps for heavy hunting and defense loads in handguns; and at 1600-2000 fps for targets and small game in rifles. In those particular ranges, good cast bullets easily equal the performance of far more costly jacketed types.

With all that in mind, except when actually shooting big game, or at long range or in bench rest competition, a great many of us could do most of our shooting with lead bullets, and get along very nicely, indeed. And we'd save a tidy sum of money in the process. No, I'm not against jacketed bullets, not at all. I'm just saying we don't *need*

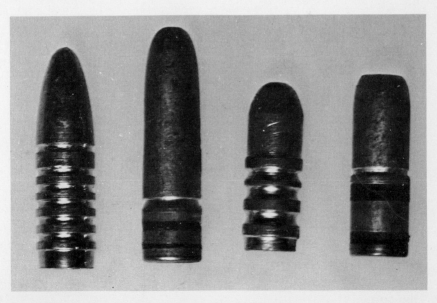

Different types of cast bullets. First and third from left have not been lubricated or had gas checks installed.

or *have* to use them as we seem to think. And right now is a good time to give that subject a little thought, because in time of war, copper for jackets often gets very dear. Remember twenty-odd years ago when you couldn't buy a single bullet, even if you offered a buck apiece?

Maybe now is the time to give a little thought to making some of your own projectiles out of that bucket of wheel weights you picked up a year or so back. Remember, as long as you have something that behaves like lead, and a mould, you can make bullets.

A lot of today's crop of handloaders have never cracked open a mould and watched a glistening, perfect, and blistering hot bullet come tumbling out. It's a hot, dirty, smelly job, they say. Who wants to cast bullets that will only travel a bit over 1,500 fps (feet per second) when it's so easy to buy a couple hundred shiny, jacketed ones that will *really* move out?

Who does? Me, that's who does. I want to when I see those six-cent slugs disappearing down range to clobber a varmint or punch holes in paper. And I darn well want to when the local shopkeeper sadly shakes his head and points at empty shelves.

What do you need to make cast bullets? A pot of some sort. Preferably cast iron, to hold lead; a heat source to melt that lead; a dipper to pour it with (unless your budget allows for a bottom-draw electric pot); a bullet mould; a stick or plastic hammer; a towel or blanket and a place to work. Add to that some means of filling bullet grooves with grease and squeezing them round and to size.

Let's take handgun bullets first, for the short guns are less critical to minor qualitative and quantitive variations and are far more tolerant than rifles of lead temper (hardness) and bullet shape. And it's easier to cast good ones. In spite of all this built-in forgiveness, perfect bullets are no less desirable.

Whether you're using an old Lyman pot on Mama's gas range or one of the big electric outfits, lead must be melted and heated to a temperature where it will flow as freely as water. As it becomes fluid, dross (dirt, grit and impurities) float to the top and must be skimmed off. However, tin and antimony alloyed with the lead also tend to float to the surface and be inadvertently skimmed off with the dross. This is eliminated by stirring and *fluxing* the mixture.

When the lead is hot enough to char a dry wood splinter without it bursting into flame (or at about 700 degrees F. with a controllable heat pot), drop in a bullet-size lump of grease. Much has been said about special fluxes and greases, but the lubricant you'll be using on the finished bullets works about as well as anything I've tried.

This will boil and bubble, smoke a lot, and stink up the whole house, so don't do it while the lady of the house is entertaining her Kaffeeklatsch. The uproar won't last long, so a fan and an open window will clear things up in a hurry. Or you can carry the pot outdoors for fluxing if you've a tyrant for a housemaid. If the kitchen range has an exhaust hood and you can work there, you're in real luck.

As soon as the smoke has cleared away and you've assured the hysterical neighbors that there's really no fire, stir the mixture well, wait a few seconds, and skim the dross off. An old, large serving spoon with a few small holes punched or drilled in its bottom works quite well for this. Don't make the holes too big.

While the lead was melting you should have washed all grease and oil out of your mould in solvent or gasoline. If the latter, I hope you didn't do it indoors. And if you had set the mould on the edge of the pot to warm up, you'd be a few minutes ahead of the game. But no matter, it will heat quickly enough as casting progresses. Better make a last-minute check to insure no foreign material is on the block faces thus keeping them from joining completely together.

Hold the mould in one hand (left, for a right-hander), with cavities *parallel* to the floor, cut off the plate toward the pot. Take the Lyman dipper (the one with the integral pouring spout) and immerse it in the lead for a few seconds to heat up. Unless the dipper is hot, lead will "freeze" in the spout. Lift dipper and bring its spout into firm contact with the sprue hole in the mould cut-off plate. Holding dipper and mould firmly together, rotate them to an upright position (counterclockwise for a right-hander) and hold steady. The lead will flow from dipper to mould and the excess lead in the dipper will exert pressure on that in the mould, forcing out air and filling all nooks and crannies.

Separate dipper from mould, rolling the two apart so that no lead is spilled. A small puddle of molten lead should remain in the sprue hole, and will harden in two or three seconds. You'll see it rapidly change color and texture as it solidifies.

In the event you are using a bottom-draw electric pot for your first bullets, we'd better cover that somewhat different procedure before continuing. The spout valve on the bottom of such pots is opened and closed by a handle connected to a rod dropping down through the molten lead. An adjustable stop controls the amount the handle will move, thus the amount of lead that will flow out the spout.

The spout appears to be made to insert in the sprue hole of the mould, and in truth, that is the way the makers intend it to be used. By all means try it that way, but in my own experience, it seldom works well. Better results are obtained (usually) by letting lead run freely into the sprue hole. Build up a support that will hold the mould with sprue hole directly under the spout and about a half-inch below it. Scrap wood works fine and so, sometimes, does a common brick.

If using wood, wrap a piece of aluminum foil over it to prevent spilled (yes, you'll spill some) lead from adhering to it. Wood has the added advantage in that you can drive a few nails into its top, forming a sort of pocket for the mould, aligning it with the pot spout. This will save a lot of time getting the mould in the right place each time.

Slide the mould under the spout and raise the valve handle very carefully. The stream of lead should enter the sprue hole deadcenter. If lead splashes over the side, the stream is too big and must be reduced by cutting down the valve opening. Adjust the valve handle stop.

As soon as lead wells up in the sprue hole, close the valve by letting the handle drop. Fast action is called for here, or lead will overflow and "freeze" on the mould, preventing the blocks from being opened. Cutting and prying off such spillage is a chore, so avoid it. Fast action is called for with the valve handle, and you'll no doubt invent a few new cuss words before you've learned to avoid overflow. A big help at this stage is a Lyman Mould Guide, which clamps to the pot uprights and provides a fully adjustable track and stop for the mould. It will fit other makes of pots, as well as the Lyman.

But back to the pot and dipper method. The dipper should be immediately returned to the pot so it will stay hot. When the sprue (that puddle of lead) has hardened, it's time to open the mould. First, the cut-off plate must be swung aside, cutting the sprue from the body of the bullet. This is best done by a sharp, light blow on the cut-off plate arm.

Never use a metal bar or hammer; the plate may be bent or warped. Instead, use a stick of hardwood (an old hickory hammer works well) or a plastic mallet. Never strike up or downward at the plate. This also will deform it so it will not lie flat on top of the mould blocks, causing lead to leak from under it. With a little practice you can cut off the sprue and have it fall directly into an old sardine can for eventual return to the pot.

Open the mould by spreading the handles apart. If the bullet doesn't fall out readily, rap the top of the mould hinge joint with the same instrument used on the cut-off plate (never a metal object). This will jar the bullet out. As the bullet comes from the mould, it is still quite hot and soft enough that dropping on a hard surface will knock it

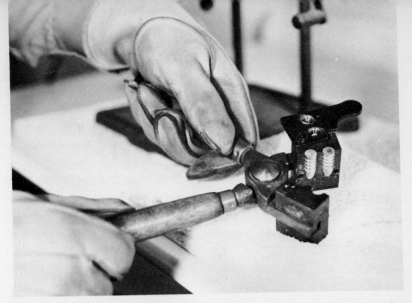

Once mold and metal are at proper temperature, neat, sharp bullets like these will be produced.

out of shape. Fold an old towel or piece of blanket into a pad on which bullets can be dropped and left until cool.

Man, that first bullet you just cast sure is a mess, isn't it? Full of wrinkles, rounded edges and voids. What's wrong? Nothing. The mould won't make perfect bullets until it has been warmed up. And if it's a new mould, it won't turn out good ones until it has been broken in by casting a hundred or so slugs.

After becoming hot enough to cause a drop of water to dance and sizzle audibly (the old-timers spit on them to check this), the mould should turn out well-formed bullets. If it doesn't, the *lead* probably isn't hot enough, so turn the heat up a bit, and keep trying. Eventually, you'll get good, if not perfect, bullets. It may take a few hours' experimentation with mould handling and lead temperature. Adding more flux will often make the difference between bad and good bullets.

Soon, though, you'll get the hang of it and be able to turn out a pair of bullets per minute, even with only a single cavity mould. You may find that after quite a few have been cast that they come out of the mould with a frosted appearance, and the sprue is taking a longer time to harden. Either the mould, or both lead and mould are too hot.

First try slowing down the casting, and if that doesn't do the trick, turn down the pot heat a bit. If this seems to slow production too much, then immerse the mould, *closed, with bullet and sprue in place.* in a bucket of lukewarm water for a few seconds. This will cool it off quickly and allow casting to continue faster. Some people I know regularly cast so fast they dunk the mould every half-dozen bullets.

By now you're getting pretty good bullets (or at least you think you are), and you've decided there really isn't much to this casting bit. But you're wrong. There's a lot more to it. Take a few of those bullets and look them over closely under a magnifying glass. Are the edges of the bands and grooves as sharp and clean as they appear in the mould cavity? Are the bases clean and square, with no "tail" or ragged cavity where the sprue was cut away? Do you find any holes when probing with a pin where the sprue was cut? Split a few bullets lengthwise — do you find air holes and voids? Weigh a dozen bullets carefully (if you don't have an accurate scale, call on your friendly pharmacist for help. Are they uniform in height? Borrow a micrometer (you really should have a good one, you know) and check a dozen bullets for diameter. Are they as uniform as you thought?

Running through this list of tests, you'll find enough wrong with your new bullets to deflate your ego a bit, maybe even to make you think the equipment isn't any good. But don't throw anything away. That setup will give you perfect bullets virtually every time it's heated up, once you learn the tricks of the trade. Don't be discouraged, for there are many things you do to eliminate the defects you discovered. And when you can do these tricks, you'll make bullets that have a natural-born X-ring tendency.

The simplest way to handle the defects you'll encounter is to build up a check list here, effect and cause, so to speak.

Wrinkled irregular surface: Check first to make certain that no oil or grease is present in the mould cavity. Check temperature of both mould and bullet metal. If either or both are too cold, this will produce this defect. If you have just begun to cast, continue casting to bring the mould up to proper temperature. If casting fifteen to twenty bullets does not produce any significant improvement, increase the temperature of the lead. These factors often require considerable juggling.

Bands, lubricating grooves, and other sharp corners are not filled out cleanly in exact duplication of the mould cavity: If using the hand-dipper, keep a larger quantity of lead in it and keep in contact with mould for a longer period of time. The weight of additional metal in the dipper, applied for a longer period of time to the fluid metal in the mould, will force lead into the mould's nooks and crannies. If using a bottom-discharge furnace, open the valve more, and hold it open for a longer period of time.

Next, try additional fluxing of the metal in the pot. Some alloys may require more frequent fluxing than others to flow well. If the condition persists, there may be insufficient tin in the alloy. At least

2½ or 3 per cent tin, by weight, is required to insure that the lead will flow freely into the mould. If the tin content of the metal you are using is not known, this may be the fault. Adding more tin may clear up the problem.

If all else has failed, it may be that your bullet metal has been contaminated by zinc, aluminum, or one of the other materials that act to greatly increase the surface tension of molten lead alloy. This contamination can easily exist when scrap lead from the local junkyard is used. Particularly, soldered joints found on some cable sheathing contains such impurities. Once so contaminated, the metal is of no further use for casting "good" bullets. You may as well use it to make vise jaws or door stops.

Air trapped in the mould can also prevent filling out all the sharp corners. Modern moulds have the blocks machined on their matching faces with a series of regular, shallow grooves which allow air to escape from the cavity as lead flows in. These grooves can become clogged, trapping air. Cleaning with solvent and an old toothbrush will usually do the trick.

Older moulds were made with smooth faces, and often required "venting." This consists of very carefully filing a minute groove from the offending portion of a mould to the edge of the block, thus providing an escape passage for trapped air.

Edge of bullet base not fully filled out: This results also from trapped air, and is usually corrected by slightly loosening the cut-off plate screw. The plate must swing freely and not be held so tightly against the face of the blocks that it prevents the escape of air. Grease and oil between the plate and the blocks can also contribute to the problem.

Holes or cavities in base of bullet: This can be caused by either trapped air, insufficient weight of metal in the dipper, or a combination of both. Loosen the cut-off plate and use more lead in the dipper.

Ragged cavity in the base of the bullet where sprue was cut off: Usually caused by striking the cut-off plate before the sprue has fully hardened, and is sometimes accompanied by smearing of lead across the cut-off plate and surface of mould blocks. Make certain sprue has actually solidified before striking cut-off plate.

Wedge-shaped lump left when sprue is cut off: Make certain underside of cut-off plate is perfectly flat and that edges of the sprue hole are clean and sharp. Loosen cut-off plate screw so that plate swings freely, then strike plate sufficiently hard to cut the sprue off in a single movement.

Band appears to be cracked or slightly separated from body of bullet:

Caused by dropping the bullet from the mould before it has cooled sufficiently. One half of the mould pulls free of the bullet cleanly as the mould is opened, but the bullet remains momentarily in the other and tips as it falls clear, thus the bands are slightly displaced. Allow a longer cooling period before opening mould.

Bullets have crystalline, frosty appearance: Either metal, mould, or both are too hot. Reduce heat of metal, and slow casting until bullet appearance is bright and normal. This can be a particular problem with large, heavy bullets in single cavity blocks. The large amount of metal heats the blocks very rapidly, even though lead temperature is held to a minimum. If a satisfactory production rate cannot be maintained, cool the mould by dipping it periodically in lukewarm water (with bullet in the cavity), or by using two moulds alternately.

Fins appear on bullet at line where blocks meet: Foreign material is holding blocks slightly apart, or guide pin holes are clogged and causing the same thing. It is also possible that you are relaxing your hold on the mould handles at the time the lead is entering the mould, allowing the blocks to be forced slightly apart. Clean mould block faces and guide pins and holes, and make sure you keep a good grip on the handle. If guide pins are loose in their holes, tap them in slightly deeper.

Fins on base of bullet: If fins are uniform around the perimeter of the base, the cut-off plate is too loose. Tighten the screw slightly until fins disappear. If the fin is predominately on one side of the bullet base, the cut-off plate is probably warped or bent — usually due to being struck incorrectly. If the deformation is only slight, the cut-off plate may be removed from the mould and polished flat on abrasive cloth stretched tightly across a hard, perfectly flat surface. The only other solution is a new plate. Once bent it cannot be straightened satisfactorily.

Bullets do not drop freely from the mould: Shift the position of the mould as it is opened. If this fails, strike the top of the hinge joint one or two light but firm blows with the stick used on the cut-off plate. If more effort than that is required to jar the bullet free, there are burrs in one of the mould blocks, embedding themselves in the surface of the bullet.

With the blocks clean and dry, examine the surface of the cavity under a magnifying glass. Burrs of sufficient size to cause trouble should be easily seen. Usually they are small and fragile, and can be very carefully removed with a pointed scraper. Often they are at the outer edge of a cavity, having been produced by the two blocks banging into each other. This type *may* be safely removed by extremely cautious application of a fine, Swiss needle file. Most authorities advise against any such work on a mould, but there is no other way to salvage a mould so damaged. I feel it is better to at least attempt to correct the condition than to discard the mould.

Air holes inside bullet (disclosed by weighing or sectioning): Usually the result of inadequate weight of metal when the mould is filled, resulting in excessive shrinkage with nothing to replace it. Also sometimes caused by too rapid filling of the mould which results in a splashing action that traps bubbles of air. Use more metal in the dipper and fill the mould more slowly.

Irregular, sharply defined small holes in surface of bullet: Caused by foreign material in the mould cavity, either falling in from the surface of the mould or being carried in with the molten lead. Keep mould clean, inside and out, and flux lead frequently so that all foreign material floats to the surface.

Careful attention to all those details mentioned above is not something that you'll be able to accomplish in the first, or perhaps even the tenth casting session. It requires continuous practice and effort over a period of some time. Often you may be casting perfect bullets, then suddenly one or more of the above conditions will crop up for no apparent reason. It's all part of the game, and rest assured that sooner or later it will happen to you. When it does, don't cuss the dog and kick your wife; just analyze the defect and take the appropriate corrective action. You'll be making good bullets again quicker that way.

Now, casting a bullet that looks perfect, is uniform in weight, and possesses no internal or external defects still does not accomplish the entire job. The bullet must be cast from an alloy reasonably correct for the job you intend it to do. There are fussy people who keep three, four, or even more bullet alloys available and use each one for a specific purpose. I do not find this at all necessary.

One alloy which has proved to me to be entirely suitable for *all* uses up to velocities of over 2,000 feet per second, is 4 per cent tin, 12 per cent antimony, and 84 per cent lead by weight. This, incidentally, is identical to the most common commercial Linotype metal. This alloy is hard enough to take care of any rifle load and to insure positive functioning in automatic actions, yet it will produce fine accuracy in low-velocity pistol and revolver loads. I can recommend it wholeheartedly for all uses, thus eliminating the need for keeping more than one mix on hand. Being a commercial alloy, it is easily obtained from linotype shops.

However, we all often find ourselves in possession of bullet metal whose contents are totally unknown. More often than not it will be too soft for use at the higher velocities. I reserve such windfall metal for use in low-velocity handgun loads or for muzzle loaders. Normally it is melted up in batches of twenty-five to thirty pounds, fluxed and skimmed to remove impurities, then a few sample bullets are cast from it. If the metal flows well and fills out the mould completely, I use it as is. If it gives any casting problems though, tin is simply added until it flows well enough to produce good bullets. Consequently, I usually have two different bullet alloys on hand — one of more or less unknown content which is soft and used only for low-velocity loads, and the other which is of known content as mentioned above and used for the higher velocity loads (or for everything if I run out of the softer material).

Just as this manuscript is being wrapped up for the publisher, SAECO has introduced a very practical and economical lead hardness tester. With it, you can determine the hardness of unknown alloys and, more important, alter one batch to match another.

But once cast, our bullet is far from ready to shoot. It requires yet two things: reduction to the diameter correct for your particular barrel and the application of a suitable lubricant in the grooves to prevent leading of the barrel. Bringing it to proper diameter and perfect round-ness is called "sizing" and is accomplished by forcing the bullet base-first through a die made for the purpose. Many statements to the contrary notwithstanding, cast bullets generally do their best work when sized to not more than .0015" larger than groove diameter of the barrel. Order your sizing die accordingly. The old-time recom-mendation of .003" over groove diameter no longer carries much weight.

Resizing is required because bullet moulds, like all other mechanical contrivances, must be made to certain tolerances. They generally cast a bullet somewhat larger than the desired diameter. While it has been said that resizing *always* damages a cast bullet to some degree, that damage can be held to a minimum by using the proper type die. Older dies were made with a two-diameter hole — the upper portion to accept the as-cast bullet, the lower of the size to which it was desired to reduce that bullet. These two holes were connected by a sharp shoulder or a sharply tapered area.

Such dies shear metal off of one side or the other of the bullet bands, yet do not touch the bottoms of the grooves. This means simply that a bullet so sized is more often than not thrown out of balance and may not be expected to produce best accuracy. More recently produced resizing dies connect the two holes of different diameters by a long, very gradual taper. This longer taper does not remove any metal from the bands, but rather compresses the bands in a reasonably concentric manner. Bullets sized in the newer dies are generally more concentric and thus more accurate.

A .30 caliber mould may well cast bullets measuring as much as .313-.314" in diameter. This bullet must then be squeezed down to .308-.3085" for best results. That much reduction in the old style die can easily produce a bullet so far out of balance that it is grossly inaccurate. However, the newer dies can achieve that much reduction without destroying the bullet's accuracy. It is generally conceded that the less reduction involved, the more accurate the bullet will be. For this reason, many modern moulds are made far less oversize than the older ones.

The lubricant may be applied either before or during the sizing operation. Fewer tools are required for the former. Lube may simply be rubbed into the grooves by hand (a most time-consuming process) or the bullets may be stood in molten lubricant that is then allowed to harden. After the lubricant is hardened, a fired case with the head cut off may be pressed down over the individual bullets to cut them free from the cake of lubricant, leaving the grooves well-filled.

Either way is entirely satisfactory. However, by far the simplest and most convenient method is the use of a lubricator-sizer machine of the type sold by SAECO, Lachmiller, and Lyman Gun Sight Co. This device combines lubricating and sizing into a single operation. The bullet is placed in the mouth of the sizing die and is then forced down into the die by means of a handle and nose-fitting punch. When the bullet reaches the limit of its downward travel in the die, it has been fully sized and its grooves are aligned with passages leading to a lubricant reservoir. A piston actuated by a ratchet handle then exerts pressure on the lubricant in the reservoir, forcing it into the grooves in the bullet. Reversing the movement of the handle then lifts the bullet back up out of the die and it is ready to use.

Lubricating and sizing is not so simple that it does not require considerable attention to detail. The nose-punch which forces the bullet into the die must fit the bullet, and it must also be aligned correctly with the die. If it is not, then the bullet may be forced through the die somewhat canted, resulting in the appearance shown in the accompanying photographs. Such a bullet is sized off-center and along a line not parallel to its longitudinal axis. Consequently, it may be badly out of balance and grossly inaccurate.

Choosing a lubricant isn't too difficult today. In times gone by, one could start a fight almost any time by bragging one lube over another.

All of the commercially-available bullet lubricants now produced are good for general use. Almost any grease will prevent leading, but only certain lubricants will produce good accuracy. All those that have survived produce acceptable accuracy. Recently, though, extensive tests have proved that mixtures containing Alox and beeswax produce better accuracy (especially in rifles at higher velocities) than the traditional grease/wax   lubricants and it seems likely the older types may not be with us much longer.

In a pinch, serviceable lubricant can be made from beeswax softened to workable consistency with petroleum jelly (Vaseline), with a small amount of finely-powdered graphite blended in thoroughly.

One other factor of considerable interest remains in regard to the use of cast bullets, particularly in high velocity loads. The plain, unprotected base of a lead bullet will be melted or fused by heat and pressure of the heavier powder charges. Available to prevent this is a shallow copper or alloy cup known as "gas check." This cup is pressed over the base of the bullet either prior to or in the process of sizing. It requires that there be a smaller diameter tenon on the bullet base to accept it, consequently special moulds are offered which produce bullets with this feature.

Application of gas checks is simple, but too many people attempt to speed the job up, resulting in checks that are not seated solidly and squarely. Best results are obtained if the check is first pressed over the base of the bullet by hand, then rapped smartly a time or two with a hardwood stick or plastic hammer to make certain that it contacts the base uniformly. The bullet may then be pressed through a hand-type sizing die, or fed through the sizer-lubricator in the normal fashion, with assurance that it is completely and properly seated and will remain on the bullet during its flight.

It has long been determined that a bullet which sheds its gas check during flight will be less accurate than one which retains it.

Of course, it goes without saying (or at least it should) that perfect bullets, perfectly sized and lubricated, must be kept undamaged until they are loaded and used. If lubricant is allowed to be contaminated with dust and dirt, or grit is allowed to collect on the bearing surfaces of the bullet, less than perfect results may be expected. Also, if the bullet base is damaged in any way, accuracy will be reduced. I find it best to store prepared bullets stacked closely on their bases in covered boxes. This protects them properly until use. Because of its soft nature, a lead bullet must also be seated carefully in the cartridge case, or it may become damaged in the process. Seating punches should fit the nose correctly, and case necks must be expanded properly and chamfered lightly to pevent shaving of lead as the bullet enters.

Jacketed bullets are another matter completely, and require entirely different tools and methods.

You must have a bullet jacket and a formed lead core; dies to assemble and shape them; and a heavy-duty loading press to furnish the power for that.

The press and dies are most important. Both must resist very heavy pressure without deformation, and the press must have a linkage with a high mechanical advantage to multiply your muscle power. Lightly-constructed C- and H-presses won't do the job.

The pressure required to swage some bullets will actually bend or break the frame. Some such presses are sold with a reinforcing tie-rod to be installed for swaging bullets and will do a good job with it in place. Even so, they are a poor substitute for a good heavy-duty O-press. After trying them all, I'm convinced the RCBS A-2 press now discontinued, is best of all, not only for its great strength and rigidity, but because of its compound leverage. The latter allows one to apply much greater pressure than can be developed by the simple leverage of most presses. RCBS now makes its Rock Chucker press which combines all the features of the A-2 into a smaller and more economical package. It is tops for swaging among currently produced presses. Bair also makes an O-type press with compound leverage, but it has not yet achieved much following among bullet swagers. Perhaps most significant is the fact that the majority of bullet-die makers recommend RCBS presses and have designed die sets around them.

In addition to strength and power, the swaging press must be of good workmanship, with close fitting parts and with the ram correctly aligned with die hole. Excessive die wear comes from bad alignment.

Dies come in two types. Those for half-jacket or short-jacket handgun-type bullets are quite simple as shown herein. A cylindrical die body is open at the bullet nose, closed at its base by a "base punch," which also serves as an ejector for the finished bullet. This die is — in most designs — seated in the press die hole. A separate nose punch is attached to the ram and contains a cavity to match bullet point shape. The nose punch is forced into the die, forming the bullet nose and expanding both core and jacket to fill the rest of the die cavity. The nose punch is then withdrawn and the base punch forces the completed bullet clear of the die. No other tools or dies are required. There are, of course, several variations of design, including the C-H die-press unit made only for swaging half-jacket bullets.

Rifle-type bullets are more demanding and require a good bit more tooling. While essentially the same general form as above, three different dies are usually required, though some bullets can be made satisfactorily with two, while perfectionists may use four.

The first die set simply pre-forms the lead core to desired weight, excess lead being forced out through "bleed holes" in the die body. In most rifle designs the die body forms the bullet (core) point, while the base is formed by the ram-mounted punch.

The second die seats the core solidly in the jacket, expanding the core into full jacket contact to eliminate the possibility of voids.

The third die brings the bullet to final shape and after ejection, it is ready to load and shoot.

As can be seen, half-jacket bullets are made easily and quickly, but rifle bullets take a good deal of time and a considerable tool investment. Short cuts are sometimes taken, as in eliminating *either* (but not both) core forming or core seating where maximum accuracy isn't required. Pistol bullets sometimes get more complicated when jackets extending forward over the ogive are used. Then, rifle-type dies must be used.

Cores and jackets: Cores you can make — up to a point and for some purposes — but jackets are well beyond the handloaders capabilities, with one exception. That exception is that with proper dies you may de-head (iron out the rim) of fired .22 rimfire cases, then use them as jackets. Once this was quite common, back when jackets were hard to obtain, but it isn't done much today. Frank Hemsted makes excellent dies for the purpose. All other jackets must be purchased. Sierra supplies excellent jackets to the trade, as does Speer from time to time. Several independent makes also supply jackets in a wide variety of calibers.

Cores are generally cut from lengths of lead wire made especially for this purpose. Several core-cutters are available and all work about the same way, shearing off uniform lengths of wire.

Usually pistol cores are cut just a wee bit over-weight and the excess is extruded when the bullet is formed. Rifle cores are cut the same, then either individually filed to weight, or the excess is extruded when the core is pre-formed. In any event, the most precise weight control is obtained when each core/jacket pair is weighed together, then the core is filed to bring the combination to exact weight *before* swaging the bullet. Allowing the excess to bleed off during swaging introduces slight variations due to time and pressure factors.

However, let's set up the dies (a typical design) and make some half-jacket bullets.

Wipe die cavity and nose punch clean. Insure that ejector pin/ base punch moves freely. Seat the nose punch in the press ram according to the maker's instructions, then lower the ram. Screw the die body part way into the ram progressively until the nose punch will enter ¼" without binding as it is brought to the top of its stroke. Correct punch/die alignment until the punch enters smoothly — it must *not* snag on the die mouth.

Now, weigh a jacket, then cut a core that will produce the weight bullet desired. Lubricate both *very* sparingly by rolling between thumb and finger moistened with case-sizing lubricant. Place the core in the jacket. Set the exposed part of the core in the nose punch and guide the whole works into the die. Gingerly "feel" the press handle through its stroke. If no resistance is felt, advance the die body into the press until you feel the lead begin to move at the top of the ram stroke.

Take out the bullet and look at it. It probably isn't fully formed, so the die must be advanced farther. Continue progressive die adjustment until the bullet is fully formed, but with the least effort that will do so. The first bullet may not go back into the die easily, so you'll use up several cores and jackets during adjustment. Once satisfied, lock the die body and start making bullets.

Both cores and jackets must be absolutely clean. Any grit or dirt whatever will eventually spoil the dies. Many handloaders wash cores and jackets, air-dry, then lube by tumbling on a slightly-oily cloth just before swaging.

Once all adjustments are made, you may, to save time, cut cores slightly over-weight and depend on the excess being bled off during swaging. However, for this bleed-off to produce uniform weight, press operation must be smooth and consistent, and the bullets must all remain in the die the same length of time. To slam the handle down hard, then immediately eject the bullet will produce a heavier bullet than normal handle movement and ten to fifteen seconds "rest" in the die. If you doubt this, slam the handle on an overweight core, then watch the bleed-hole carefully. You'll see that bleed-off continues *after* the ram and nose punch have come to rest. Slamming is also very hard on nose punches and often cracks them.

The bases of finished bullets should be very slightly radiused. If the edges are very sharp, too much swaging pressure is being used. Wrinkles or folds in the jacket indicate trapped air and voids brought about by bent or poorly-shaped cores. They also occur when the core itself contains a void. This doesn't happen with extruded lead wire, but is not uncommon with cast cores. Cast cores were once popular. Lyman made special multiple-cavity moulds casting simple cylindrical cores, and they were (are) subject to the same problems as cast bullets. Cast cores are more economical than wire, but not nearly so satisfactory, and require more time and equipment.

At one time special dies were made by Lakeville Arms to carry the process further and squeeze-in a crimping cannelure which was also claimed to produce a better core/jacket assembly. This was done as a final operation in a collet-type device installed separately on the press. Today much better cannelures can be rolled in on the S.A.S. Bullet Canneluring Tool.

During that same period, dies were made to produce Pro-T-X Bore swaged bullets. They were swaged in quite similar dies but without a jacket. Instead, a zinc washer was placed in the base of the die a teat of core extruded through the washer center to lock it in place.

In the absence of jackets, half-jacket dies can be used to produce gas-check bullets. The check is substituted for the jacket without any other change. Such bullets are a last-ditch affair, for they contain no lubricating grooves, yet the entire bearing surface is exposed lead. Bad bore leading almost invariably results from their use.

The process is a bit more complicated for rifle bullets which must operate under entirely different conditions.

Work begins with selection and preparation of jackets. It must be of the proper length to produce the type and weight bullet desired. Either hollow-or soft-point types may be made. All the well known makes of jackets are good and will make up into fine bullets. However, makes do differ in internal profile, meaning that the core forming die and core seating punch must be made to match the jacket. To eliminate voids, the core must fit the jacket closely, yet be a shape that can be expanded under pressure to fit tightly into the jacket. This means a core-forming die to match the jacket. Likewise, the core seating punch must fit closely enough in the jacket to prevent lead from extruding between punch and jacket. In tapered jackets, this sometimes means different diameter punches for different length cores, even with the *same* jacket.

So, you begin by obtaining dies to match the jackets to be used — or vice versa.

Straight wire will cut more uniformly into cores than as it comes off the spool. Cut wire into twelve-to eighteen-inch lengths and roll between two clean steel or glass plates to straighten. Wipe clean with solvent and inspect for dents and scratches, then pull through a lightly lubed cloth to lubricate very sparingly. Clean up one cut end and adjust the core cutter to produce slightly heavy cores.

Set up the core-forming die and punch exactly as described for handgun bullets. Carefully adjust the die to fully form the core and bleed off excess lead to exact weight. From this point onward, the core will be covered by the jacket and no further bleed-off is possible, so any errors in weight will show up in the finished bullet, Unless the punches fit precisely in the die, lead may extrude around them, and may also extrude if too much pressure is used.

Some bench-rest shooters are convinced that greater weight uniformity and dimensional stability is produced if cores are swaged again a second, or even a third time, combined with a two-or three-week aging period after each swaging, and after seating as well. I know of no definitive tests that prove this theory, but a few people swear by it.

Once formed, cores should be treated gently to avoid dents and nicks that might cause flaws during further operations. For that last iota of accuracy, some shooters remove all lubricant. This is easily done by placing cores in a collander or wire basket and pouring alcohol or similar solvent over them. Don't use a petroleum-base solvent, for it leaves its own oily residue.

Jackets must be inspected and often wiped or washed if they've been exposed to dust and dirt. Run-of-the-mill jackets produce good bullets, but the perfectionists check wall thickness uniformity and concentricity with an arbor and dial indicator. Some even use elaborate machines to bore or ream jackets to more uniform thickness and concentricity than can be obtained from factories. Only champion shooters can tell the difference, and by no means do all of them find such meticulous preparation necessary. One fellow I know spent 400 hours tool-room time to build a machine to bore out his jackets. That means a $5000 machine!

Cores are inserted into jackets by hand prior to seating. Cores should slip easily to the bottom of the jacket under finger pressure only. A short piece of dowel or rod serves to press cores all the way down. Put only the slightest trace of lube on the *outside* of the cups by rolling between oily fingers.

Set core-seating dies in the press just like the others, with special care to insure that the seating punch does not scrape the jacket mouth. Adjust the die so that you can *feel* the core expand to fill the jacket properly. You'll first feel the core begin to collapse as the punch engages it, then as all available space is filled, resistance will become quite stiff. Stop there. Excessive force is not needed. Examine the exposed end of the core to be certain lead has not extruded up around the punch.

Pointing or point forming is next, and last in the home-swaged rifle bullet. The die is set up in the press as before and is adjusted until it produces the proper point shape on the core/jacket assembly. This can be a tricky adustment, for the very small-diameter ejection punch may wedge in the bullet's hollow-point, or may pierce a soft-point and jam. So, it is best to proceed gently, making certain there is lubricant on the jacket to ease ejection. Hollow-points must be adjusted so that the point closes sufficiently to prevent the ejector pin from entering. When soft-points are to be made, not much help is available in point-ejection dies. Anyway you look at it, that exposed lead is soft, and if a bullet sticks a bit, the ejector is going to dig in. If you insist on making soft-point bullets, better stick to one of the less-common base-ejection die types such as Hollywood. Most first class dies are of point-ejection type, since the major demand is from target and varmint shooters who prefer hollow-point bullets. In addition, base-ejection die bodies must be made in two parts, a more costly method.

Regardless of the type, most uniform results are obtained when the press is operated uniformly with smooth, even handle movement and the bullet is left in the die for ten to fifteen seconds before ejection. Lead and copper alloy do not stabilize instantly when deformed. A few seconds "rest" confined in the die aids stabilization.

Up until the final pointing/forming operation both core and jacket are slightly undersize. Each operation expands the lead core which in turn expands the jacket. This is called "expanding-up" and makes for a tight assembly of core to jacket. Copper alloy "springs back" after being deformed, back toward its original shape and dimensions. Lead does not do this to nearly so great a degree, so holds its new shape and dimensions — resulting in the jacket being left in a state of tension, clinging tightly to the core. The same principle is applied in factory production of jacketed bullets.

Bullets should be inspected after final forming. Dents or wrinkles can be the fault of too much lube. Too-sharp bases are the result of too much pressure. You can weigh finished bullets if you like, but the weight won't have changed since the cores were seated. Minor scratches on the bullets will not effect accuracy, but they indicate dirty or scratched dies or jackets. Overlaps, folds, or splits in the jacket over the point may effect accuracy if very deep. Significant variations in bullet diameter indicate fairly wide variations in forming pressure.

Less pressure than you might think is required to form jacketed rifle bullets. Splitting the effort into three stages keeps pressure requirements down. With compound-leverage presses it is easy to apply too

Modern Handloading

much pressure, even enough to bulge a die. Always experiment to learn the minimum pressure that will do the job fully, then try not to exceed it. Wear and tear on both you and the dies will be a lot less, and the bullets will be better.

There is some measure of economy in swaging half-jacket pistol bullets, considering the current price of components and factory bullets, but not nearly so much as in casting. However, if you place any reasonable value on your hard-won time and labor, homebrewed bullets are the costliest. Rifle bullet cores and jackets cost more, and a full set of tools can easily run several hundred dollars. It's easy to see, then, that the advantages to be gained from swaging for rifles aren't really economic — they are in making something you can't buy in either size or performance, or in simply being able to say, "Made 'em myself."

CHAPTER 16

Commercial Loading Tools Today

TODAY THE handloader may choose from the widest variety of tools and equipment ever offered. Only Lyman offered a lubricator-sizer for cast bullets, and other items were equally dear. A catalog of pre-war items would have been mighty thin and contained few names and items. We could cover them all in detail here in a half-dozen pages.

Now, though, to catalog every item would require a large portion of this book. Consequently, we are going to list only major items by the better known makers, and also those items of unusual interest and/or utility. A complete listing of makers grouped by products will be found in the Appendix, and catalogs and brochures may be obtained by writing if you wish for more.

Bair Company

Bair Company got its start only a few years ago by obtaining the tooling of what was then the entire Pacific line. Bair manufactures both metallic and shotshell tools and equipment, as well as an extensive line of accessories.

Metallic Presses include three models and three types, beginning with the **Grizzly** Bair C-type press. It features simple leverage, detachable shell holder head, swinging primer arm, provision for auto primer feed.

Next is the **Brown** Bair H-tool with three die and shell holder stations and simple leverage, post priming, and detachable shell holder

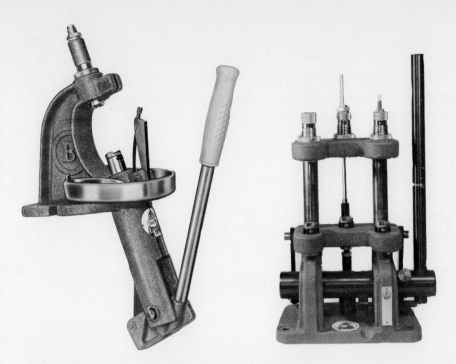

Left: Bair C-type press. *Right:* Bair 3-station H-type press for metallic loading.

heads. Top of the line is the O-type **Kodiak** with compound leverage and sliding-bar auto primer feed standard equipment. Other features are as for the other presses.

Die Sets of conventional two- and three-piece design are offered in most popular calibers. Die bodies are chromed, other parts blued.

Micro-Measure is Bair's micrometer-adjustable powder measure. It is of conventional design with a rotating drum containing the adjustable metering chamber.

Magna-Damp is Bair's magnetically damped powder scale of 510-grain capacity. Base is metal with one leveling screw. This is a conventional beam balance.

Honey Bair is the basic, single-post, shotshell loading tool. It possesses the usual five stations and the shell must be moved through them manually. An auto primer feed is available optionally. The GLACIER Bair is a more sophisticated tool of H-type with the five stations in generally a straight line, with a reservoir for expelled primers. All operations end on a positive stop. Powder and shot are handled by a single fore-and-aft charge bar. Top of the shotshell line is the POLAR Bair, a manually-indexed circular progressive tool with automatic primer feed and charge bar operation. The entire unit is housed in a

Commercial Loading Tools Today

Typical Bair 3- and 2-die sets.

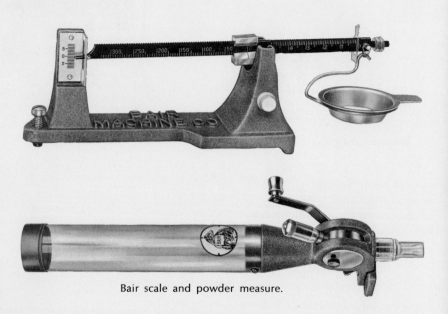
Bair scale and powder measure.

wide H-type frame and produces a completely loaded shell with each handle stroke.

Bair also offers many accessories, including the following:

Parts Kit	Primer Tube Filler
Complete Loading Kit	Powder Funnel
Pistol Powder Measure	Primer Turning Plate
Auto Primer Feed	Case Lubricant
Primer Pocket Reamer	Case Trimmer
Powder Dribbler	Bullet Puller
Chamfering Tool	Lubricant Pad
Form & Trim Die	

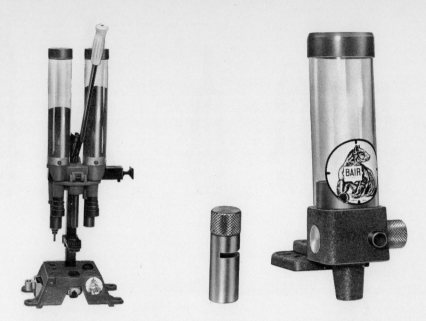

Left: Bair low-cost shotshell tool. *Right:* Bair simple fixed-charge pistol powder measure.

Bonanza Sports

Bonanza was formed only a few years ago by the owners of Gopher Shooters Supply as a separate company to produce a line of unusual (and excellent) new designs of handloading equipment. These new tools have been very well received. The entire line originated with and is built around the unique CO-AX press.

Co-Ax Press: An unusual design combining features of C- and O-types where the ram or riser rides vertically, attached to two steel rods that slide in holes at the top and bottom of a heavy cast frame. A handle pivoted to the *top* of the frame attaches to the riser by two heavy links. The riser carries a universal type shell holder composed

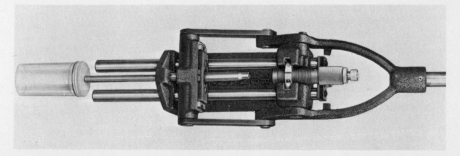

Bonanza unique CO-AX metallic press with universal semi-automatic shell holder.

Commercial Loading Tools Today   201

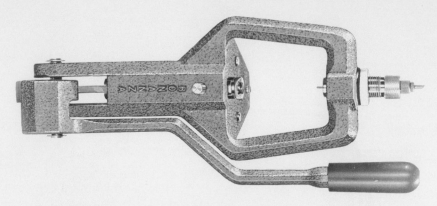
Bonanza M68 O-type press with in-ram priming system.

of opposed spring-loaded notched jaws. The jaws are cammed apart to allow the case to be removed at the bottom of the riser stroke by a tapered rod. The case is snapped into place simply by pressing it downward against the jaws or, alternatively, automatically as the case is forced into the die.

Dies are held in the CO-AX press in an unusual manner. They are slipped horizotally into a smooth vertical slot which grips only the lock ring. Conventional ⅞ x 14 dies may be used, but special Bonanza lock rings are required in most instances.

Bonanza priming tool with unusual adjustable, eccentric-disc shell holder.

Modern Handloading

Bonanza tool for measuring bullet/case runout.

Priming is accomplished by a post and shell holder setup attached to the upper ends of the vertical rods and to the top of the frame.

Bonanza also produces an excellent O-type press of conventional design, designated Model 68.

Co-Ax Primer Seater: An unusual priming tool which holds the case horizontally and feeds primers for a vertical magazine. The unique shell holder consists of three steel discs which may be rotated about their eccentric axis to accommodate any size case head. The tool base incorporates a primer tray from which magazine tubes are easily loaded.

Bench-Rest Powder Measure: A simple rotating-drum measure with a unique angled metering chamber. The adjustment plug rides in the angled handle whose axis coincides with that of the metering chamber. Adjustment is by a sliding vernier scale on the handle rather than the usual threaded stem.

Co-Ax Dies: Conventional ⅞ x 14 dies that will fit most presses, except that in the bottle-neck calibers the E-Z OUT expander design is used. This consists of the decapping stem threaded most of its length, allowing the expander to be moved upward to a point where it just clears the case shoulder. Thus, the case is pulled over the expander early in ram travel while mechanical advantage is great, thus reducing effort required. Bonanza also offers its BENCH-REST bullet seating die in which case and bullet are supported and correctly aligned by a spring-loaded inner sleeve throughout seating. The inner sleeve moves with the case through all seating travel.

Commercial Loading Tools Today   203

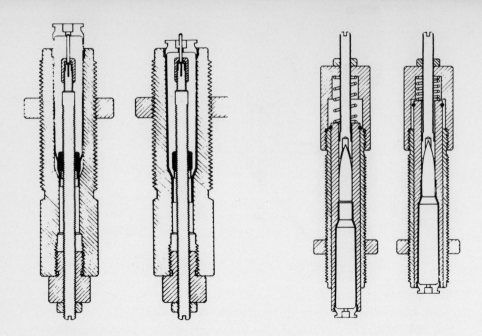

Left: Bonanza die section view of unusual adjustable expander ball which eases operation. *Right:* Bonanza Bench-Rest seating dies in section, shown action of guiding sleeve.

Bonanza also offers a wide line of accessories:

Powder Funnel	Die Storage Box
Long Drop Tube	Sizing Lube
Lube Pad	Case Conditioner Kit
Deburring Tool	Bullet Puller
Gunsmith Screwdrivers	Case Trimmer
Bullet Alignment Indicator	Powder Scale

C-H Tool & Die Corporation

C-H has been up and down and was at its peak about 1960, but encountered lots of problems after the death of its founder, Charles Heckman. C-H tools had been off the market for several years until

Bonanza bullet puller — the simplest on the market.

C-H 3-station H-press set up for rifle caliber.

a short time ago. Thus far, only a few items of the once very extensive line are again available.

No. 205 Press: A conventional O-type press of very heavy construction with simple leverage. Fitted with lugs for auto primer feed; detachable head shell holder; swinging universal-type primer arm; and ⅞ x 14 die hole. Excellent for swaging bullets.

No. 333 Press: A conventional three-station H-type press with detachable shell holder heads and post-style priming.

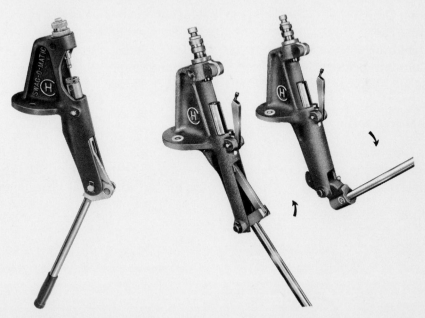

Left: C-H Swag-O-Matic bullet making press. *Right:* C-H typical C-type metallic press.

Commercial Loading Tools Today

Left: C-H bullet swaging dies for ¾-jacket designs. *Right:* C-H hexagonal body dies.

No. 203 Swag-O-Matic: An integrated press/die combination for production of swaged half-jacket bullets and not suitable for any other purpose. This is the simplest and most economical setup for making such bullets and was instrumental in making them popular over a decade ago. Interchangeable die sets available in most popular calibers.

No. 204 Press: A conventional cast-iron C-type press with simple leverage; detachable shell holder; swinging primer arm; lug for auto primer feed; and ⅞ x 14 die hole.

C-H bench primer seating tool with auto feed.

Modern Handloading

C-H tool for canneluring bullets or cases.

Bullet Swaging Unit: A simple die set for one-operation swaging of half-jacket bullets in most popular calibers. Designed to be used in most ⅞ x 14-thread O- and C-type presses. Bullet ejection is by separate hand lever.

C-H also offers the following accessories:

Powder Dipper	Powder Funnel
Powder Measure	Loading Blocks
Powder Measure Stand	Primer Catcher
Bullet Puller	Primer Pocket Swager
Split Lock Rings	Resizing Lube
Case Trimmer	Deburring Tool
Two- and Three-Die Sets	Powder Scale

English

Bill English produces an excellent portable loading tool called **Pak-Tool.** It consists of a body to which is assembled a handle and toggle link and a ram. Short neck-sizing, expanding, and seating dies screw into the ram and are forced by the handle over the case placed in the lower part of the body. Decapping is performed similarly, while priming is done by a long punch screwed into the ram while the case is reversed in an external shell holder. The Pak-Tool performs all operations except resizing and is small and light enough to carry in one's pocket. It is excellent for on-range loading and comes in a plastic carrying case.

Hensley & Gibbs

This firm has been offering fine multiple-cavity bullet moulds since the 1930's. Not a high-production item, H&G moulds probably have

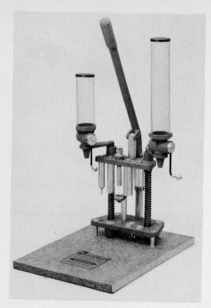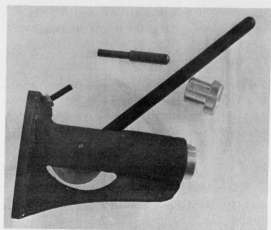

Left: Lachmiller Super Jet shotshell loader. *Right:* Lachmiller bench priming tool.

the best reputation of all makes. They are genally hand-made in a small shop. An extremely wide variety of bullets — exceeded only by Lyman — is offered, but today only moulds of four or more cavities are made, and only in pistol calibers. A very high percentage of large-volume bullet casters use H&G moulds exclusively.

LEC (Lachmiller Engineering Co.)

Another one of the relatively old-time companies operated on a rather small scale since WWII, LEC has recently undergone considerable expansion and is in process of adding many new products. Interestingly, just recently this plant burned completely to the ground, and yet was in near full production again in less than one year.

M400 Press: A very heavily constructed C-type press of conventional design and characteristics, but not provided with a priming device. Frame threaded 1¼ x 18 for shotshell dies and fitted with a ⅞ x 14 adapter for standard metallic dies. It is supplement by the M707 press, a conventional ⅞ x 14 O-type with swinging primer arm.

Super Jet Shotshell Loader: A H-type, 5-stations-in-line tool differing from most in its two widely-spaced posts and its separate rotating-drum measures for shot and powder.

Priming Tool: A simple cam-operated, manual, bench-type priming tool with removable shell holder head. It provides excellent feel in seating primers as a separate operation.

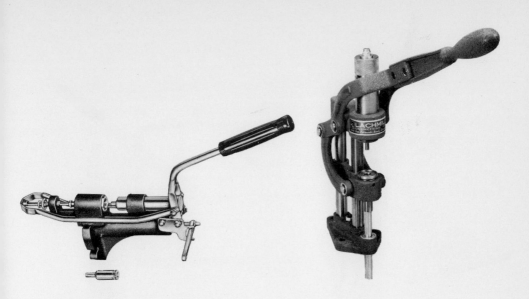

Left: Lachmiller bullet lubricator-sizer for cast bullets. *Right:* Lachmiller shotgun reconditioner.

Ultra Powder Measure: A fairly conventional rotating-drum measure featuring a one-pound capacity reservoir and large metering chamber that will throw charges up to eighty-seven grains of 4350. The companion PISTOL measure uses fixed-size inserts in its rotating drum to throw charges from 2.0 to 9.0 grains Bullseye by ½ grain steps; nine to twenty-four 2400 by 1 grain steps.

Bullet Mould: LEC now offers a line of conventional four-cavity moulds in most popular pistol bullet weights and shapes.

Lubricator-Sizer: A new design based on the principles of the old Lyman lubri-sizer, but fitted with a ratchet device to maintain constant pressure on lubricant in the reservoir. Available in all popular calibers and diameters.

Bullet Swaging Dies: Simple, conventional dies for making half-jacket handgun bullets. Designed especially for use with the M400 press. Auto ejection of finished bullets.

LEC also makes a variety of accessories for both metallic and shotshell loading.

Loading Dies
Berdan Decapper

Shotshell Reconditioner
Lead Melting Pot

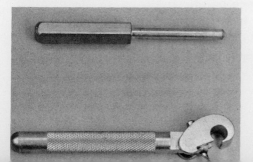

Lachmiller Berdan decapping tool.

Left: Lyman Spar-T turret press with powder measure installed. *Right:* Lyman All-American turret press for metallics.

Lyman Gunsight Company

This is the oldest producer of reloading tools still in business. Tools have been produced continuously since the 1880's, first under the name IDEAL. For many years Lyman was the only quantity producer of bullet moulds and the only maker of lubricating/sizing tools for cast bullets. Lyman is also known for its scopes and metallic sights for rifles. Lyman covers the entire metallic and shotshell loading field with the single exception that it makes no bullet swaging equipment.

310 Tong-Tool: Unchanged in basic design since the 1880's, this is a compact, self-contained, pliers-type unit that will perform all operations except full-length resizing. For leisurely loading of small quantities, it is still about the handiest unit available.

Spartan Press: A basic, simple, aluminum-alloy, C-type press. Not for bullet swaging. Possesses lug for auto primer feed; detachable shell holder head; swinging universal-type primer arm; simple leverage; ⅞ x 14 thread.

Spar-T Press: A variation of the Spartan with a six-die turret. Uses a special auto primer feed mounted on primer arm.

All-American Turret Press: A heavy-duty C/post-type press with a four-hole turret. It requires special shell holders and is available also with a push-button auto primer feed. Post-type priming.

Universal Trimmer: A hand-cranked case trimmer conventional in design except for its unique clamp-type shell holder which accepts all sizes from .22 Hornet upwards.

No. 55 Powder Measure: Like the Tong-Tool, this measure is essentially unchanged since the 1880's. It uses a rotating drum, but metering chamber is varied by horizontal movement of wedge-shape segments with micrometer adjustment. One of the few measures today fitted with a "knocker." Screw-clamp mounting.

D-7 Powder Scale: A conventional beam balance of good quality. Magnetically dampened; riser to take load off bearings when not in use; 505 grain capacity.

All-American Dies: Two- and three-die sets of conventional design, ⅞ x 14 thread, hexagonal locknuts, supplied with a wrench. Two-step expanding die available for cast bullets. The P-A (positive alignment) seating die available separately uses a floating bushing to hold bullet aligned with case neck as seating begins.

Bullet Casting Equipment: Lyman produces the largest line of bullet moulds in this country, including many for obsolete cartridges and for muzzle loaders. One-two- and four-cavity moulds are available, as are hollow-point, and hollow-base types. The traditional cast-iron pot and dipper for melting and handling lead is available as is the modern Mould Master electric bottom-discharge furnace and Mold Guide. The No. 450 Sizer and Lubricator, though recently redesigned, is a variation of the traditional No. 45. Dies are offered for virtually any bullet shape or diameter.

Lyman bullet casting furnace with mold guide installed.

Lyman Easy-Loader shotshell tool.

Easy Shotshell Loader: A fairly conventional five-station, straight-line tool containing separate charge bars for powder and shot. Supplied in all popular gauges for paper and plastic shells with star crimp. Can be ordered to produce any safe load, or optional micrometer-adjustable charge bars may be had to simplify load changes. A separate model is offered to produce the old-fashioned roll crimp in most gauges. Gauge conversion kits available for both models.

Lyman also produces a large variety of accessories, including the following:

Wad Fingers	Buckshot Moulds
Powder & Shot Dipper	Shotgun Slug Moulds
Roll Crimp Head	Shotshell Case Trimmer
Wad Cutter	Resizing Lubricant
Gas Checks	Cake Cutter
Bullet Lubricant	Tong-Tool Bullet Sizing Die
Shotshell Handbook	Powder Dribbler
Reloading Handbook	Powder Funnel
Inertia Bullet Puller	Case Lube Kit
Hand-type Resizing Die	Chamfering Reamer
Primer Pocket Reamer	Tungsten-Carbide Dies
Drill Press Case Trimmer	

MEC (Mayville Engineering Co.)

MEC got into shotshell loading tools since the war, and has adhered strictly to that field, producing no metallics or other accessories. From

Left: Lyman Reloading Manual, in publication longest of all. *Middle:* Lyman power case trimmer for use with drill press. *Right:* Lyman lubricator-sizer for cast bullets.

the beginning, MEC Tools have typified the all-in-one shotshell loader that is now the accepted standard. MEC pioneered many of the design and construction features now found on other similar tools.

MEC 400: This may be considered the basic MEC single-station tool. It is a simple and sturdy tool with a circular die head carrying five stations and a single charge bar handling both shot and powder on alternate strokes. The shell is passed manually through all five stations. Die head is spring-returned to upper position. Convertible to all

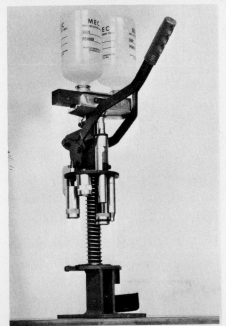

Typical low-cost MEC shotshell loading tool.

Commercial Loading Tools Today

Left: New Ohaus dual-purpose powder measure. *Right:* Ohaus bullet mold.

popular gauges and loads. Charge bar uses interchangeable bushings. The 250 is a trimmed-down and less costly version of the 400, but functions in the same way. The 600 Jr. and 700 are more sophisticated, yet reasonably-priced single-stage tools that feature greater convenience of operation.

MEC 650: This is the basic MEC progressive loader featuring a manually-indexed rotary table carrying six shells under the various die stations. When the table is filled, each handle stroke produces a fully-loaded shell, and one operation on each other shell in the table. Primer feed is automatic, and a mechanical cycling device also operates the powder/shot charge bar. The only manual operations are indexing, case insertion, wad insertion, and of course, handle movement. The Super 600 is the same, but lacks the charge bar cycling device. At the top of the line is the HYDRAMEC, which consists of the 650/600 series linked to a self-contained hydraulic power unit operated by a foot control so that hands are free for handling components. An astute operator can easily load over a case of shells per hour with ease on the Hydramec.

MEC produces few accessories, mainly a shell conditioner, which resizes metal case heads and reseats base wads; wad dispensing bins; a packaging guide; and a primer-magazine-filling device.

Ohaus: The Company has long been a producer of a wide variety of scales and weighing devices for industry. For quite some time. it manufactured the powder scales sold by Lyman, but has recently placed a new line of handloading scales on the market under its own name. As this is written, Ohaus has shown prototypes of other handloading items, including powder measures and bullet moulds, but they are not yet in production.

5-0-5 Scale: A low cost, magnetically damped beam balance of conventional design and 505 grains capacity. Plastic base, one leveling screw.

10-0-5 Scale: A higher grade scale of 1005-grain capacity. Base combines with a separate plastic cover to form a fitted storage case for the entire scale. Other features as for the 5-0-5.

Dial-O-Grain 304 Scale: The most sophisticated scale offered regularly to handloaders. It is a large scale resembling laboratory types more than the usual small beam balance, and has a capacity of 3100 grains. It features a large calibrated, easy to read dial by which all fine adjustments are made. All increments below 100 grains must be read off the dial. It is magentically damped and is unique in that it is furnished with a powder trickler located so as to feed directly into the scale pan.

The D-O-G 314 is essentially the same unit without the dial adjustment and with a third poise weight and scale added to take its place.

Pacific

Pacific Gun Sight Company is one of our oldest makers of handloading gear, continuously in operation since the 1920's. Now com-

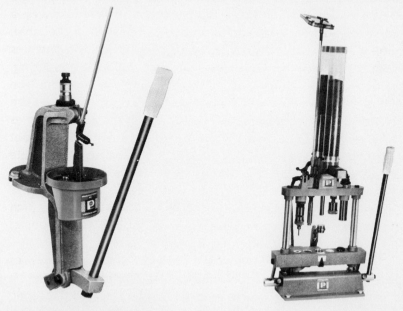

Left: Typical Pacific C-type press. *Right:* Middle of Pacific shotshell tool range is represented by this DL 266.

Commercial Loading Tools Today

bined with Hornady Mfg. Co., Pacific introduced the C-type press and ⅞ x 14 dies in the late '20's and also originated the now-standard swinging primer arm. Many innovations have come from Pacific over the years. Its line of metallic and shotshell loading items is one of the largest.

Power-C Press: A typical conventional C-type loading press with detachable shell holder, ⅞ x 14 die hole, auto primer feed lug, swinging primer arm, and simple leverage.

Multa-Power Press: This is essentially an O-type modification of the Power-C, but fitted with an RCBS-type compound leverage to supply greater power for case-forming bullet swaging.

DL-155 Shotshell Loader: A basic medium cost shotshell tool of single-post design, carrying the usual five stations. Available with optional auto primer feed. Large capacity detachable powder and shot reservoirs; dual charge bar.

DL-266 Shotshell Loader: A large H-type loader with the five stations in a straight line. Bar carrying shell holders moves up to meet fixed die head. Auto primer feed actuated by the charge bar.

DL-366 Shotshell Loader: A circular progressive loader on the base and frame of the DL-266. Primers, powder, and shot are fed automatically, but case and wad insertion, and indexing are manual. Will easily produce a case of shells per hour with an experienced operator.

Pakit: A screw-operated, hand-type loading tool for metallic car-

Left: Top of Pacific line is DL 366. *Right:* Economy Mesur-Kit screws to powder can; throws adequately accurate charges.

Pacific Pakit is screw-operated, pocket-size reloading tool for metallics.

tridges, intended for use in the field. A companion item is the MESUR-KIT which is a single adjustable powder measure which screws on the mouth of a standard powder can.

Pacific also produces a large variety of accessory items, including the following:

Powder Scale	Powder Measure
Case Trimmer	Pistol Powder Measure
Powder Funnel	Deburring Tool
Primer Pocket Reamer	Case Lubricator
Case Cleaner	Auto Primer Feed
Primer Plate	Primer Tube Filler
Loading Manuals	Shotshell Pouch
Shotshell Wads	Loading Tool Sets

Pacific loading manuals are available for both metallic and shotshell.

Left: Big Ponsness Size-O-Matic shotshell tool is without peer in its field. *Right:* Ponsness Du-O-Matic functions same as big tool, but slower and at much less cost.

Ponsness-Warren

Ponsness-Warren is another relatively new entry, formed only a few years ago to produce and market Lloyd Ponsness's excellent designs. It produces only two models of shotshell loaders, not even any accessories. The P-W tools have come to be very highly regarded and produce some of the finest handloads I've seen.

Size-O-Matic: The basic P-W tool is a large circular progressive tool with automatic primer feed, indexing, and powder/shot charging. The operator need only feed in empty cases and wads, and work the handle, producing a fully loaded shell with each stroke of the handle. The S-O-M is unique in that the case is resized full length at the first station and remains in that same resizing die throughout all other operations. In order to accomplish this, each of the eight stations on the shell table contains its own resizing die. By this method, the shell is restrained from bulging or wrinkling, and is spared the working caused by moving in and out of several different dies as in some other tools. This is truly a production tool and with practice one operator can easily crank out as many as 600-700 shells per hour. Operation requires less effort than many comparable tools and is likely due primarily to good workmanship and design.

Modern Handloading

Ponsness feature is individual resizing die in which case stays for all operations.

Du-O-Matic: A single-station tool using the basic principle of the S-O-M. It contains a single resizing die on a pivoted carrier; thus the die can be swung under each of the loading stations after the empty case is seated. Though much slower of operation, and requiring everything to be done manually, it produces loads equal in quality to those of the S-O-M- at much less cost.

RCBS, Inc.

This firm was formed by Fred Huntington who still operates it, and growing from a one-man garage operation, it has become one of the most highly regarded in the industry. It pioneered 3- and 4-die handgun caliber sets, interchangeable shell holder heads, and compound-leverage presses. It also produced the first practical O-type presses designed especially for bullet swaging. RCBS produces only metallic loading presses, dies, and accessories. No shotshell tools, bullet moulds, bullet-swages, etc.

Rock Chucker Press: This is a compact, powerful O-type unit using compound leverage and suited to bullet swaging. It possesses a seating lug for auto primer feed; swinging primer arm of universal type; detachable shell holder head; large mounting flange; ⅞ x 14 die thread; hardened pins throughout. An excellent heavy-duty press.

RCBS Jr. Press: A medium-duty aluminum alloy O-type press suitable for all but bullet swaging. It has all the other features of the Rock Chucker *except* compound leverage, instead it uses a conventional toggle-joint.

"Precisioneered" Dies: Conventional type ⅞ x 14-thread dies in natural hardened steel finish, cavities polished after hardening. 3- and 4-die sets available in straight-case calibers. An extensive line of special calibers and case-forming dies is available, and custom dies will be made to order at reasonable prices. Tungsten carbide sizing dies are offered in appropriate calibers.

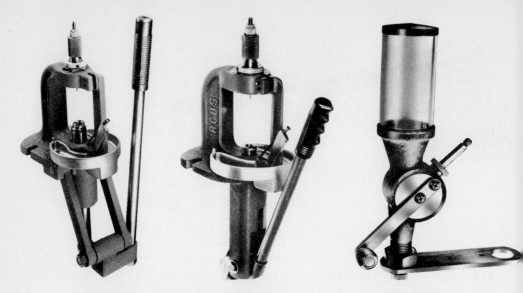

Left: RCBS Rock Chucker; top of the press line. *Middle:* RCBS Jr.; medium price press. *Right:* REBS Uniflow powder measure.

RCBS Powder Measure: An excellent, well-proven design of rotary-drum volumetric measure for both rifle and pistol use. Excellent charge uniformity. Housing threaded ⅞ x 14 to permit installation on turret presses or press bracket supplied. Bench stand also offered.

RCBS Powder Scale: A typical beam balance of good quality, magnetically dampened. Cast metal base with one leveling screw, 550-grain capacity.

RCBS Priming Tool: New as this is written, this is a bench-type priming tool with a lever-operated seating punch and removable shell holder head. Fitted with a swinging auto primer feed that works very well. Readily switched to either size primers. Gives excellent feed.

RCBS pioneered 3- and 4-die pistol caliber reloading die sets.

RCBS powder and bullet scale.

RCBS also offers the following accessories:

Case Lube Kit	Powder Spoon
Primer Pocket Swage	Sizing Lubricant
Primer Pocket Cleaner	Auto Primer Feed
Case Brushes	Parts Kit
Case Extractor	Neck Reamer
Form & Trim Die	Bullet Puller
Powder Trickler	Burring Tool
Reloading Guide	Powder Funnel

Left: Simple RCBS primer pocket swager avoids case rim damage. *Right:* Stuck case extractor.

Commercial Loading Tools Today

Redding Hunter

Commonly called just "Redding," this firm built its reputation on excellent drum-type powder measure and conventional beam balance powder scales noted for accuracy and consistency. In recent years it has diversified and entered both metallic and shotshell loading fields with fairly conventional tools and accessories.

Model 24 Press: A basic, conventional C-type press with auto primer feed lug; ⅞ x 14 die hole; detachable shell holder head; simple leverage; swinging universal type primer arm.

Model 25 Press: A beefed-up version of the 24 but with a four- or six-station rotating turret to accept ⅞ x 14 dies. All else as the 24.

No. 3 Master Powder Measure: A conventional rotating-drum measure with micrometer-adjustable metering chamber. A smaller pistol chamber is offered optionally. This measure has been made for many years and has been widely copied. It gives excellent results.

No. 4 Standard Powder Measure: An economical separate-chamber measure. The screw-adjustable metering chamber is hand-held while the reservoir is swung horizontally across and back to fill the chamber. Powder is then poured from the chamber directly into the case. Ideal for the casual handloader on a low budget.

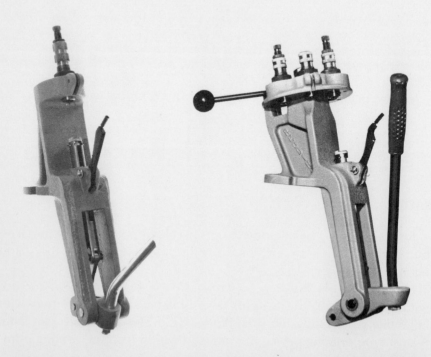

Left: Redding Hunter Model 24 press. *Right:* Redding Hunter Model 25 press.

No. 1 Standard Scale: The basic scale upon which much of the R-H reputation was built. Excellent workmanship and performance. 380-grain capacity; cast base with one leveling screw, oil damped.

No. 2 Master Scale: Similar to No. 1, but magnetically damped, heavier construction, and 505-grain capacity.

No. 16 Shotshell Loader: A fairly conventional single-station tool of twin-post type. Five die stations through which the case must be moved manually. Single manual charge bar alternately feeding powder and shot.

Redding-Hunter produces also a number of accessories, including:

Sizing Lube   Powder Measure Stand
Powder Dipper   Auto Primer Feed
Conventional Loading Dies

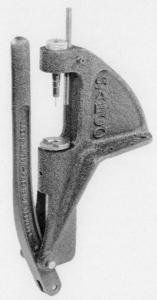

Unusual SAECO Match-Precision loading press with adjustable ram stop for use with stubby dies.

SAECO

SAECO has long been known for its excellent bullet-casting furnaces, but does produce an extensive line of metallic loading tools and equipment as well.

Match-Precision Press: An unusual C-type press designed for use with short STUBBY neck sizing and bullet-seating dies. Fitted with large diameter removable shell holder head and adjustable ram stop. A post-type priming device is contained within the ram, making prim-

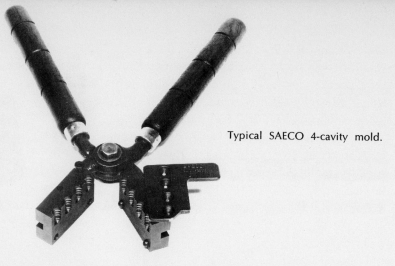

Typical SAECO 4-cavity mold.

ing a separate operation *after* decapping on the same press. Conventional dies may also be used.

Stubby Dies: Unusually short dies for seating and neck sizing/decapping/expansion which accommodate only the case neck. They are threaded ⅞ x 14 to fit standard presses, but require an adjustable ram stop on the press for proper operation. One size suffices for all cartridges of that bullet diameter.

Lubri-Sizer: A heavy-duty lubricating and sizing tool for lead bullets. It presses the bullet into the sizing die, forces lube into the grooves,

Left: SAECO lubricator-sizer. *Right:* best known SAECO item is excellent M-24 bottom-draw bullet casting furnace.

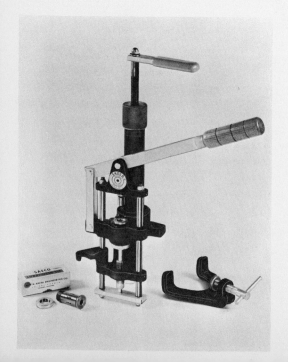

SAECO lead tester allows duplication of alloy hardness.   SAECO stubby dies.

then pushes the bullet back out the way it entered. Will seat gas checks. Dies available in most popular diameters.

Bullet Moulds: These are the original Cramer designs obtained by SAECO some years ago. Pistol bullets are offered in the popular weights and profiles, but rifle bullets all include the long bore-riding pilot essential to good accuracy. Plain and gas check styles are available. Moulds are offered in two- three- and four-cavity sizes with large blocks and heavy cutoff plate. Blocks are interchangeable on handles.

Melting Furnace: A heavy-duty, thermostatically-controlled, bottom-discharge furnace for casting lead bullets. This unit holds roughly eleven pounds of lead and has a most excellent reputation for durability and long life. Capacity is adequate for four- and six-cavity moulds.

T31 Furnace: A heavy-duty utility furnace for melting and blending bullet metals, or for casting when a separate ladle or dipper is used. Thermostatically controlled; twenty-four-pound capacity; available for either 115V or 220V current.

Powder Measure: A large rotating drum measure with angled metering chamber and micrometer-click adjustment stem protruding at an angle from the right end of the drum. Will throw charges from 2.0 grains of Bullseye to 95 grains of 4350. Must be used with optional measure stand.

STAR MACHINE WORKS

Since the 1930's Star has produced a single basic reloading tool in several versions that have been improved over the years. It is known simply as the STAR PROGRESSIVE tool and was once aimed primarily at the police and custom loader market. It is furnished only for short, pistol-type calibers up to .44 Magnum, .44/40 and the like. It may be had for similar rifle calibers that do not exceed those in length and diameter.

The Progressive is a circular progressive tool that produces a loaded cartridge with every handle stroke. In the basic model, cases and bullets must be inserted manually; primers and powder are fed automatically; shell plate indexing is manual. The more costly models include a magazine type automatic case feed and permit 600-800 rounds per hour to be loaded by a single operator; 1000 or more by two operators.

Many accessories have been produced by other firms for the Star Progressive, including auto bullet feeders, auto indexing devices, etc. With or without them, the Progressive remains the most popular large-production tool among handgun reloaders.

TEXAN

Texan is another firm making an extensive line of both shotshell and metallic loading tools, though it is better known for the former. Texan's name and fortunes have changed a few times, but it still produces good quality equipment, some of which has become quite popular.

Model M-II-A Autoloader: The mainstay and the tool that built

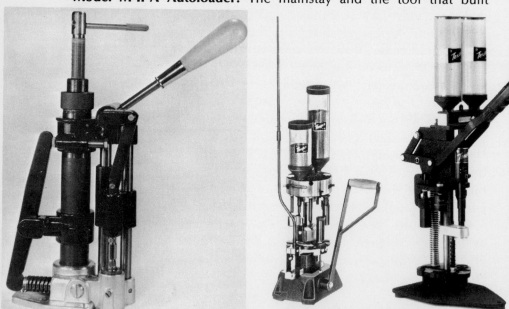

Left: Star push-through, semi-automatic lubricator-sizer (Star Progressive tool is shown elsewhere). *Middle:* Texan M-11-A semi-automatic shotshell loader — one of the best available. *Right:* Texan Model DP-1 low-cost shotshell tool.

Modern Handloading

Left: unusual Texan powder measure with built-in powder trickler. *Middle:* Texan shotshell reconditioner for reforming heads and base wads. *Right:* Texan Model 101-T-11 for metallic loading.

Texan's reputation among shotgunners. At the time of its introduction, it was the only reasonably priced auto-indexing progressive type. A rotating shell plate carries cases through all the operations so that, once filled, each pull of the handle produces a fully loaded shell. Cases and wads must be fed manually; all other operations are automatic. Available in all popular gauges.

Texan offers the DP-II, FW, and A-II single station loaders at lower cost, the latter being one of the least costly on the market. All three are conventional post-type priming and an auto primer feed is available optionally. The turret has seven stations, providing for two three-die sets and a powder measure. Also offered is the 301-H3, a perfectly conventional three-station H-press, and the No. 256, a simple C-press.

Powder Measure: A unique measure with a square metering chamber in a drum rotating about its vertical axis, and with a powder trickler built into the drum. Threaded ⅞ x 14 to mount on loading presses. A bench stand is also offered.

Powder Scale: Texan produces a conventional beam balance with oil or magnetic dampening. Capacity is 500 grains.

Texan also produces a line of accessories including the following:

Powder Funnel Bracket	Two-and-three-Die Sets
Loading Kits	Bullet Puller
Sizing Lubricant	Case Trimmer
Deburring Tool	Shotshell Base Sizer
Auto Primer Feed	Shotshell Reconditioner

Commercial Loading Tools Today

CHAPTER 17

Powder Measures and Scales

THE POWDER measure or scale is simply a necessary means for insuring that the correct amount of powder is placed in each cartridge before the bullet is seated. Unless you are working with a known, unvarying charge, for which you may buy a measure already set up, the intelligent and safe use of *any* powder measure requires the coordinated use of an accurate scale. Conversely, though, a scale is sufficient in itself. This is so because the scale is capable of measuring varying weight (in grains avoirdupois), while the measure deals only in *volume*.

Propellant powders, being a solid cut in granular form and of many widely-varying sizes, do not all have the same weight/volume relationship. Nor, does even a given powder have the same weight/volume relationship under different conditions. For this reason, only the value of *weight* may be considered constant. One-hundred grains of any powder will weigh 100 grains any time, any place. On the other hand, one cubic inch of powder will weigh more or less, depending upon conditions and upon the particular powder involved.

Traditionally, powder scales are simple beam balances provided with sliding weights, usually a "coarse" (5-grain increments) and a "fine" (1/10-grain increments) will a pointer over a graduated scale at one end of the beam, and a weighing pan at the other. All mounted on a sturdy metal or plastic base or frame. Some have beam movement

Typical beam-balance powder scale made by Bonanza.

damped by means of a paddle moving in a container of oil, others use a magnetic damper, but many have no damping system whatever.

Given that basic design, accuracy and sensitivity depend almost exclusively upon quality of workmanship and, of course, on the operator. Such quality means carefully balanced components, accurate weights and graduations, and friction-free (as nearly as possible) bearing surfaces. Most scales on the market today are described as being "accurate within one-tenth grain."

This isn't quite true. Extensive independent tests conducted several years ago for *Gun Digest* indicated that most current scales will produce an average error of 1 per cent above 20 grains. A few will produce 1 per cent above 10 grains, but a few can barely manage 1½ per cent above 20 grains. This is a rather far cry from 1/10 grain. In fact, in a 50-grain charge it runs to ½ grain or more. This doesn't mean all charges will vary that much. Very careful and detailed attention to proper operating procedures will keep the error down. The key to minimum error is absolute uniformity in operation and in conditions of the scale itself.

Certainly greater accuracy can be had from more expensive laboratory-type balances, but it isn't necessary for typical handloading. Even bench-rest shooters have not been able to justify better charge accuracy and uniformity than can be obtained by handloading scales and measures.

As for sensitivity, nearly all scales currently offered will deflect visibly to a single kernel of IMR 4350 powder, and that weighs only .03 (3/100) grain. Certainly no greater sensitivity is really needed. In fact, greater sensitivity makes a scale difficult to use.

Powder Measures and Scales

Operating a scale properly is not difficult—it just requires proper preparation and attention to detail.

Remove the dust cover and set the scale on a *level* surface.

Even a minor slope of an ¼ inch per foot or so (not readily visible) can cause up to ½ grain error. This level surface cannot be assumed. Check it with a small machinist's level in two directions. Since it can be quite difficult to level a table or bench or shelf, much time and effort can be saved by making a scale base. Get a 8″ or 10″ square of ¼″ steel plate with one ground surface. Drill and tap two adjacent corners for 1″ x ¼″ machine screws; drill and tap the opposite edge for one screw. Run the screws in from the smooth side and set the plate on the protruding screw-legs, smooth side up. Set the plate on your work area, then level it with the three screws and a level. Level in two directions at right angles. Lock the screws with jam nuts and recheck level. Now you *know* you have a level weighing surface and a major source of error is removed.

Inspect the scale to make certain that, 1) the ends of the knife-edges do not contact the frame; 2) the knife-edges are in their proper notches and that both surfaces are clean and dry; 3) that the pointer does not rub against the scale; 4) that there is no interference with the damping device (if present); 5) that pan and hanger are located properly; 6) that all weights move freely and seat properly in their notches on the beam.

Level the scale with the screw provided in its base so that the pointer indicates zero with all weights at zero. Turn the scale through 180° and recheck the pointer. If it does not indicate zero, the adjustable zeroing nuts on the end of the beam must be shifted until zero is indicated 180° apart on the leveled plate.

You may now proceed to weigh charges. Always double-check weight settings, then recheck them after each charge. On many scales the shock of setting the pan in its hanger a bit roughly can cause weights to bounce and shift. Always replace the pan in the same relative position.

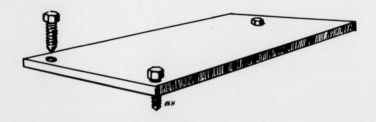

Leveling plate for powder scale.

It serves no useful purpose to bring the pointer completely to rest. In reality, more accurate results are obtained if the pointer swings freely an equal amount on both sides of zero. When the pointer swings between, say, four graduations above and below zero, the charge is right on the button.

Don't operate the scale in the draft of a fan, open window, air-conditioner, heating vent, etc. Strong air currents will deflect the beam. Don't jiggle the table or bench and don't breathe heavily on the scale. Above all, don't let anyone interfere or interrupt, and don't try to weigh charges while watching TV or any other entertainment. You'll screw things up for sure, at least eventually, if such distractions are allowed.

I knew one expert handloader/rifleman years ago who insisted there was only one way to weigh powder charges accurately. One night each week he sent the wife and kids out, locked the doors, took the phone off the hook, and then retired *alone* to a half-ton bench bolted to deep concrete footings (for stability, not strength) in the basement and weighed his charges. No amount of horn-honking or door-banging by potential visitors could get through to him. He'd simply stop work and wait until the interruption subsided before proceeding. The wife and kids couldn't come in until he turned on the door-light, indicating he'd finished "powdering" as he called it. Many's the night mama *und Kinder* sat in the car an hour out front, waiting for the all-clear light to come on.

Often lower-priced scales will be found which are adequately accurate insofar as uniformity is concerned, but may not quite give correct weight — being a half-grain or so off when checked with a precise set of master weights. This happens when balancing wasn't carried out quite right at the factory or when for one reason or another the balancing weights or nuts have worked loose. When you have reason to suspect that this condition exists, perhaps you can have your local pharmacist check the scale against his own set of weights which is costly and accurate and periodically certified correct by a government agency. Once the error is known, you can allow for it.

As with most other mechanical devices, you may expect longer and more accurate service from the more costly models. There are numerous low-cost scales available in the ten- to twenty-dollar range, with others ranging up to the Ohaus "Dial-O-Grain." The latter costs a bit over $50 as this is written and probably represents the maximum of sophistication that can be justified by a handloader.

Damping devices have been mentioned. They are handy in that they cause the pointer to come to rest more quickly. In doing so they reduce sensitivity and can also hide other errors. For example, if the pointer comes to rest quickly, how can one be certain the beam quit swinging because of the damper, or because the knife-edges are rubbing against the frame. Exactly this has happened to me several times. If no damper is present, then you *know* any quick stop is the result of some error creeping in.

Friction or liquid (paddle in oil bottle) are most likely to cause trouble. The more recent magnetic dampers eliminate the errors of the others and are, technically, error free, except for a *very* slight effect on sensitivity. About the only problem with magnetic dampers is that the arm attached to the beam — and which passes through the magnetic field — may rub where it passes through the magnet. This is easily corrected, once it's been identified, by bending or filing.

Naturally, scales will produce the best results when kept clean and handled gently. The first bit means that you simply *must* keep some sort of a dust-proof cover over the unit when it is not in use. Simply throwing a dirty wiping rag over it is not the answer. By far the simplest, most economical, and most effective way of protecting your scale is to slip it inside a plastic bag and tie up the mouth with a twister wire. After each use, before the bag, wipe the bearing surfaces, those V-shaped notches or round notched rods on which the knife edge is borne, and all moving parts free of dust, oil, or dirt. Pay particular attention to the pan and pan hanger which often collect enough dirt and grit to make your charges a tenth or two light. The bearing surfaces are quite hard, but this does not relieve you of the responsibility of making certain the beam doesn't get banged around to drop on them. A blind grab for the pan could easily result in one end of the beam being inadvertently lifted to drop down with a crash on the bearing surfaces, a sure way to mar them and reduce the scale's accuracy and uniformity. In the event the knife edges or the bearing surfaces do become nicked or damaged, the better makers are set up to re-grind those areas and restore their accuracy.

Careless dumping of an excess amount of powder into the pan will cause the beam to bottom and bounce around, which can eventually chew up bearing surfaces. Less than the required amount of powder should always be placed in the pan first, then add more very slowly until the beam pointer swings up to the zero mark, indicating the proper charge weight has been reached. The ability to add only two or three kernels of powder at a time comes only with practice. I find it convenient to do this with a thin, light teaspoon, which is tapped very gently with the fingertip to cause individual powder kernels to vibrate toward the lip and fall off into the pan. Other people find it

Typical powder "dribbler" or "trickler."

more convenient to use an empty cartridge case of fairly large caliber, say, .45 Colt or .45/70 dipping the case half full of powder, then rotating it while held nearly level to feed one or two kernels at a time over the lip. Whichever you are accustomed to will work well. Alternatively, at relatively little additional cost, we have the powder "trickler" or "dribbler" offered by several manufacturers. This little gadget passes a hollow tube through a small reservoir of powder. When the tube is rotated, holes allow powder to pass into its interior, and very shallow threads inside the tube cause the powder to move out to the open end where individual kernels can be fed out at will. Some handloaders find it convenient to throw an underweight charge on the scale pan by means of a measure, then feed on the additional powder needed to bring the charge up to weight by means of a trickler installed on the bench or on a bracket so that the mouth of its tube falls over the pan. In any event, the one operation that must be mastered to avoid frustration while weighing powder charges is that of adding varied minute amounts to bring the charge up to weight. The rest of it is easy.

For all practical purposes, the only disadvantage possessed by a good scale is that it is slow of operation. An experienced operator can throw twenty charges with a measure in the length of time it takes to weigh out one with the average scale. For that reason, I personally use a measure for all but the very hot loads, and also when loading small quantities where the time spent in setting up the measure is not really justified. Unless you intend going all out for speed, then a single powder scale will meet all of your needs.

As we've said, a powder measure will meter out whatever *volume* of powder it is set to produce. The actual weight of this volume of powder doesn't even enter into the measure's functioning. By looking backward from loading data tables, it is up to you, the operator, to determine what volume of the powder involved will weigh the correct amount, then set up the measure to throw that particular charge. The measure will then throw that charge repetitively forever without caring the least what it weighs. That's your problem.

Except in certain instances where a very great many of the same powder charge must be thrown, a fixed-charge measure is not practical. Therefore, the most common measures on the market have adjustable metering chambers. Then, with a single measure, it is possible to adjust and lock the chamber to throw any reasonable weight of any powder available.

Adjustable measures come generally in two styles. The one uses a rotating drum containing a hole within which is an adjustable plug. This drum is beneath the column of powder in the reservoir and is rotated so that powder flows by gravity into the chamber, then the drum is rotated roughly 180°, sealing off the reservoir and dumping the powder out of the chamber and into your case or measure or scale pan through a drop tube. This type is characterized by having a constantly-changing "head" of powder over the measuring chamber, and this is often said to cause undue variation in charge weight. The

Typical adjustable drum-type powder measure by Lachmiller.

other type of adjustable measure is typified by the very old Belding and Mull design which utilizes a separate horizontally-sliding chamber to transfer powder from the reservoir to a constant-volume chamber from which it may flow into an adjustable metering chamber or tube. The metering tube is generally held in the hands and after it is filled, the powder is poured manually into the case. This type is typified by a constant head of powder over the metering chamber and is generally thought to give superior accuracy and uniformity, though it is substantially slower of use.

The reason for the alleged differences in charge uniformity between the two types is simply that the weight of a tall column of powder tends to compact powder in the metering chamber, and, as this weight varies, from powder being consumed, the degree of compaction will also vary. In theory, the last few charges from such a measure when nearly empty will weigh less than the first few thrown when the reservoir was full. This makes sense when one considers that as much as a full pound of powder may be contained in some reservoirs and this pressure may be compared with only an ounce or so when the reservoir is nearly empty. On the other hand, the sliding transfer chamber of the Belding and Mull type measure always has the same head of powder, about an ounce or two at most. The pressure it exerts on the powder in the metering chamber will always be reasonably uniform.

In recent years, though, baffles have been introduced for the rotary-drum type measures. In some instances, these have consisted simply of a metal washer or perforated plastic disc fixed a couple inches above the bottom of the reservoir. More recently, Jim Normington of Rathburn, Idaho, developed thin aluminum "inverted-V" baffles which may be fitted into almost any measure reservoir. With both types, powder above the baffle flows into the lower portion of the reservoir, but the piled-up powder blocks off the small space through which it flows *before* the secondary chamber can be completely filled. Consequently, when a charge is thrown, an amount of powder approximately equal to it flows into the secondary chamber from the reservoir and thus maintains a constant head of powder for every charge thrown. When rotary-drum type measures are baffled in this manner, I've been able to determine no measurable difference in the accuracy they produce and that of the Belding and Mull type.

It goes without saying that moving parts of powder measures must be very closely fitted. For example, if the rotating drum is not fitted with virtually no clearance into the housing, fine-kernel powders (such as Bullseye) may work their way into the space between drum and housing and cause vibration and difficult operation. Difficult movement of the drum and/or vibration will cause variations in the powder charge thrown. This is simple enough to understand when we consider that any such outside influence will cause the powder granules to settle or compact in the metering chamber. When this occurs, more kernels are contained in the chamber, thus the charge thrown is heavier. It is essential, then, that one be able to operate the measure very smoothly and uniformly. It has been proven many times that the uniformity of charges thrown by a measure is much more dependent upon the skill and uniformity of the operator than upon the measure's design.

All measures are equipped with stops which control the movement of the metering chamber, bringing it to a halt at the proper position for filling from the reservoir and for discharging the metered charge into the case. Easing the handle up against the stop on the filling stroke will cause a light charge to be thrown, while slamming it vigorously against the stop will produce a heavy charge. The impact and vibration of the handle striking the stop will cause the powder to compact or settle in the metering chamber, therefore allowing more powder to be contained.

Irregular movement and vibration of the measure caused by unstable mounting will have the same effect. When small-kernel powders such as Bullseye or ball-type powders are being measured, the drum may be rotated with relative freedom. However, when large-kernel extruded powders such as IMR 4831 are being measured, one or several individual kernels will occasionally be caught between the lip of the metering chamber and the opposing lip in the bottom of the reservoir. This makes considerable additional force necessary to cut the offending grains, and the vibration and impact resulting can cause charge variation. It is my personal practice to return all such charges to the reservoir rather than dump them into the case. Of course, if you are simply throwing underweight charges into the scale pan, this effect will be negligible. Because of the effect that impact, vibration, and movement have on charge weight, it was once more or less standard practice to attach a "knocker" to powder measures. The venerable Lyman adjustable measure (which is still on the market and nearly twice as old as I am) was so fitted. The principle of the knocker was simply that one compensated for variations introduced by other factors by giving one, two, or three sharp raps on the measure housing with the knocker while the metering chamber was picking up its charge. In this fashion, a deliberate attempt was made to secure uniform settling and compaction of the powder in the metering chamber before

Modern Handloading

cutting it off, and thus offset other lesser variations. Today, however, knockers are not in vogue and I know of no other measure on the market that comes so equipped.

Boiled down, the following conditions must be met to obtain maximum uniformity with a rotating-drum type measure and, to a lesser degree, with the Belding and Mull type: Measure must be rigidly mounted so as not to shift or vibrate while in operation; a constant head of powder should be maintained, usually with a baffle; all operating parts must fit closely and be clean and dry and permit free movement; the handle must be moved smoothly and evenly with equal velocity and brought to a stop at both ends of the stroke with equal force; and, the adjustable metering chamber must be very accurately set in the beginning and locked securely so that it cannot come loose and shift during operation. In addition to that, the operator must take the time to practice with the individual measure and to carefully weigh practice charges to determine what effect variations in his technique of operation will produce. Once the most accurate technique is determined for a particular measure, then it should always be used with that measure.

Some elaboration is in order on the adjustment of metering chambers and on their design. With the notable exception of the Lyman among rotary-drum measures, the metering chamber consists of a round hole drilled radially into the drum proper. A relatively large diameter hole is used for large charges of coarse powders, while at the other end of the scale a small-diameter hole is used for the finer pistol and shotgun powders. The receiving portion of the hole is smooth-walled and sharp-edged. The balance of the hole, or, in some instances, a smaller-diameter extension of it, is threaded to accept a threaded stem on a movable plug. The plug is a close sliding fit in the smooth portion of the hole, and turning the threaded stem in this hole moves the plug and thus varies the volume of the metering chamber. A jam nut of some sort is supplied to fix the adjusting stem in position once the desired setting has been achieved. A more sophisticated variation on this theme is the addition of a micrometer-type thumb nut and graduated scale so that numbered settings may be established and recorded for future use. In some measures, the metering chamber and adjustment stem passes through the drum at right angles to its longitudinal axis. In others, it may be angled to one side or the other, simply to provide clearance for the housing. One significant variation from this basic design is found in the Bonanza Bench-Rest powder measure which has its metering chamber drilled completely through the drum at a sharp angle. A sliding plug varies

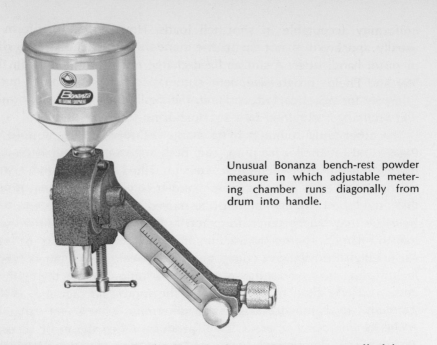

Unusual Bonanza bench-rest powder measure in which adjustable metering chamber runs diagonally from drum into handle.

metering chamber volume, but the plug position is controlled by a knob sliding in the hollow drum handle and fitted with a vernier-type numbered scale. An exception to this round hole/sliding plug design is the old Lyman measure which is still very popular and widely used. It, too, uses a rotating drum but the drum has a V-shaped recess (just like a wedge of pie) machine to accept a series of three horizontally-sliding matching plugs. The plugs are stacked one atop the other, and are moved as a single unit for large charges, while two or one may be moved for the smallest charges. Graduated scales are provided for the purpose already mentioned above. This measure produces a wedge-shaped metering chamber, a shape sometimes claimed to be less accurate than round — but when it is handled with equal skill, it seems to produce results fully up to the standards of later designs.

Fixed-charge measures do have their place as already mentioned. The most commonly encountered type is found in shotshell loading tools where a simple straight hole is drilled vertically through a sliding "charge bar." The bar is moved first under the reservoir where powder flows into it by gravity, and is then moved back over the drop tube where the powder falls into the case. Tools of more recent manufacture are not totally fixed in regard to powder charge, but are supplied with replaceable fixed-charge bushings machined to throw standard charges of the more popular shotshell powders. Changes in charge weights can be made only by replacing bushings. This type works quite well when accurately fitted, and produces a degree of uniformity acceptable in shotshell loads. However, in the form it usually appears it is not up to the standard demanded of the rifle or pistol handloader. A similar fixed-charge measure is found on the Star and Phelps progressive, semi-automatic loading machines made primarily for pistol cartridges. Here, the desired degree of uniformity and accuracy is obtained by very close hand-fitting of parts.

The other fairly common fixed-charge measure is the type sold by Pacific and Lachmiller for pistol use. Both are rotating-drum measures quite similar to their adjustable cousins. The Pacific is offered with the drum drilled with a metering chamber to contain any one of certain established charges of the popular powders. Charge changes may be made only by replacing drums or by drilling or reaming the existing hole larger. The Lachmiller functions in exactly the same fashion except that the metering chamber is drilled over-deep and different length plugs may be used to vary the metering chamber volume for certain specific charges. Thus, if one orders a measure set up to throw 2.7 grains of Bullseye, the plug supplied may be knocked out and replaced with another which cause it to throw six grains of Unique, etc. The only disadvantage here is that extreme care must be taken to insure that the plug is fully and properly seated or the full charge weight identified with that particular plug will not be thrown.

Graduated charge cups are handy for loading only a few rounds of some odd caliber — less time-consuming than setting up a powder measure.

Powder Measures and Scales

It might appear that charge cups of the type furnished with the Lee Loader and other similar tools are not really measures. Yet, they are by function. Those simple little plastic cups function in exactly the same manner as the most sophisticated powder measure in that powder flows into the metering chamber (cup) and the excess is then struck off, and the amount remaining in the cup is the proper charge. A skillful operator who pays attention to detail and uniformity can throw charges with Lee dippers (or any similar type) which differ very little in uniformity from those thrown from $20 to $30 measures. It does take a good bit of practice and it is easy to goof — but it can be done. Those who doubt the accuracy of charge cups should consider that much factory-loaded ammunition is "plate-loaded." In that process, a steel plate drilled with individual cavities for fifty or more powder charges is set upon a smooth steel plate, then the powder is thrown across the top with a scoop, filling every cavity. The operator then wipes off the excess powder with a device somewhat like a squeegee, the entire unit is set over a plate full of cases, the bottom plate withdrawn, and the powder flows into the cases. That "charge plate" is nothing more than a series of charge cups fastened together. If it works well enough for the best of factory ammunition, there is no reason it can't work equally well for you if you learn how to handle it.

When setting up an adjustable powder measure, regardless of type, for a specific charge of a particular powder, too many handloaders make the mistake of juggling the adjustment stem until they get one or two charges which weigh out right on the nose. They then lock the adjustment and proceed happily on their way. This isn't enough to insure precise and uniform charges. It is not good policy to attempt to base charge adjustment on single charges. Over the years I have developed a procedure which is very nearly as fast and yet gives me a far better picture of exactly what is happening. A rough adjustment is first made, based on memory and/or notes and the graduated scale on the measure. Several charges (at least five) are thrown and returned to the reservoir to settle the powder. A single charge is then thrown and quickly weighed on the scale, merely to determine whether my eye-ball adjustment is anywhere in the ballpark. If the charge weight is way off, I make a compensating adjustment and throw and weigh a second charge. This second charge is usually pretty close, at least close enough that I can crank in another minor adjustment after which I throw five consecutive charges directly into the scale pan, paying particular attention to uniformity of operation. Those five charges are then weighed together and the average charge weight is quickly calculated. It will still be probably at least one-half grain off, so a final adjustment is made — and I can usually recall about how many turns or fractions of a turn of the adjustment stem that particular measure requires to produce that much change with that particular powder. A second series of five charges is thrown, weighed and averaged — and the average weight is usually right on the nose. If not, almost invariably one more very slight correction will make it so. Then, as a final check, a ten-charge series is thrown and weighed as a unit, and the average is computed. This is what I go with. It is a far more valid and representative of the average charge that will be entering the cartridge cases than any single or two or three charges I might throw and check. If the charge is for a load intended to produce extreme accuracy, or very high pressures, I will weigh those last ten consecutive charges individually and determine the extreme spread between the lightest and the heavist. However, if the measure being used is one with which I am accustomed to working, I will already know approximately what extreme spread is expected from it with a particular powder, and this additional weighing is really not necessary.

If one simply begins adjusting the measure and weighing single charges and then quits trying when one charge comes in very close to the desired weight, accuracy cannot be assured. That *one* charge may have been at either end of the extreme spread for that particular measure/powder/operator combination. This can easily result in the average charge weight for that particular batch of ammunition being above or below the desired weight by the amount of the extreme spread. With some powders and measures and less-than-expert operators, this could amount to a difference of three or four grains of powder. Add that combination to a load already near the excess pressure threshold, and one might well be at least skirting with trouble.

Powder measures and scales are actually relatively simple mechanical devices. Mechanically, they should hold no secrets from anyone. We all work daily with other items far more sophisticated. However, it behooves one not to take their functioning and accuracy for granted. A few hours spent in becoming familiar with your particular tools and in learning their idiosyncrasies and the effect of your own method of operation will pay big dividends.

CHAPTER 18

Military
Surplus
Ammunition

FOR THE past quarter of a century, military surplus ammunitions of all types have had a great deal of significance, not only to the hand-loader, but to the non-reloading shooter. Certainly, the reasons for the tremendous quantities of surplus ammunition used in past years by civilian shooters are many. By far the most popular reason has been pure economics. At a time when a box of twenty 8x57mm Mauser cartridges currently made for sporting use cost $5.25, one could buy 100 rounds of military surplus ball 8x57mm ammunition for as little as $4.50 to $6.00, depending upon the quantity purchased. Even as this is written, one must pay $5.75 for a box of twenty .30/06 cartridges of new commercial manufacture, while the surplus vendors will sell a case of 1,000 rounds of the military counterpart for a mere $65.00. In short, the commercial cartridge costs roughly twenty-nine cents, while the military surplus costs less than seven cents. Certainly, where the shooter could buy four rounds of surplus ammunition for the price of one commercial cartridge, he would be delighted to do so.

One other major reason has been simply that domestically-produced commercial ammunition has not been available in some calibers for which many thousands of rifles are in circulation. In such instances, military surplus ammunition is the only means of shooting such guns unless one gets deeply involved in case-forming and handloading for that specific purpose.

Naturally, since far more shooters don't than do handload, this has been significant. For example, at a time when the major domestic ammunition manufacturers had all but determined to discontinue the .303 British cartridge, over 100,000 .303 caliber Lee-Enfield rifles were imported and eventually sold in this country. The limited stocks of U.S. .303 ammunition were immediately swallowed up, and those thousands of Lee-Enfield purchasers had no place to turn but to the vendors who had fortuitously imported several million rounds of military surplus .303 ammunition. Then, there were the 6.5mm and 7.35mm Italian rifles and the 6.5mm Mauser and Krag rifles (to name only a few) for which ammunition has never been manufactured in this country. Hundreds of thousands of the rifles were imported, and buyers who wished to shoot them could only turn to the military surplus ammunition. And, it was the same story over and over, as various nations phased out their WWII and earlier rifles in favor of modern assault weapons and disposed of obsolete guns and ammunition at extremely low prices.

The presence of millions of rounds of low-cost surplus ammunition presented (presents) mixed blessings. Initially, manufacturers feared that such great quantities of cheap ammunition would seriously reduce their sales. Such has not proven to be the case, for in the intervening years, domestic commercial ammunition sales have continued to increase at hitherto-unexperienced and unexpected rates. During 1968 and 1969, for example, commercial ammunition sales reached all-time highs in spite of many millions of rounds of military surplus ammunition being sold during the same period. Consequently, we can only assume that the presence of this cheap ammunition greatly increased shooting in general throughout the nation — and of course, the vast quantity of surplus guns brought in at the same time contributed greatly to this.

In another area, the presence of cheap ammunition encouraged poor shooting and hunting practices to some extent. Many people who had never before owned a "deer rifle" bought a $15 Lee-Enfield and one hundred rounds of ammunition. Those same people in many instances did a lot of unsafe shooting and also took that sharply pointed, full-jacketed military ammunition hunting, and we all know that far more game is wounded than bagged by such cartridges.

All in all, it is this scribe's opinion, though, that the presence of military arms and cheap surplus ammunition contributed greatly to the tremendous growth of shooting that this country has experienced in the past score of years.

Two very important factors must be considered in the use of military surplus ammunition, be it for rifle or pistol. First is the reliability. Naturally, ammunition intended by a sovereign nation to be used in fighting its wars is originally manufactured to a very high standard of quality. Such ammunition is eventually declared surplus and sold for a fraction of its individual original cost — but, in the intervening period (which might well be twenty-five or thirty years), it may very well deteriorate simply from age or from damage or exposure to extreme conditions of climate. Though normally quite stable in storage for as much as twenty years or more, ammunition is by no means imperishable, especially when exposed to high or low temperatures and moisture. Added to that is the fact that much of the surplus ammunition was produced under the press of wartime conditions where quality standards were lowered because of urgency, resulting in both propellants and primers becoming more subject to deterioration from age. In other instances poor-quality materials and processes are sometimes used in wartime.

Generally speaking, the more reputable surplus dealers and distributors conduct tests and classify their surplus ammunition and readily admit which lots or batches do not produce 100 per cent reliability. Advertisements will be found reading "100 per cent sure-fire" and "Misfires to be expected." However, "100 per cent sure-fire" doesn't mean that the ballistic performance of the ammunition is up to its original standards. Extreme spreads in the velocity, pressure, and other performance factors may be encountered. "Sure-fire" just means every round will go "Bang."

Ammunition manufactured many years ago was also without the benefit of recent advances in technology. Some surplus ammunition is actually seventy-five years old. Consequently, it is not unusual to encounter split necks, even in sure-fire lots. Splits are safe to fire, so long as they hold the bullet in position well enough for feeding and chambering, and will not normally damage the gun. However, excessive firing of split necks may well produce some errosion of the chamber neck and that in turn can produce extraction difficulties with even good ammunition. Some lots of older ammunition are assembled in cases made from a relatively hard brass, and the cases may sometimes split upon firing. This in itself isn't dangerous, but a fair number of splits occurring at the same point in the chamber may eventually produce some roughening thereof.

Primer leaks may also be encountered even in relatively low-pressure loadings where primer cups have become embrittled over the years. Primer leaks can be dangerous to both gun and shooter, since in their more advanced form they allow high-pressure gas to

Modern Handloading

squirt back through the bolt toward the shooter's face. Older gun designs sometimes leave a clear path right to the shooter's eyes for such escaping gas.

Important to those who are too lazy to clean their guns thoroughly after each shooting session is the fact that virtually all military ammunition produced prior to the early 1950's is loaded with so-called "corrosive" chlorate-type primers. In addition, some military ammunition, notably that of French manufacture, was loaded with mercuric-type primers which have a deleterious effect on the cartridge case. The importance and handling of these factors is discussed in detail in the chapter pertaining to primers. Also, there will be found detailed information on identifying the type primer with which a particular lot of military surplus ammunition may have been loaded.

Generally speaking, the bulk of the military surplus ammunition sold in this country has been of the type loosely described as "ball." This means simply that it is loaded with a full-jacket, lead-core bullet. Minor variations of ball exist, some containing aluminum or fibre as a portion of the core, and others containing a soft steel core. Bullets of this type generally have no deleterious effect on gun barrels and may be used in quantity without encountering any problems. Lesser quantities of armor-piercing and tracer ammunition in standard rifle calibers have been offered. Armor-piercing is generally acceptable for use in anything except bench-rest grade barrels. In most instances, it has been found to be no harder on barrels than ball ammunition, and will, in some instances, even give accuracy superior to that of ball. Tracer, on the other hand, is to be avoided like the plague. First, it creates a tremendous fire hazard wherever used. Certainly no one in his right mind would fire tracers promiscuously into the countryside where they will almost without exception start brush and timber fires. This scribe — and I'm sure many other ex-military people — has spent countless hours and days fighting brush and grass fires started just that way during military training operations. In fact, many states and most ranges prohibit by law and regulation the firing of tracer bullets. Additionally, some tracers leave an abrasive and corrosive residue in the bore. Only a few rounds of some types of tracer bullets will result in abrasions on the bore's surface from the gritty residue which is caused to act like sandpaper by subsequent bullets passing over it or carrying it down the bore. Others may leave an ash residue which attracts moisture from the air and promotes rusting in only a few hours. Either one can ruin a fine bore in short order.

Vendors usually try to weed out other types of hazardous military ammunition, such as incendiary and explosive. However, because of the manner in which some such ammunition is obtained, this cannot always be accomplished. It is not unusual for a large quantity of ammunition to be simply dumped loosely into large shipping containers by the selling government without much regard to type. Consequently, nothing less than 100 per cent individual visual inspection can insure that an occasional round of explosive-bullet or incendiary-bullet does not get out to a shooter with his order for ball. Consequently, when buying loose-pack surplus ammunition, one should inspect closely for bullets with unusual shapes or unusual colored markings. When such are found, they should be set aside and by no means fired. Positive identification can usually be made by referring to Volumes I, II, and III, of *Cartridges for Collectors,* by Fred Datig.

Some incendiary and explosive bullets are not "bore-safe." This means simply that they *could* detonate or ignite in the barrel under the forces exerted upon them by firing, and also that the amount and type of pressure involved in pulling those bullets might well set them off. Consequently, under no circumstances should an attempt be made to salvage case and powder by pulling such bullets.

Of what value is surplus military ammunition to the handloader? Well, it invariably has *some* value, aside from simple shooting. Fired cases first. Steel cases, readily recognizable by their lacquer or plated finish (usually gray or green in color), and most often found in 7.92mm German and 7.62mm Russian ammunition can also be positively identified with a magnet. They have virtually no value to the handloader. They can be reloaded, but the time and effort involved certainly can't be justified except in rare instances. Brass cases that survive firing without splits or separations can be reloaded. Of course, the bulk of such ammunition of foreign manufacture is assembled with Berdan-type primers with their attendant difficulties. Usually, primers of the proper size can be obtained if one wishes to go to the trouble, but I personally do not ordinarily recommend attempting to reload Berdan-primed military cases. It is simply just too much trouble. However, if you feel so inclined, the necessary procedures are outlined in the chapter on priming. On the other hand, Boxer-type cases are readily reloadable with standard domestic primers and tools. Such military surplus brass is fully as useful as any new domestic brass you might buy across the counter so long as it is in good condition.

The only difficulty encountered in the re-use of fired military cases is the primer crimp which is almost invariably present in one form or another. It may take the form of a complete 360° *ring* crimp in which the mouth of the primer pocket is forced over the radius of the primer by a hollow punch; the form of a *stab* crimp where two or three or

Modern Handloading

more very short segments are punched in over the primer radius; or may be a segmented crimp where longer sections of the mouth of the primer pocket are pushed in over the primer. In any event, these crimps serve to prevent the primer from backing out of the case head when fired in automatic weapons whose headspace may be slightly excessive or may have grown excessive because of heat-induced elongation of barrel and other parts. In most instances, decapping presents no great problems, though a decapping pin may occasionally push through the center of a primer cup without removing it. Repriming cannot be accomplished until the crimp is either removed or turned back out of the way — and the procedures for this operation are described in the chapter on priming.

Generally speaking, U.S. and most foreign makes of military brass of recent manufacture are considered to be at least equal and sometimes superior to commercial cartridge cases. Consequently, such military cases may be used for re-forming as well as conventional reloading.

Often, it is more practical or more worthwhile to break down military surplus ammunition into its individual components rather than to attempt to fire it and then re-use the cases. Bullets can usually be salvaged, even though the balance of the cartridge is completely unserviceable. Over the years, there have been numerous instances of large quantities of ammunition with split necks, defective primers, or deteriorated powder charges being sold at sufficiently low prices that salvaging the bullets was an economically sound operation. At one time some splitneck 7mm and .30 ammunition could be purchased for less than $.01 per round — far less than the value of the bullet alone.

Often, even though the case may be damaged or the primer defective, powder as well as bullet may be salvaged. When the powder type is known, as it usually is or can readily be determined in the case of U.S.-produced ammunition, it may be salvaged and re-used. This is not practical where only a few hundred rounds of ammunition are involved. One-thousand rounds of .30/06 military ammunition will contain roughly eight pounds of powder — which I would consider the minimum quantity worth cleaning and salvaging and developing loading data for it. Such salvage powder will contain a certain amount of dust and graphite and will also have mixed in with it flakes and rings of water-proofing compound that was originally placed inside the case mouth to seal the loaded cartridge. Small pieces of brass are also sometimes encountered, punchings from the flash holes. This material won't necessarily affect the burning of the powder or its utility, but will surely gum up a powder measure. It is best removed by trickling the powder slowly from one container to another through the gentle breeze from an electric fan. The light dust, water-proofing material, and chips and fragments of individual powder granules will be blown out of the falling stream of powder. Once thus cleaned, the salvaged powder should be placed in light-tight and water-tight containers and sealed. More detailed instructions for the use of such salvage powder whose type is *known* is contained in the chapter on smokeless powders.

Often, however, it is not possible to identify the powder salvaged from broken-down military surplus ammunition. This is *usually* true of foreign ammunition. Even then, after cleaning, the salvage powder can be of some use if there is a sufficient quantity to warrant developing loading data for it. However, one must be especially cautious breaking down any foreign ammunition because of the fact different powders were used in cartridges of the same caliber. For example, literally dozens of different powders were used in 8mm Mauser cartridges loaded in different ways by different arsenals in different countries. When a mixed bag of such ammunition is broken down, one has no way of knowing how many powders are involved unless they are physically different in appearance and can be segregated. Consequently, I cannot recommend attempts at using such powder — except where it is known that the ammunition being broken is of a single lot from a single manufacturer and all of the same loading. This can be determined only when the ammunition is in the original packaging and still carries the original markings and lot number. The mere fact that it is all the same type load and carries the same head-stamp does not necessarily assure that the entire lot is loaded with the same powder type and charge.

At one time I became involved in the breaking down of 100,000 rounds of Israeli-manufactured 7.92mm Mauser ammunition loaded for machine-gun use. It was in perfect condition, all in the original packaging and all clearly identified by lot number. Careful examination of the powder in several cartridges out of each case indicated the same type and charge weight of powder had been used throughout. Consequently, the entire lot of salvage powder was first cleaned, then blended, and used to load 7.62mm NATO ammunition, but only after extensive tests to develop a charge weight producing the desired pressure and velocity. In another instance, the break-down of a large quantity of 7.35mm Italian Carcano ammunition was involved. Again, the ammunition was all of the same lot number and was still packed in its original sealed metal containers. Nevertheless, it was found that

Modern Handloading

three different powders had been used, each with a slightly different charge weight, in loading that batch of ammunition. Two of the powders were of typical dark gray, graphited, single-perforation, extruded type but differed slightly in granule dimensions. The third was also single-perforation extruded, but in much shorter and thicker granules and of a reddish color. Consequently, no attempt was made to utilize this salvage product — it was eventually destroyed by burning. These two incidents have been described to show clearly that under no circumstances may one simply *assume* that a particular batch of surplus ammunition will produce powder that can be salvaged and used.

Primers cannot be salvaged for re-use, except in that if they are in good condition and function reliably, the primed cases may be reloaded after bullets and powder are removed. Insofar as Boxer-primed cases are concerned, removal of defective or partially defective primers presents no problem. No mechanical problem, that is. For example, I once obtained several hundred rounds of quite old .30/40 military ammunition which gave a very high percentage of misfires. Powder, bullets, and cases were in good condition — so the ammunition was broken down, the cases decapped, and fresh primers used to reload the original components. The depriming operation presented only one hazard — the possibility that an overly-sensitive primer might be detonated by the decapping punch as it was forced out of the case. Any danger of injury was guarded against simply by rotating the ram of the press being used for decapping 180° so that the slot through which ejected primers fell was facing directly away from the operator. Thus, if a primer detonated (and none did, though it has happened on other similar projects), the flash and fragments of the primer would be thrown away in a harmless direction.

Berdan-primed ammunition is not so easily handled. There is no practical mechanical way in which live or partially live Berdan-primers can be removed. If it should become absolutely necessary to salvage cartridge cases containing defective Berdan-primers, then the cases must be submerged in a hot oil solution for several days or boiled vigorously for an hour or so in water to completely destroy the detonating qualities of the priming mix. After that, the cases may be de-primed using the same various methods described in the chapter on Priming. Alternatively, such cases may be de-primed hydraulically, as described in the chapter on priming.

Particularly where salvaging cases of *old* military surplus ammunition, reloading life may be greatly extended by annealing. Much older ammunition was made with case necks far harder than is the custom of today, and those hard necks may split upon the first firing after being reloaded. This can be avoided by simply annealing the mouth and shoulder area of the case before attempting any resizing or other reloading operations. A detailed description of annealing procedures will be found elsewhere in this volume, and it is unnecessary to say that you should never attempt to anneal live primed cases.

From the foregoing, it should be evident that military surplus ammunition definitely does have a place in the activities of the ardent handloader. Often it is possible to obtain components or shootable ammunition from surplus vendors that could not be had in any other way. A typical example of this was the 6.5mm Italian Carcano surplus ammunition before Norma began producing it commercially. Prior to that time, nothing but surplus ammunition was available for use in the thousands of Italian rifles circulating in this country. Another example is the use of 7.35mm Italian service ammunition broken down and re-formed into cases for loading the 7.62 x 39mm Soviet M43 cartridge which is generally not available in this country, for use in the SKS carbines being brought back as war trophies from Vietnam.

CHAPTER 19

Case Forming and Alteration

BACK AROUND the turn of the century, Dr. Mann worked slowly and laboriously through his many detailed experiments. He was determined to produce one-hole accuracy from a rifled barrel. Mann designed the cartridges for this work, and in producing cases for them, he was probably the first man to "reform" cases as we understand the term today. That also may classify him as our first "wildcatter." Of course, even before that, the arms producers did much of their new-caliber experimental work by re-shaping existing cases.

Mann used the .30/40 Krag case for much of his work. When a 7mm or .25 caliber was needed, he tapered, necked, and shortened the Krag case to suit him. Many examples of this work are shown in his fine work *The Bullet's Flight*.

Not until WWII does it seem that many persons other than outside the wildcatting clan put Mann's ideas to work. During the big fuss, though, ammunition for sporting use was virtually impossible to obtain. Yet, many shooters had bushels of fired cases they had saved for reloading. Especially plentiful was .30/06 military brass.

Probably simultaneously, hundreds of reloaders discovered they could make .270 Winchester cases by simply running '06 brass into a full length resizing die. The ammunition shortage no longer kept .270 fanciers from shooting. An influx of war souvenir Mausers in 8 x 57mm caliber prompted other lads to run the old, reliable '06 case into a 8 x 57mm die. It came out just right to fit the Mauser chamber, after trimming to length. This '06 to 8mm conversion became vastly popular. I ought to know, for in the middle 40's, I financed many a new gun and an occasional bacchanalian spree by selling re-formed 8mm cases for around three cents each. It was a hell of a lot of work, with an old Pacific press, die, tubing cutter, and a file as my only tools — but a corporal will do a lot for beer money.

From those beginnings, case forming has become an integral part of the reloading scene. Several reloading shop operators specialize in making obsolete and hard-to-get or expensive cases from readily available brass. At one time I did this myself, and even wrote a book on the subject called *Cartridge Conversions,* which is, fortunately, now in its second, revised, edition. And it is, if I do say so myself, a pretty good reference on the subject.

Indicative of the popularity of case forming is the fact that most major makers of reloading tools and dies produce special forming-die sets. Most outfits will also make dies on special order for pratically any re-forming job. Just tell them what you want to make and what from. They'll oblige with the correct dies, though the cost is pretty steep in some instances.

All over the U.S., their voices now still from long dis-use, lie count-less thousands of fine old rifles, mostly of pre-1900 vintage, the disappearance from the shelves of one caliber ammunition after another having spelled their doom. Most of the loads for the old '76, '86, and '95 Winchesters have been dropped. The revered Sharps and Ballards, too, are without fodder except in .32/40, .38/55, and .45/70 calibers — those three, let's hope, will be with us for many years to come.

These venerable relics of our Nation's formative frontier days hang silent on the walls, perhaps never again to belch forth those fragrantly pungent clouds of spark-shot white smoke as they did when a new nation was being carved from the wilderness.

They can be rescued, though, for few indeed are the obsolete American calibers for which it is not possible to make up a supply of cases from some currently available brass. The informed handloader, using the correct basic case, can make up just about anything he needs. Many present-day cartridges, simply logical developments of earlier loads, retain the basic characteristics and dimensions of the old-timers. The .45/70, for instance, is identical except for length and bullet diameter to a dozen or more of yesterday's discontinued numbers. Similar situations exist with respect to other, more modern cases. That

Modern Handloading

RCBS forming die set to make .250-3000 from .30-06 brass. A set of loading dies is still required.

Use of Forster power trimmer to remove large amounts of excess quickly and easily.

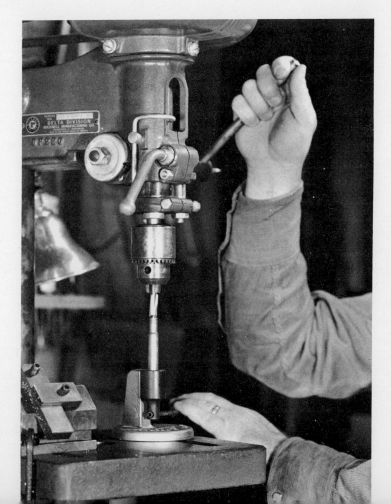

old workhorse, the .30/06, can be used to form more current and obsolete metric calibers than you can shake a ramrod at.

Now, let's get down to cases (no pun intended) on what you can do to make Great-Gran'pappy's old buffalo gun or that ornate Deutsche sporter bellow again.

Little more than what is already to be found in the average hand-loader's shop will be required for most forming jobs. Perfectly good, safe cases can be (and have been) turned out with nothing more than a hand-type sizing die, a file, and a good bench vise. For speed, efficiency, and comfort, though, a heavy-duty press, case trimmer, and a large assortment of dies will save a lot of trouble.

With the exception of tong-tools and a few light-duty presses, practically all presses currently available will do a good job of case forming. The operator must do his part, though. Presses can't think. Presses with the greater mechanical advantages will, of course, make the job easier. The discontinued RCBS Model A-2 is ideally suited to this sort of thing and after using one for over a decade, I consider it indispensable. Its replacement, the Rock Chucker is nearly as good. The Super Turret that Hollywood Gun Shop used to make is also excellent, as is the excellent Pacific "Pro" Deluxe. Case-forming does place more of a stress and strain on tools than normal reloading.

Selection of the correct basic case for forming to the desired caliber requires particular care. Most important is to insure that head and rim dimensions are as close as possible to those of the case to be formed. The head of a case receives the least support during firing and, there-fore, is most likely to let go if all is not well. This is especially so in the older guns. If you must use a case a few thousandths large at this point, it can be reduced by swaging, which we'll get to later. Con-versely, a case somewhat small at the head (no more than .015" smaller) may be used, fire-forming it with a light load to expand it before using full charge loads. Remember, only a few thousandths of an inch of soft brass stands between your face and white hot gas at a pressure of many tens of thousands of pounds!

Normally, only rimmed basic cases may be used to form rimmed types, and vice versa. There are, however, a few exceptions to the rule — take a rimmed, *solid head*, turn a new extraction groove cut, and we have a rimless case. Thousands of 9mm Luger cases made in this fashion from .38 Special brass were used by the writer back in '46 and '47; they were satisfactory with light loads, and these so-formed cases should be limited to this type load.

Conversely, a rimless case can be reduced in diameter ahead of the extraction groove while the rim proper is left full diameter. Depending

Modern Handloading

on the amount of reduction, it may be converted to a semi-rimmed or rimmed case. Thousands of 6.5mm Japanese semi-rimmed cases have been made up this way from .30/06, .300 Savage, and 7.62mm NATO brass.

The case alterations just described are extreme, last ditch resorts, to be used only when there is no other way to get what you need. Such work should never be attempted on anything but late-manufacture, solid head cases. The earlier folded-head and balloon-head cases simply do not have enough metal in this critical area to withstand the strain. To make certain your cases are sufficiently strong after such a job, section the first few before going further. Clamp the formed case horizontally in your vise; then use hacksaw and file to cut it down to the midpoint. You can then easily see just how much brass is left and determine whether it will hold.

Case condition is also quite important. Odds and ends picked up on the range or donated by well-meaning friends should be viewed with suspicion. There is no way of knowing how many times they were fired, or with what. Mercuric primers, repeated firings, or excessive pressures may have damaged them. And, such damage may not be evident until firing, and then it is too late. Be safe. Stick to cases that you know are new or once-fired with sensible loads. Any sign whatsoever of leaks, splits, or cracks is ample reason for rejection. Throw 'em away. In any event, you'll be able to select a basic case to meet your particular needs by examining the accompanying dimensional chart. On it, locate the case you want to make, and then look for another available caliber that comes close enough to it dimensionally to do the job.

Resizing

This will be familiar to anyone with a bit of loading experience and is covered elsewhere in this volume, but a few points are worth additional emphasis. Lubrication is vital. Those of you have wrestled a case stuck in a die know how the lack of a bit of lubricant can louse things up. Forming cases *works* them much more than normal full length sizing; thus, they're more likely to stick in the die if not adequately lubricated. Any of the prepared lubricants furnished by the loading tool makers are excellent. Anhydrous lanolin from the corner drugstore is equally good, as is "STP" and similar auto engine oil additives. Apply a thin, even coat with your fingers or a cloth, but do it sparingly. Any excess will be trapped at the junction of neck and shoulder as the case is forced home in the die to produce oil dents.

Minor dents do no harm, other than to make a sloppy looking job, but the larger ones can split or crack the brass. Incidentally, if you lack any other lubricant, moisten the fingers and rub them over a cake of toilet soap and apply the soap in a very thin film over the case. It works fine. Don't use light household or machine oils — they lack the film strength necessary for this kind of work.

When full length dies are used, adjust them so that the full-sized case chambers in the gun with *slight* resistance. Make certain that the force required to press the case into the die does not push the die away from the shell holder. With a case in the die and the holder at the top of its stroke, holder and die will usually make firm contact. If they do not, be not concerned so long as the resulting case chambers properly.

In some instances, it may be desirable to use intermediate stage dies. Less effort is required to form 7 x 57mm cases from .30/06 brass if the shoulder is first pushed back to the correct position by running the case into a .308 Winchester die. The case will then go into the 7mm full-length die much easier. Two easy operations rather than a single hard one. Some jobs make the use of intermediate dies a necessity rather than a convenience, as in forming .30 and .35 Newton cases from Norma cylindrical belted brass. Attempts to run the basic case directly into the full-length die will result in crumpled cases like that shown. Adjustment of intermediate dies is by trial end error until the desired results are obtained. One can, however, record the distances between shell holder top and die mouth after the correct setting has been obtained. This will enable you to get it right next time on the first try.

Occasionally, it will be possible to form a case without using the die of that particular caliber. The 6.5 x 55mm Mauser-Krag cases can be made up this way. First size .30/06 brass full length in a 7 x 57mm die, and then trim to 2.156". Now run the case into a .250-3000 or .22-250 die until the bolt will close on it with just the faintest trace of drag. Next, expand the neck to hold .264" bullets and fire-form. This method is of particular value when the number of cases to be made up does not justify the cost of a new set of dies. Judicious comparison of case shapes and sizes will enable you to select proper dies for this sort of operation.

Trimming

Most basic cases will not be of the correct length and will require trimming from a few thousandths to upwards of a half-inch. When only

Modern Handloading

a small amount is to be cut away any of the commercial trimmers do a fine job. They're all rather slow, though, and a better method is required if you have lots of brass to remove or many cases to process. Forester-Appelt makes a fine tool for this purpose. It consists of a spot facing tool and pilot to fit the case neck, intended to be chucked in a drill press. It is used with a collet type case holder (like those used in their bench type trimmer) that is bolted to the press table. Once set up and adjusted to the correct case length, any amount of brass can be trimmed with one smooth stroke of the handle. With a little practice, several hundred cases per hour can be handled with ease.

Large amounts of brass can be removed with a common, hardware-store, tubing cutter, though the burr and crimp left will have to be removed before a bullet can be seated. Run the case over an expander plug of the correct diameter, and then chamfer the mouth with a knife or with the Wilson chamfer tool. Several strokes with a fine cut mill file will remove any feather edge that remains.

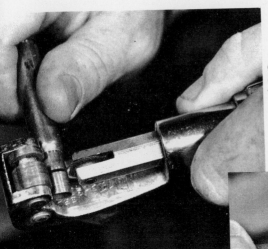

Common tubing cutter is suitable for cutting off large amounts of brass, but very slow and tedious.

Form and trim die makes both operations simple and is the only extra tool needed for many forming jobs.

Generally, final trimming should be done after any required neck expansion and sizing are completed. These operations tend to drag one side of the neck down a bit, leaving the case mouth out of square. Trimming last eliminates this. The finished case length should be a few thousandths of an inch or so less than specified. This will allow it to stretch during subsequent firing, obviating the need to trim again after only a few loadings. Don't however, cut them *too* short if you want to crimp them in a standard seating die. Strike a happy medium; short enough to stretch, long enough to crimp. Remember that a good, solid crimp is essential to proper magazine functioning in many of the older arms.

Neck Expansion

Many jobs will require that the neck of the basic case be opened up to accept bigger bullets. When this increase is slight, it may be done during resizing by having the correct diameter expanding plug in place. Some plugs have a sharp shoulder that will shave brass or even crumple necks when much expansion takes place. These should be ground to a smooth taper and polished; then no further trouble will be encountered. Tapered plugs are available in some calibers and are well worth the nominal extra cost. Such a plug will open .30/06 necks up in one pass to take .357" bullets with no trouble at all.

When a really great amount of neck increase is required, such as opening up .30/40 necks to take .403" bullets (for the straight Sharps cases), it should be done as a separate operation. Several stages may be required to avoid splitting the necks. The RCBS expander die, ideal for this purpose, fits the press frame, and is bored out to accept a large variety of cases. Expander rods, in a wide range of diameters, are threaded into this die from the top. If in doubt about the amount of expansion that a case will stand without splitting, go up by .020" steps. The amount that can safely be done in one pass will vary from one lot of brass to another. Some recent arsenal .30/06's will stand as much as .060" to .075" at one pass without splitting or crumpling.

Fire-Forming

Body and shoulder dimensions of a basic case may be less than required, even though other dimensions are okay. The only easy and practical way to correct this is to go ahead with other forming operations, then fire the case in the chamber for which it is intended. This is known as fire-forming. The expanding powder gases push the soft brass case walls out until they are stopped by the chamber walls, thus,

Modern Handloading

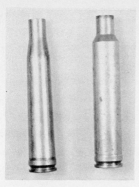

Typical fire-forming operation. Case at left was fired in "improved" chamber to produce result at right.

the case is reshaped to fit the chamber to perfection. Except where a great amount of expansion is to take place, medium-to-stiff loads should be used. Too-light loads won't finish the job in one shot.

An example of full load fire-forming is found in making 8.15 x 46R cases from .32/40 brass. The .32 is a straight taper case while the 8.15 has a very slight bottleneck. Only a few thousandths expansion is required to produce the latter's shoulder. Here a full charge load may be safely used. Thirteen grains of Hercules 2400 powder behind a 150- to 180-grain cast bullet will produce perfectly formed cases every time, and deliver full velocity.

An alternative to the use of bulleted loads for forming is the filler load. It consists of a light charge of fast-burning pistol powder, such as Bullseye, behind a case full of granular filler such as corn meal or Cream of Wheat. Powder and filler should be separated by a thin wad. A quarter sheet of bathroom tissue works well for this. IPCO grease wads will prevent spillage during handling; but if that isn't available, just stick the case mouth into a stick of bullet lube or cake of soap, give it a twist and pull it out. This leaves a plug of grease or soap in the neck. Such loads, happily, require no range facilities for firing. Nothing but dust-like filler emerges from the muzzle; its energy dissipates within a very short distance, it is perfectly safe to use in your base-ment — assuming mama will hold still for the not-inconsiderable noise.

It is possible, however, to develop dangerous pressures with filler loads unless caution is observed. Since no published data is available, start with a charge of Bullseye that fills approximately 10 per cent of the case volume. Work up from this until satisfactory expansion is obtained. Be alert for any signs of excessive pressure and back off immediately from the load that caused them. Thick, heavy cases may not always blow out completely with this type of load. If not, don't worry about it. They will be so close to final shape that the first firing with a full charge will finish the job in fine order. A filler load may also be substituted for

Case Forming and Alteration   **259**

other operations, such as neck expansion. The .30/06 may be fire-formed directly to .35 and .400 Whelen without any preliminary expansion by fifteen grains of Bullseye plus corn meal. This also produces a nice square neck and mouth, not always possible with mechanical expansion.

Virtually all of the increased capacity wildcat calibers require that the body and shoulder of the parent case be expanded anywhere from a little to a lot. In addition, they often require moving the shoulder considerably forward of its original position, as in the Gibbs' series of cartridges.

Fire-forming is the most common method of accomplishing such reshaping. It is generally satisfactory, though where large amounts of radial expansion are involved, cases often split near the shoulder/body junction area and are ruined. Even so, there is no danger to gun or shooter so long as the case is properly headspaced in the chamber in which it is fired, and the head is of correct diameter.

Shoving the shoulder forward is another matter. With rimmed and belted cases, fire-forming does the job neatly and with dispatch — and equally important — without danger. Rimless cases are not so simply handled. To avoid severe weakening of the case near the head (even complete separations occur sometimes), a separate auxiliary shoulder must be formed prior to fire-forming. Sure, I know that many shooters of the .30 Gibbs and similar calibers simply fire factory loads in their stretched-out chambers, relying on the extractor to hold the round in place. I also know I've sectioned many such cases and found them seriously weakened just ahead of the web. I definitely do not approve of such shenannigans.

There is a far simpler, faster, and cheaper method of forming the many blown-out "improved" cases. Do it hydraulically. If the original case (either new or fired) is placed in a die, then filled with fluid to which force is applied, the case will expand to fill the die cavity without any such damage. And, it takes far less force than you might expect to get the job done.

After starting this book, I stumbled (practically by accident) on a very practical method of hydraulic case forming. I was reaming the necks of some well-used .30 Gibbs cases. It suddenly occurred to me that the RCBS reaming die being used would be perfect for the job if a plunger was substituted for the reamer, and proper force was applied to it.

A piece of drill rod was lapped to fit the reamer pilot hole. A fired .30/06 case, with primer still in place, was filled with tap water and run into the reamer die. The drill rod was greased to prevent escape of water past it, and dropped into the pilot hole. The first two blows of an 8 oz. machinist's hammer expanded the case pretty well, but the third popped the primer out, spraying water round and about.

While not fully formed, the case was expanded enough to hold full charges of powder. Best of all, the shoulder had moved forward and assumed enough of its new shape to headspace properly in my .30 Gibbs Mauser.

A plug was soldered in the central hole of a spare shell holder. Repeating the process, the plug kept the primer in place and three hefty belts with a one-lb. hammer did a perfect job of expanding the case to fill the die cavity fully. Only a slight radius at shoulder/body junction gave any indication the case had not been fire-formed. All this without the expense of powder, primer, and bullet, and without the trouble of going to the range to do the job. More important yet, the cases were not weakened by the stretching action that so often causes incipient ruptures when the job is done by fire-forming.

Turning Rim and Head

On occasion, one will be forced to use a basic case that has a rim too great in thickness and/or diameter. This is easily corrected in a lathe. Set the case up in a tapered collet or chuck and turn the rim to desired thickness and diameter. Always remove the excess metal from the front of the rim, not from the rear face where the headstamp appears. The latter course will reduce the depth of the primer pocket, allowing it to protrude. A premature firing could result from the use of such a case. Of course, the pocket could be deepened with a drill or end mill, but why add another operation to the job? Also, deepening the pocket will reduce web thickness, possibly to the danger point. Do it the safe and easy way — from the front.

A tool bit ground to the shape shown can be used to reduce both diameter and thickness in one pass. While dimensions do vary from case to case, this tool will produce a serviceable rim in nearly all instances. Just make sure it is set up to give the correct thickness, thus, headspacing the new case correctly. Using this tool, it's relatively simple to reduce .30/40 rims to form the rimmed Mauser 57mm series. If only a few cases are to be made up, the job can be done with a safe-edge file, but it's a lot of work.

Cases may be safely reduced a *slight* amount just ahead of the extraction groove. This is practical only in thick-walled, solid-head modern cases. The older folded and balloon head types simply do not have any metal to spare at this critical point. Any reduction here is achieved

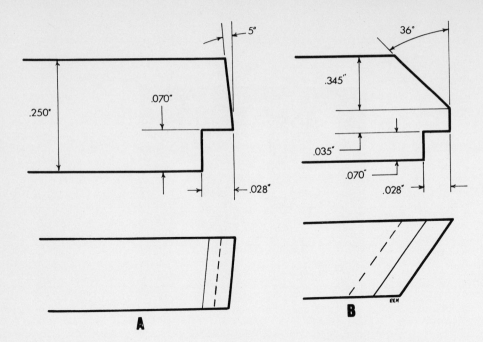

Tool bit ground to reduce head and rim and recut extractor groove.

at the expense of wall thickness and strength. This, too, can be done with a file if you have no lathe at hand. Go easy on this sort of thing and section the first few you make up, as described earlier.

Some turning may be required on rimless cases, usually when the head diameter has been reduced by swaging or other means and the extraction groove must be deepened or recut. This can be done with the same tool bit already mentioned. It is shaped to reproduce approximately the groove of the .30/06 case. While not all cases have exactly this same groove, they are close enough that this shape will function quite well. The extended angle will also permit the same tool to be used for turning the reduced head type of case. It will carry the forward slope of the groove on out to the edge of the head.

Swaging

Cases may be reduced at the head by swaging as much as .040'' by using dies of the proper diameter. Considerable power is needed so a hydraulic or arbor press is a convenient item to have. Lacking either, a heavy bench vise or the more substantial loading presses will do the job. Simple dies (as shown) can be made up for this job. The drawing shows a threaded die for use in a loading press but no threads are needed for use in a vise or other type press. The interior should be

Modern Handloading

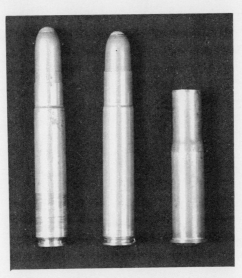

Two cases at left began as Norma cylindrical belted magnum; belts and rims were turned down, and extractor groove deepened to make 11.2 x 68 and 11.2 x 73 cases.

reamed and lapped to size, then hardened. If more than .005″ reduction is desired, a set of dies (each one of .005″ reduction) will be needed, unless you have a very powerful press. For use in a loading press, the standard shell holder should be faced off as shown — case rims often won't stand the strain of pulling them out of this type of die. A rod is inserted in the top of the die and the case is driven or pressed out. Care must be used or the head will be deformed and the primer pocket distorted. A smooth, even push is best.

Swaging .30/06 case heads will enable one to make up 6.5mm Japanese, 6.5mm Carcano, 6mm Lee-Navy, 7.35 Terni and numerous other hard-to-find calibers. Even the long 9.3 x 72R can be made up by combining swaging with redrawing to get the length required. The writer has used many .45/90 cases drawn from .45/70 brass by Robert Pomeroy of Waterbury, Conn. He does a good job and the cases have held up well, even when loaded heavily with 53 grains of 4895 behind a 405-grain bullet.

Redrawing

This is mechanically simple but the equipment required is beyond the means of the average loader. Precision-made dies and punches are used, just as in original case manufacture. It is a quite advanced job requiring complete machine shop and heat-treating services.

Using some procedures already covered, let's take up a couple of typical forming jobs and see just how much work is involved. The first, quite simple, will produce nice shiny new brass for that .40/60 Winchester '76 you've been wanting to shoot. If new cases are used, the result will be more durable than any of the original cases you might find around these days.

While the .40/60 case head is almost identical to that of the readily available .45/70, the latter is 2.10″ long as opposed to the 1.88″ of the .40/60. Run the .45/70 case into a .40/60 full-length sizing die, removing the expander rod if need be, to allow the longer case to go all the way in. If a hand-type die is used, support the neck end on a large nut or a piece of metal with a half-inch hole in it. This will allow the excess case length to protrude from the die without crumpling against the vise. After sizing, trim to 1.87″ and chamfer the mouth lightly to aid

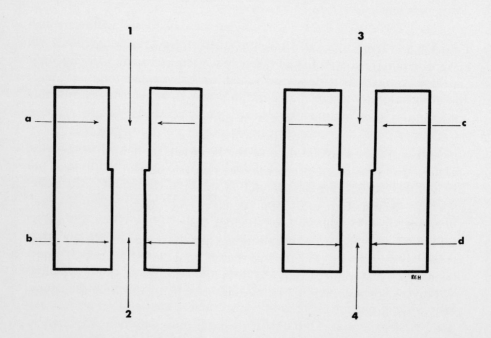

Two-die set for reducing case head diameter. Diameters A, B, C and D are progressively smaller, and case is pressed into them in sequence to produce reduction desired.

Modern Handloading

bullet seating. Now run the case over the expander plug, if necessary, to bring the neck to correct inside diameter. All done, and you've made a new .40/60 case at least equal in quality to the original factory product and, by your own efforts, returned to service a fine old vintage rifle.

The second example, the .45/75 Winchester, is a bit more complicated, but still well within the capabilities of the average loader. The only large rimmed case that comes reasonably close to it, the .348 Winchester, is a bit small at the head and rim, but its heavy construction will easily stand the strain of expansion at this point. The rim is still large enough for the extractor to get a good bite. The .348 is 2.25" long compared to 1.88" for the .45/75, so it will do nicely for length when trimmed. First expand the neck to accept .457" bullets. This is best done in two stages, to about .400" first, then up to .457". Now size the new case full length in a .45/75 die, then trim to 1.879" or a wee bit less, and chamfer the mouth. Now your new case needs only a small amount of fire-forming to fill the .45/75 chamber fully — 22 grains of Hercules 2400 behind a 300-grain cast bullet will do the trick. Now you've got a case that will probably last twice as long as any originals you might find.

The foregoing examples show what can be done without much trouble or equipment. For your convenience, we show a chart that will let you quickly select the proper basic case for many forming jobs, without comparing and measuring several dozen specimens.

Whichever centerfire metallic rifle case you need, there's probably a solution. A little thought and research, the use of our table, a few tools, and you'll have that old buffalo buster a-bellering again.

"SBO," meaning "seat bullet out," will be seen in the remarks on the case chart. This is necessary when the basic case used will not finish out to the full length of the case being formed. In magazine arms, the bullet must be seated far enough out to produce an overall length that will function correctly through the magazine. To do so, seat the bullet progressively deeper into an unloaded case, trying it in the gun until proper functioning of the magazine and feed mechanism is obtained. Now, record this dimension for future use. In single-shot arms, bullets should be so seated that they lack about 1/16" of touching the rifling. Seat a bullet progressively until it just touches the lands, then turn the seating stem down 1/16". Many single-shot shooters, the benchresters among 'em, feel that this seating depth gives the best accuracy.

Dimensional data in this chart comes from many sources. Much of it is from personal measurement and examination of the cartridges and weapons concerned, some from standard reference works. The rest comes from friends and associates who, handy with micrometer and caliper, checked specimens from their collections. These sources are considered reliable, but not infallible. In some instances, only one or two specimens of a particular caliber were measured. In others, references were relied on in their entirety. In view of this, dimensions of your cases or cartridges may not fully coincide with the data given here. This does not mean that your cases are wrong or that my figures are wrong. Both are probably within the rather generous tolerances allowed in years gone by in the manufacture of the old-timers. Should any cases formed in accordance with these instructions fail to chamber correctly in your gun, or if you are seeking a case not on our chart, drop me a note. We'll try to work out something that will do the job for you.

One perennial objection to wildcat cartridges is the lack of an identifying headstamp. In many instances, there are other wildcats very similar in size and shape, probably formed from the same caliber brass and bearing the same markings. This can (and does) lead to confusion, inconvenience, and on occasion, danger.

Since one seldom makes up more than a hundred or so cases for a wildcat, why not exert just a wee bit more effort and give them distinctive identifying marks? It can be the correct caliber designation, your own coded identification, or even your name or initials. Not much is needed in the way of equipment — access to a drill press is necessary, though, a lathe will do the job with less fuss and bother. Also needed is a means of holding the cases, head up, a set of small (1/16" is what we use) number and letter stamps, an old twist drill from which to make a fly cutter, and a few pieces of scrap material.

The case holder can be easily made from the chamber end of an old rifle barrel, though even a hardwood block will do. I have two, one made from a Krag barrel (for rimmed cases) and the other a shot-out M-1 Garand tube. They will handle most jobs that come up. Cut 3-4" of the chamber end of barrel off, clamp it solidly in a vise and hacksaw a two-inch slot down the middle. Drill the slot terminating holes first and saw down to the holes. Simplest way to secure this holding fixture to the drill press table is to weld it to a small plate and clamp it to the press table. For securing cases solidly in this rig, an "O" clamp is easiest and simplest to make. Rimmed cases rest on their rims in this fixture so may be recessed without preparation as long as they will enter the old chamber. Rimless brass seats on the shoulder and must be uniformly resized before the cutting is done. If this is not done, variations in headspace will cause groove depth to vary from case to case.

This holder is secured to the drill press table, carefully aligned with the chuck. An old ½" twist drill is then ground to the shape shown, producing a single point fly cutter 3/32" in width and properly located so as to describe a circle about 1/32" inside the case rim. This cutting edge should be ground so that it will just cut away the original headstamp.

A case is dropped in the holding fixture and the clamp screw tightened just enough to keep the case from rotating. Check the cutter and adjust depth of quill stroke so that the cutting edge will go just deep enough to remove the old headstamp. Run press at its lowest speed and apply water to the cutting edge by means of a squirt-can as it contacts the brass. Feed with moderate pressure until cut is completed. Sounds like a long drawn-out procedure, but it actually requires only a few seconds per case.

Once the original headstamp is removed, it is a simple matter to take 1/16" number and letter stamps and apply the markings desired in the machined groove. Take care to hold the stamps vertical and strike uniform blows. A little care in positioning the stamps will produce a neat and attractive marking. Stamping is best done before removing case from the holding fixture. If not done this way, another holding fixture

Cutter made from twist drill to remove original headstamp from case.

REH

must be devised. Cases must be rigidly supported while stamping. For this purpose, a hardwood block can be drilled to accept the case, then slotted and clamped in your vise. Cheap and simple.

Of course, there are some among you who are horrified, or at best skeptical, at the removal of brass from the case head. I have encountered no indications of impaired head strength, even with loads in the 50,000 psi range. In the years following World War II, RWS produced a large amount of modern sporting ammunition with recessed heads like those described above. I shot hundreds of rounds of those cartridges in high intensity calibers such as 7 x 64mm Brenneke and 8 x 68S with not a single indication of case failure. Some of those cartridges were loaded to pressures well over 50,000 psi. Considering that performance, I feel perfectly at ease with modern solid head cases altered as described above. If it was good enough for RWS, it seems okay for me.

Re-headstamping cases to properly identify wildcat calibers and reformed obsolete numbers is entirely practical and not at all difficult. It eliminates confusion among calibers and lends a touch of distinction to your ammunition. Give it a try and watch the look on your shooting buddy's face when he picks up one of your rounds marked ".266 Lederhosen Special."

Abbreviations and Footnotes to Case Interchangeability Chart

[1]Most large caliber Newton chambers accept the belt of the 300 Magnum without alteration. If yours won't, turn or file the belt down until it enters the chamber freely. Do *not* reduce rim diameter. You can now size the case full length, trim it and finish up by fire-forming.

[2]The smaller British belted cases can be formed from 30-06 brass by swaging a belt on the latter. In a block of steel about 2½" long, drill and ream a .448"-.450" hole full length. Counterbore both ends to .470", to a depth of .220". Lightly chamfer the bottom of one counterbore, but leave the bottom of the other clean and sharp. Press your 30-06 case into the chamferred end first, then into the other. Be sure the case goes all the way in so that case head is flush with the face of the die. This produces a belt .470" in diameter and .220" long. Size this belted '06 full length in the proper die and trimming to length will give you the small belted British series.

[3]Reduce the '06 neck in a 250-3000 or similar die, then trim it to 2.40". Now size progressively deeper in a 22-250 die until the bolt will close on it. Ream the neck, expand to hold .227" bullets and trim to the length shown. To last more than a couple of firings, these cases must be annealed at the neck and shoulder. After annealing, fire-form and the job is finished.

[4]Turn or file the belt down to the head diameter shown. Now reduce the rim to the diameter shown, deepening the extractor groove if necessary to give the extractor a good bite. Size full length and trim. Fire-form if necessary.

[5]To make a .512" (or less) diameter case from the belted magnums, first turn the belt down flush with the body. Swage the case to required diameter as outlined in the text under "swaging." The rim must now be turned to the diameter given and the extractor groove deepened at the same time. When all this is done, size full length and trim to length.

[6]Cases for low pressure calibers may be formed by sweating an extension on to an existing case of the correct head dimensions.

For the 9.3x72R, expand the neck of a 30-30 case up to an outside diameter of about .358-.360" and tin the first quarter-inch thoroughly. Now cut the head off a 38 Special case and tin a corresponding length of it on the inside. Press this extension over the 30-30 neck. Stand this composite case in a half-inch of water and apply enough heat to flow the solder. When cool, clean off excess solder and trim to length. Sizing full length will finish the job. Examine such built-up cases carefully after each firing to see that the joint is holding. Keep loads moderate and pressures low.

BSA — Birmingham Small Arms
BN — Bottleneck
BP — Black powder
BPE — Black Powder Express
DEG — Deepen extractor groove
EL — Extra Long
EN — Expand neck
Exp — Express
Extr — Extractor
FF — Fire-form
Fl — Flanged
G&H — Griffin & Howe
HP — High-Power
Mann — Mannlicher
M-S — Mannlicher-Schoenauer
Mau — Mauser
May — Maynard
NE — Nitro Express
Nor — Norwegian
NS — Neck Size
NTE — Nonte-Taylor Enterprises
(a) NTE case No. 1 may be used by sizing and trimming.
(b) NTE case No. 2 may be used by sizing and trimming.
(c) NTE case No. 3 may be used by sizing and trimming.
R — Ream neck
Rem — Remington
Rem-Hep — Remington-Hepburn
Russ — Russian
S — Size
Sav — Savage
SBO — Seat bullet out
SFL — Size full length
Shps — Sharps
Sht — Short
SL — Self Loading
SS — Single shot
St — Straight
T — Trim
WCF — Winchester Center Fire
Win — Winchester
WR — Westley Richards

CASE INTERCHANGEABILITY CHART, METRIC AND BRITISH CALIBERS

Old case	Head dia.	Rim dia.	Length	Bullet dia.	New case	Head dia.	Rim dia.	Length	Forming data
METRIC CALIBERS									
5.6x61 VomHofe	.476	.479	2.40	.227	30-06	.466	.469	2.494	See footnote 3.
6.5x50 Jap.	.451	.474	1.968	.263	30-06	.466	.469	2.494	Swage head to .451, SFL, T to 1.967, R.
6.5x55 Mau.	.476	.476	2.157	.263	30-06	.466	.469	2.494	SFL, T to 2.156, R.
6.5x68	.520	.509	2.661	.265	300 Mag.	.512	.532	2.850	See footnote 4.
280 Halger Mag.	.529	.554	2.598	.283	300 Mag.	.512	.532	2.850	SFL in 280 Ross die, T to 2.587, FF.
7.5x53.5 Swiss	.496	.496	2.177	.304	7.62 Russ.	.485	.564	2.107	Turn off rim, cut new extr. groove, SFL, FF.
7.62 Lahti	.480	.481	1.375	.310	7.62 Russ.	.485	.564	2.107	Turn off rim, cut new extr. groove, SFL, T to 1.374.
7.7x58 Jap.	.472	.473	2.271	.300	30-06	.466	.469	2.494	SFL, T to 2.270, FF.
7.92 Kurz	.470	.470	1.299	.318	30-06	.466	.469	2.494	SFL, T to 1.298, R.
7.92x60.8 Nor.	.488	.469	2.396	.322	300 Mag.	.512	.532	2.850	See footnote 5, then SFL, T to 2.395.
8x51R Lebel	.541	.627	1.99	.327	348 Win.	.546	.605	2.40	SFL, T to 1.989.
8x57 Mau.	.467	.470	2.237	.323	30-06	.466	.469	2.494	SFL, T to 2.236.
8x60S	.470	.460	2.338	.323	30-06	.466	.469	2.494	SFL, T to 2.337.
8x68S	.522	.510	2.657	.323	300 Mag.	.512	.532	2.850	Turn belt to .522, DEG, turn rim to .510, SFL, T to 2.656.
8x72R	.429	.483	2.838	.324	9.3x72R	.427	.481	2.835	SFL.
8.15x46R	.421	.506	1.815	.316	30-30 Win.	.418	.497	2.044	SFL, T to 1.814.
9.3x62	.467	.466	2.480	.366	30-06	.466	.469	2.494	T to 2.479, EN for .366 bullets, FF.
9.3x72R	.427	.481	2.835	.369	30-30 Win.	.418	.497	2.044	See footnote 6, then SFL.
10.75x68	.492	.488	2.67	.424	375 Mag.	.512	.532	2.850	See footnote 5, then SFL, T to 2.669, FF.
11.2x72 Schuler	.536	.467	2.815	.439	300 Mag. Norma spec. case	.512	.532	2.850	T to 2.815, NS for .439 bullets until breech will close, FF.
BRITISH CALIBERS									
.240 Apex	.448	.467	2.49	.245	30-06	.466	.469	2.494	See footnote 2, then SFL, T to 2.49.
.350-7mm Rigby	.470	.528	2.490	.280	9.3x74R	.465	.524	2.925	T to 2.489, SFL in 7x57 die until breech closes, FF.
.303 British	.452	.528	2.209	.311	30-40 Krag	.457	.545	2.314	SFL, T to 2.208.
.310 Cadet	.354	.407	1.075	.316	32-20 Win.	.353	.405	1.315	SFL, T to 1.074.
.333 R'less NE	.540	.542	2.450	.333	348 Win.	.546	.605	2.40	Turn rim to .542, cut new extr. groove, SFL.
.450-400 2⅜ B.P. Exp.	.548	.615	2.37	.406	348 Win.	.546	.605	2.40	EN for .406 bullets, SFL, T to 2.369, FF, (c).
.416 Rigby	.589	.589	2.90	.416	378 Wea.	.584	.606	2.92	Turn belt to .589, SFL, DEG, EN for .416 bullets.
.577-450	.668	.746	2.335	.454	577 NE 3"	.660	.739	3.00	SFL, T to 2.334.
.500-465 NE	.572	.646	3.23	.468	500 NE 3"	.571	.641	2.99	SFL, SBO, (c).
.470 NE	.571	.650	3.28	.468	500 NE 3"	.571	.641	2.99	SFL, SBO, (c).

CASE INTERCHANGEABILITY CHART, U.S. CALIBERS

Old case	Head dia.	Rim dia.	Length	Bullet dia.	New case	Head dia.	Rim dia.	Length	Forming data
6mm Lee-Navy	.443	.440	2.36	.243	.30-06	.466	.469	2.494	Swage head to .440, recut extr. groove, SFL, T, R.
.25-36 Marlin	.417	.500	2.12	.256	.30-30 Win.	.418	.497	2.044	SFL, T.
.25 Rem.	.417	.419	2.05	.256	.30 Rem.	.418	.417	2.05	SFL, T.
.256 Newton	.468	.468	2.457	.264	.30-06	.466	.469	2.494	SFL, T.
.30-30 Wesson	.380	.442	1.66	.306	.357 Mag.	.375	.432	1.28	NS for .306 bullets, SBO, FF.
.30 Newton	.525	.520	2.497	.308	.300 Mag.	.512	.532	2.850	See footnote 1.
.32 Ideal	.350	.404	1.75	...	.32-20 Win.	.353	.405	1.40	SBO, FF.
.32-40 Rem-Hep.	.455	.534	2.110	.307	.30-40 Krag	.457	.545	2.314	SFL, T to 2.109, (b)
.32-40 Bullard	.450	.508	1.84	...	.30-40 Krag	.457	.545	2.314	Turn rim to .508, SFL, T to 1.839, (b).
.33 Win.	.505	.608	2.105	.338	.45-70	.501	.602	2.10	SFL, T to 2.104, (a).
.33 Newton	.522	.520	2.495	.333	.300 Mag.	.512	.532	2.850	See footnote 1, then EN for .333 bullets.
.351 Win. SL	.380	.410	1.38	.352	.357 Mag.	.375	.432	1.28	Turn rim to .410, DEG, SBO.
.35 Rem. Auto.	.435	.454	1.920	.359	.30-06	.466	.469	2.494	Swage head to .453, SFL, T to 1.919.
.35 Newton	.520	.525	2.495	.357	.300 Mag.	.512	.532	2.850	See footnote 1, then EN for .357 bullets.
.38-40 Rem-Hep.	.454	.537	1.772	.372	.30-40 Krag	.457	.545	2.314	EN for .372 bullets, T to 1.771, SFL, FF, (b).
.38-40 Ballard	.421	.504	1.807	.396	.30-30 Win.	.418	.497	2.044	EN for .396 bullets, T to 1.806, FF.
.38-56 Win.	.501	.603	2.11	.376	.45-70	.501	.602	2.10	SFL.
.38-72 Win.	.460	.519	2.57	.375	9.3x74R	.465	.524	2.925	SFL, T to 2.569, (b).
.40-50 Shps. St.	.451	.551	1.855	.401	.30-40 Krag	.457	.545	2.314	EN for .401 bullets, T to 1.854, FF, (b).
.40-60 Win.	.503	.618	1.87	.408	.45-70	.501	.602	2.10	SFL, T to 1.869, (a).
.40-65 Win.	.500	.601	2.10	.403	.45-70	.501	.602	2.10	SFL, (a).
.40-82 Win.	.508	.610	2.390	.408	.45-70	.501	.602	2.10	SFL in .40-65 Win. die, SBO, FF, (a).
.40-85 Ballard	.478	.556	2.93	.403	9.3x74R	.465	.524	2.925	EN for .40 bullets, FF, (b).
.40-90 Shps. St.	.477	.540	.325	.403	9.3x74R	.465	.524	2.925	EN for .403 bullets, SBO, FF, (b).
.40-110 Win. Exp.	.544	.651	.325	...	.450-400-3¼	.542	.618	3.25	SFL.
.44-90 Shps. 2⅝	.517	.633	2.626	.443	.45-90 Win.	.504	.603	2.395	SFL, SBO, FF, (a).
.45-90 Win.	.504	.603	2.395	.454	.45-70	.501	.602	2.10	SBO, (a).
.45-120 Shps.	.506	.610	2.875	.454	.45-90 Win.	.504	.603	2.395	SBO far as possible, (a).
.45-100 Ballard	.498	.487	2.812	.454	.45-90 Win.	.504	.603	2.395	Swage head to .498, SBO, (a).
.45-3¼ Shps.	.505	.601	3.25	.457	.45-3¼ Nitro	.544	.612	3.25	Swage head to .505, SFL, (a).
.45-125 Win.	.533	.601	3.246	.456	.45-3¼ Nitro	.544	.612	3.25	Use as is, rim may need reduction, (c).
.50-70 Gov't	.563	.660	1.75	.510	.348 Win.	.546	.605	2.40	EN for .510 bullets, T to 1.749, FF, (c).
.50-95 Win.	.562	.628	1.913	.499	.348 Win.	.546	.605	2.40	EN for .499 bullets, T to 1.912, FF, (c).
.50-100 Win.	.552	.602	2.407	.499	.348 Win.	.546	.605	2.40	EN by FF in chamber with filler load, SBO, (c).

CHAPTER 20

Reduced Loads

THE TERM "reduced load" means different things to different people. To the revolver shooter it might mean a bullet of about half the usual weight and a hardly-visible powder charge for short-range indoor plinking; to the big-bore rifle hunter it might mean the standard jacketed bullet loaded down to about 1500 fps for taking small game for camp meat in the field; and to the target rifleman it might mean a 2300/2400 fps load of finest accuracy for 300-meter shooting.

About the only valid generalization that may be made about reduced loads is that they produce significantly less pressure and velocity than typical factory or arsenal loads in the same caliber. Most powerful of the lot is what used to be called "mid-range" rifle loads with velocities in the neighborhood of 2000-2300 fps in modern bottle-neck calibers. The name came from their use at intermediate or middle ranges in the .30 caliber rifle matches popular in those days. The boys felt they didn't really need full-charge ammunition for 200- and 300-yard shooting where ranges and sight settings were known precisely. In addition these lighter loads lessened recoil and aided recovery in rapid fire. Consequently, they worked up loads that were just potent enough to give the accuracy they needed for the "mid" ranges, and at less cost than full charges. When good jacketed bullets could be had or afforded, they were used — but many a fine rifleman shot his 200- and

300-yard scores with carefully-cast, lead, gas-check bullets. As mentioned elsewhere, *good* cast bullets will deliver excellent accuracy at 2000-2400 fps.

Mid-range loads with properly-chosen jacketed bullets will kill game reliably — but bullet selection is critical. For example, the .30/06 normally drives a 150-grain bullet 2800-2900 fps and the bullet is designed to expand at that velocity. A mid-range load needs a bullet designed to expand down in the 2000-2300 fps range — and that can be found in the .30/30 150-grain bullet. So, use .30/30 bullets in the .30/06 if you want normal expansion in game. The same rule is to be followed in any caliber and select a bullet *intended* for the reduced velocity.

In the larger calibers, say, .33 and upward, blunt-nose cast bullets in mid-range loads will kill deer and similar game well. Because of their large diameter, they produce large wound channels even though they won't expand reliably when cast hard enough for good accuracy. Naturally, the larger the caliber, the better it will kill. For this reason, the various .35's (.350 Rem., 9mm Mauser, .35 Rem.), the .375 H&H, .444 Marlin and .458 Winchester are the best of the lot. Even in rifles the big-bore pistol calibers can't produce true mid-range velocities, nor can the venerable .45/70 because of gun strength limitations (except in the 1886 Winchester and fine single-shot rifles), so aren't considered here.

In reality, mid-range loads will serve for a very large percentage of the average rifleman's shooting. Remember that back in the 1930's many experienced and match-winning riflemen used them for everything except the largest game. And, they are considerably cheaper than even handloaded full-charge ammunition.

One step down we have what most people think of as "reduced" loads; those in the 1500-1800 fps range and *usually* assembled with cast lead or, sometimes, swaged half-jacket bullets. Here, too, excellent accuracy can be produced with a bit of care and attention to detail — more than adequate for taking small game, target shooting to a bit over 100 yards, and even for close range varmint hunting. After all, until relatively recently, many old-time chuck hunters had nothing better, and they hit regularly out to 300 yards.

Even at these velocities, cast bullets must be quite hard (type metal equivalent), so even when hollow-pointed, cannot be relied upon to expand to any degree — though when the cavity is deep and large enough, the bullet may break up to some degree. When expansion is required, half-jacket bullets with pure lead cores may be used — but they are usually prone to produce considerable leading. The exposed core sets back and rubs on the bore. A better bet for expansion and accuracy are high performance handgun bullets with ¾-length jackets. The jacket extends far enough forward over the ogive to prevent leading. Unfortunately, such bullets are available in relatively few large (.35 and upward) calibers. There are a few factory handgun bullets in the smaller calibers which will serve to some degree in this velocity range, i.e., .22 Jet, .256 Winchester, .22 WRFM, but there aren't many choices.

Proper powder charges to produce these velocities in modern cartridges occupy only a small portion of the available powder space. This permits the charge to shift willy-nilly about the case, according to the position of the cartridge and the handling it receives. Depending on the position of the rifle, the powder may be bunched up against the bullet base (far from the primer flash) or back against the case head and in direct contact with the primer flash, or in any position between those extremes. It has long been known that such shifts cause variations in ignition and, consequently, in velocity. Bad for accuracy.

This can be avoided by placing a high-volume, lightweight, inert filler on top of the powder charge to hold it against the case head and insure uniform ignition. Many materials have been used with some degree of success, including cornmeal, sawdust, rolled-up toilet tissue, even granulated Teflon and other space-age materials. Probably the best combination of effectiveness and availability (doesn't matter how good it is if you can't get it) is found in Kapok fluff. It is easily obtained by pulling apart Kapok pads available from upholsterers and many of the larger mail order houses. Or, old furniture and cushions can provide a cheap and ample source.

The Kapok must be shredded finely, then rolled into loose balls slightly larger than the diameter of the case. Then a ball is poked through the case neck and down on the powder where it will expand and hold the charge in place. It will, that is, if you've made the ball large enough. One ball is enough; don't try to "improve" on the method by packing the case full. That will raise pressures.

Incidentally, cornmeal and Cream of Wheat were once popular for this use. Such granular fillers must completely fill the balance of the case and this greatly increases loading density and chamber pressure. One fellow I know found that those fillers actually pulled off case necks and carried them down the bore.

The best cast bullets for light loads are those of moderate to heavy weight for the caliber (170-180 grains weight in .30 caliber, for example) and with a long, bore-diameter, cylindrical section ahead of the bearing surface. I feel that the ideal bullet should have a bearing surface about 1½ calibers long, then a 2-caliber front section capped by a fairly blunt or rounded nose. This "pilot" section should fit snugly on top of the lands, thus aligning the entire bullet perfectly with the rifling before firing. Bearing surface diameter should be from groove diameter to .0015" more.

Bullet seating and factors surrounding it are discussed in the chapter devoted to seating.

Many shooters prefer to only neck size cases used for light loads. This is fine so long as the cases are paired to a particular rifle. However, this does not necessarily improve bullet alignment in the bore unless cases are marked and oriented in the chamber for every shot. As pointed out elsewhere in the volume, cases expand off-center and this can actually *cause* poor bullet alignment unless cases are oriented. If you don't orient, try full-length sizing before condemning it.

Bottom of the line in reduced loads are what have been called "squib" or "ultra-light" loadings of lightweight lead bullets. Generally they are not quite as accurate as the more powerful types, and they are suitable only for very short range plinking and target shooting. They will, of course, kill small game, but not as well as more powerful loads.

Squib load velocities seldom run over 1000 fps, and may go as low as only 500 fps in some instances. Plain-base cast lead bullets are normally used, and they may be of relatively soft alloys — though I've personally had best results with linotype metal. Bullets are usually light for the caliber, like, say, 90-120 grains for the .30/06, 60-75 grains in .257 Roberts. Bullets so short cannot have a long pilot as described earlier, so seating alignment is very important; and seating depth should be quite shallow so the bullet needn't travel too far before engaging the rifling.

Though not so common these days, round balls were once popular for very light loads, especially in those calibers which ready-made buckshot will fit. No resizing at all is required; simply press the ball half its diameter into the case mouth over a grease or graphite wad. If the latter isn't available, a smear of soft bullet lube over the exposed portion of the ball will do the job fairly well. No. 0 buckshot will fit .30 and .32 barrels; 00 will do for .35's; 000 fits .375. Other sizes must be cast since they aren't available as prepared shot.

Squib loads require the case necks be resized to insure adequate grip on the bullet except as just noted. The old-time theory of much-oversize bullets allowed them to be snug in unsized necks, but that won't work with bullets sized just over groove diameter, except in undersize chambers.

The quite low chamber pressures do not generally expand cases very much, so eccentric expansion presents no problems. However, this very low pressure often is not sufficient to overcome the initial primer thrust which sometimes drives the case forward forcibly enough to push the shoulder rearward. Unless sufficient chamber pressure is then developed to drive the case rearward and fire-form the shoulder, the case remains foreshortened, producing a condition of excess headspace. Such cases should be restricted to use with squib loads thereafter. The tendency to foreshorten can be reduced by using pistol primers whose thrust is less than that of rifle primers. Because small quantities of easily-ignited powders are used, pistol primers produce entirely satisfactory ignition and, in some instances, even better accuracy than rifle types.

All reduced rifle loads require careful attention to powder selection. Often one is tempted to simply use less of the same type suitable for full charge loads. Only occasionally is this successful, and there is evidence that it can be dangerous — the latter a matter we'll discuss further later. Generally, mid-range loads require a powder slightly faster burning than full-charge loads for the same bullet weight. For example, a full-charge 180-grain load for the .30/06 is 56.0 grains of IMR 4350 powder, producing 2750 fps and about 49-50,000 psi chamber pressure. A well-balanced and efficient load. Reducing that charge to about 50.0 grains produces 2300-2400 fps and drops pressures down to about 35-36,000 psi, a range where 4350 does not perform efficiently. A far more suitable powder is faster-burning IMR 4895 which does perform efficiently at that pressure level, where approximately 42.0 grains will give mid-range performance.

Reducing velocity further to the 1500-1800 fps range requires pressures in the 25-30,000 psi range where even IMR 4895 does not perform well. Hercules 2400 performs well there and 20.0 grains will produce approximately 1700 fps at somewhat under 25,000 psi, substituting a cast lead bullet of equal weight for the jacketed type.

Moving downward to squib loads we are dealing with pressures of less than 15,000 psi where not even 2400 will burn efficiently. Purely pistol powders such as Unique, AL-5, and Bullseye are required. Charges as light as 2.5 grains of Bullseye will burn cleanly in .308 and .30/06 with bullets in the 80-110-grain weight range, producing velocities well under 1000 fps.

So, generally speaking, mid-range loads require the faster IMR-type powders (4895, 3031, 4198); light loads, the small-case rifle powders such as DuPont 4227 and Hercules 2400; squib loads, the faster pistol and shotshell powders.

Reduced Loads

In recent years there has been much controversy over so-called "pressure excursions" or "secondary explosion effect" which *sometimes* results when smaller than normal charges of the slowest-burning powders are used with light bullets in large-capacity bottle-neck cases.

There are documented instances of propellant detonations which destroyed rifles and injured shooters where those powders were used in less quantities than had been proven entirely satisfactory as full-charge loads with the same component combination in the same gun. Generally the victims reported that full-charge loads in cases like the .25/06 had functioned normally, but when the charge weight was cut 10 to 15 per cent for various reasons (usually to reduce barrel erosion), a detonation occurred, completely destroying the rifle.

These detonations are not as yet predictably reproducible in the laboratory, though some experimenters claim an *occasional* reproduction. Consequently, the exact cause or causes are not yet known. Professor Lloyd Brownell has published in *Handloader Magazine* an extensive series of articles presenting theoretical analysis of the phenomenon which he calls S.E.E. Even this data has not made predictable reproduction of detonations possible.

The best advice that can be given is simply to avoid use of the slower powders such as 4350, 4831, N205, etc. in anything but full charges in large-capacity, sharply bottle-necked cases similar to the .25/06.

Reduced loads in handguns present a somewhat different set of conditions. Standard full-charge factory loads in all but magnum calibers *already* use the fastest powders available and already work at fairly low pressures. The more or less standard powders such as Bullseye and Unique work quite well at reduced-load pressures. Experience has proven that Bullseye is generally the best available for reduced loads in straight-case handgun cartridges such as the .38 Special and .45 Colt.

To some extent, reduced loads can be produced by simply cutting down the powder charge while retaining standard bullet weight. However, due to the relatively slow rifling twists of many handguns, long, heavy bullets won't be well stabilized at very low velocities. As velocity is reduced, accuracy goes to pot.

Shorter and lighter bullets *will* be stabilized at lower velocities, so they offer a solution. Wadcutter-shaped bullets — both solid-base and hollow-base — will stabilize at lower velocities than other conventional shapes. The hollow base, with its center of gravity well forward, will stabilize at a lower velocity than any other shape of equal weight.

About the lightest conventional bullet that will deliver good accuracy is Lyman's number 358101 in .38 Special. It weighs only 77 grains and is actually less than one caliber in length — that is, it's shorter than it is wide. Comparable bullets are made by Lyman in .44 and .45 caliber, but nothing of this sort is available in .32 or .41.

In developing a reduced load to suit your needs, decide whether you can settle for standard-weight bullets. If so, just start cutting back on a full charge of Bullseye ½ grain at a time until the recoil and velocity level desired is reached. If accuracy deteriorates, work back up in 0.2 grain increments until it improves. If this doesn't reduce recoil and velocity enough to suit you, then a lighter bullet is necessary.

With the lighter bullets, reverse the load development procedure. Begin at the low end with a very small charge of Bullseye: ½ grain in .32 cal.; 1 grain in .38; 2 grains in the bigger bores. From there work upward in 0.2 grain increments until the performance level you want is reached. Very small powder charges will burn erratically, producing wide velocity errors that spoil accuracy, usually in the form of vertical stringing on target.

The absolute minimum bullet for low-power handgun loads is the round ball. Moulds are available in all sizes, but factory swaged 0 buckshot fits .32's, 00 the .357/38's, and .45 caliber balls are available for the various .45's. Balls should be chosen a few thousandths of an inch larger than barrel groove diameter and lubricated by a grease wad seated in the case mouth. Balls may be seated half their diameter into the case mouth with larger powder charges; but the smallest charges will burn better if the ball is pushed down into the case against the powder (or as near there as case wall taper will allow). A short piece of dowel works well for seating thus.

Lubricant for reduced load handgun bullets can be *any* acceptable variety. However, when shooting is to be indoors at home, then you'll find some types produce far less smoke than others. The smokey types can be quite irritating in the typical poorly-ventilated basement or garage range, so it will pay you to try several and choose the least-smokey.

Smoke is much less a factor on properly designed commercial indoor ranges. They have powerful ventilation and exhaust systems which gobble up the smoke with ease, though a *really* smokey load will still get a dirty look from shooters on adjacent firing points.

CHAPTER 21

Special-
Purpose
Loads

EVENTUALLY THE practicing handloader is going to decide that he simply must have some special loads that are either not available commercially or for which most loading manuals and other sources don't supply much data. I know that in my office I receive many queries for this sort of thing. The queries range all the way from people wanting black powder loading data for use in their ancient Damascus-barreled shotguns (which it is generally agreed are not entirely safe for use with modern smokeless powders) up through snake loads (shot) in handguns and also to include special-effects blanks. One fellow wrote me quite extensively stating that he was an amateur film producer and wanted instructions from which he could prepare several thousand "noisy and bright" blanks to be fired through a Thompson .45 submachine gun. In any event, all manner of oddball loads that would be nice to have will occur as you develop more handloading experience.

Generally speaking, I think the demand for handgun shotshell loads far exceeds all the others, followed by black powder loads for obsolete guns in second place, and blank loadings bring up the rear. Others, like multiple-bullet loads, tracers, armor-piercers, etc., crop up from time to time. We'll touch upon them all in the next few pages.

First, let's take a look at handgun shotshell loads. Many years ago the major ammunition manufacturers offered shot loads in the more popular handgun cartridges, the .44/40, .38/40, .45 ACP, .38 S&W, etc.

In almost all instances, these loads consisted of the standard case and a light powder charge topped off with a paper or hollow wood container filled with shot. This shot container extended beyond the mouth of the case and was more or less of bullet shape and gave the same overall length as the standard bulleted load. Generally, such special shot cartridges carried only a few shot and were none too spectacular in performance. A few such loads used cases extended to cartridge length, doing away with the separate shot container.

Doubtless someone else tried it first, but my first contact with handgun shotshell loads dates back to the middle 1930's when I read Bud Dalrymple's comments on his experiments. He used the old .45 caliber Colt single-action revolver almost exclusively and by hand laboriously smooth-bored and choked barrels for same to obtain optimum shot performance. Dalrymple used the standard .45 Colt case with card and fibre wads of the period and all the Number 9 or Number 12 shot the case would hold. I've never had the opportunity to try the loads he developed in a proper smooth-bored and choked gun, but he reported excellent patterns and results on game such as jackrabbits and partridge at ranges out to twenty yards or so.

Prior to WW II, Remington produced a .45 ACP "Riot Shot Load" specifically for use in Thompson submachine guns. This consisted of the standard case with a bullet-like metal shot container crimped in place. I've never been able to lay hands on enough of these to try, but authorities of the day didn't get too worked up over them. When WW II came along, the demand for a "Survival Cartridge" to be used in the standard service sidearm resulted first in paper shot containers being loaded into .45 ACP cases. These proved less than completely satisfactory, and were eventually replaced by a longer brass case which contained the shot charge completely in itself and was closed at the mouth by a thin card wad. Because the .45 ACP headspaces upon the mouth of the case, this longer case (made to the length of the loaded .45 ACP ball cartridge) was slightly necked at the proper point to produce a squarish headspacing shoulder. Those cases are excellent for reloading if you can find them, but in my experience they are mighty few and far between. They can, incidentally, be duplicated by turning a *rimless* head on .45 Colt cases, then trimming and necking accordingly.

In the early 1950's, perhaps before, the late Jim Harvey (who owned and operated Lakeville Arms) produced the first *truly* practical shot revolvers and ammunition. In searching for the ultimate in shot performance in a woods gun, he finally evolved a system which hasn't yet been surpassed for performance. It involved taking a modern revolver

U.S. military .45 ACP shotshell used full-length brass case necked down to produce a headspacing shoulder.

Left to right: .41 Magnum, .44 Special, and .44 Magnum, each with cylinder-length cases ready for fire-forming in unaltered chambers, after the Harvey manner.

in .44 or .45 caliber (he preferred to work with the M-frame S&W in .44 Special) and reaming the rifling from the bore; polishing it mirror-bright; and finishing with an integral choke of the dimensions he'd found best.

The chambers of the cylinder were then reamed to straight holes clear through; eliminating the bullet-diameter throats. Cases were then cut to cylinder length from .30/40 Krag brass expanded to fit the .44; .405 Winchester or similar to fit .45's. Today the .44 Marlin would do beautifully for the latter. These cases were loaded with zinc O-P and O-S wads and at least ¼" of felt filler wad and No. 9 shot. As finally developed, these loads would match the ½ ounce, 2½", .410 shotshell load in .44 and .45 caliber revolvers. Loaded to Harvey's specifications and used in guns he'd reworked, performance was all that one could ask.

Unfortunately, Harvey ran afoul of the federal law in the middle 1950's. For some reason, as yet unknown to knowledgeable people within the shooting fraternity, the old ATTD which enforced the Federal and National Firearms Acts in those days decided that Harvey's smooth-bored shot revolver was to be classified as "Sawed-Off Shotgun." In short, such guns fell under the registration and transfer tax provisions of the law at the time. The ATTD ruling was eventually upheld by the courts, in spite of the fact that only a very limited

Modern Handloading

number of the guns had ever been produced and that even until *today* there has not yet been a *single record instance* of a shot revolver ever having been used in a crime of violence.

This unwarranted action by the Federal Government put the skids to any further development of Harvey's shot revolvers and ammunition. The ATTD ruling did permit alteration of the chambers to take Harvey's special loads, but required that the barrel of the revolver remain *rifled*. Unfortunately, as most everyone who tried it knew, the rifling in the bore prevented achievement of acceptable pattern performance with the components available in those days. Today, the rifling effects can be reduced somewhat by using a plastic shot sleeve, but those weren't available at the time.

Quite logically, interest in shot loads was pretty slim for a few years because no one wanted to place himself in a position where he might be busted by the Feds.

Today, even under GCA '68, the original ATTD ruling still applies to shot handguns. They are in violation unless they retain a fully rifled barrel. However, the Feds have relented to some degree in that they do not regard the attachment of a smooth-bore choke device to the muzzle as being in violation of the law. Technically, then, you may add a tapered choke tube to your favorite big-bore revolver, then brew up cylinder-length loads which will match the performance of the standard 2½" .410-gauge load. Performance with a screwed-on choke tube isn't up to the standard of Harvey's and Dalrymple's smooth-bored revolvers, but with modern plastic-protected shot charges, it comes mighty close.

At the time people were talking about Harvey's shot revolver developments, I was cooking up my own version — admittedly based on Harvey's accomplishments — simply because I couldn't afford the price of one of those Lakeville conversions. It was something like $55.00, plus the gun. The methods and techniques I devised then for both gun and ammunition are what I still employ today and I think they give the best results that can be had within the law.

Left: .45 ACP and cylinder-length case for shot loads in unaltered .45 M 1917 revolvers. Note slight headspacing shoulder formed by RCBS sizing die. This case has been fireformed and reloaded with ½ ounce of shot. *Right:* .45 Colt and cylinder-length shot case for same, formed from Norma 8 x 57 JR brass. Note long case is necked slightly so as to enter unaltered chamber freely.

Special-Purpose Loads

Any large revolver caliber may be used, however, I consider the .44 Special the smallest that throws a sufficiently large shot charge to be worthwhile. Probably the best gun and caliber to be used, considering both economy and shot capacity, is the M1917. 45 military revolver by either Colt or Smith & Wesson. I prefer the latter because of its lighter weight. To use the guns as is, one first trims .30/06 or .308 Winchester cases to a length of 1.70''. Those cases should enter the chamber of the revolver freely, but if they do not, they must be necked down ever so slightly at a point .90'' from the case head to allow them to slip past the headspacing shoulder in the chamber. While special dies are available from RCBS, Inc. for this, it is easily done by dropping a piece of pipe or washer with an inside diameter of .450'' over each case and tapping it down to the proper point. This will squeeze the case in sufficiently to allow it to chamber for fire forming.

These cases should then be loaded with from five to six grains of Hercules Unique powder topped with a fairly thick card wad solidly seated on the powder. Because of the relatively small diameter of the case mouth, wads will be distorted somewhat in seating, but for the initial fire forming, that presents no problem. The case is then filled to within 1/16'' of the mouth with small shot, usually Number 9, which is then secured in place by a card wad either crimped or cemented in place. These loads are then placed in the three-shot half-moon clips intended for use with .45 ACP ammunition in M1917 revolvers.

The loads are then fired and the cases will fire-form to fit the chambers perfectly. In fact, you may find when you try to extract them that they fit too perfectly. They may appear to be stuck tightly in the cylinder. This is generally caused by burrs turned up on the inner edge of the headspacing shoulder inside the chambers. This results from previous extensive firing of steel-case military ball ammunition in the gun without the half-moon clips. Absence of the clips causes the complete cartridge to be driven sharply forward against the headspacing shoulder, and over a period of time, such repeated impacts peen the shoulder inward. When your first loads are fire-formed in those chambers, then, the burrs bite into the brass and hold the cases tightly. If it does happen, just tap the cases out gently with a rod or dowel inserted into the chamber throats and then polish or scrape the burrs out of existence.

After fire-forming, the cases are ready for loading with standard charges. Fired cases should always be resized full length in order to insure that they will chamber freely. While RCBS makes a special die

Modern Handloading

set for this load and which includes a full length sizing die, it is possible to get by without it. Simply first resize the rear portion of the case in a standard .45 ACP or .45 AR die. Follow this by forcing the forward necked-down part of the case into a .44 Special or .44 Russian die or into a .45 caliber bullet resizing die. Either will reduce the forward portion of the case sufficiently for it to chamber freely. Even so, it's a good idea to try the resized cases afterward in the gun to make sure you've got the job done properly.

After the reformed cases have been decapped and primed in the usual fashion, charge them with 7 to 7.5 grains of Hercules Unique (for a full-charge load) and seat a .45 caliber card wad tightly on the powder. The wad will be squeezed down somewhat in going through the case neck, but won't be deformed enough to prevent it from expanding to cling to the sides of the case. I have had best results from 1/16" thick cardboard over-powder wads, or two thinner wads.

Follow this with a 3/16" to 1/4" thick composition wad rammed snugly into place, then pour the case full of shot to within 1/16" of the mouth. Press a .44 caliber gas check, skirt up, into the case mouth and crimp securely in place. Again, the special RCBS die simplifies this job, but it can be accomplished easily enough by simply pressing the case into a conical hole in a block of steel or an old die body, while using a piece of dowel to hold the gas check in place. Loaded in this fashion, the cylinder-length case will hold just under one-half ounce of Number 9 shot. More shot can be gotten in by leaving out the 3/16" to 1/4" of filler wad, but the shot will be much less deformed by acceleration if the wad is in place. Enough shot will be deformed by the rifling that we don't want to make the situation any worse.

The sharp-edged rifling in the barrel will badly deform the outer layer of shot and will also tend to give the entire shot charge a spinning motion which causes it to scatter badly. This can be reduced, though not completely eliminated, by using a simple polyethylene shot wrapper. To accomplish this, simply cut strips of thin garment-bag polyethylene that can be rolled up and placed in the case mouth on top of the filler wad. The polyethylene should stop about 1/16" short of the case mouth so that it does not interfere with seating and crimping the overshot gas check. As for length, the strip of plastic should be long enough to wrap completely around the shot charge with only a very slight overlap. If the plastic available to you is unusually thin, it may be wise to use two layers.

The foregoing load, used with a shot wrapper, is capable of taking small game and breaking clay targets consistently out to about forty feet. To improve performance beyond that range, it will be necessary

Special-Purpose Loads

to fit a choke tube to the muzzle. A dimensioned drawing of a typical choke tube in .45 caliber is shown here, and it's construction should be self-explanatory. Attachment by threads on the barrel is shown, but there is no reason that a bayonet-type attachment fitting around the base of the front sight couldn't be constructed if you wish to avoid threads on the muzzle of your gun. I suggest that if you do use threads, that you make a small protective nut to screw over them when the choke tube is not in place.

The methods just described above may also be used to prepare cylinder-length cases and loads in .44 Special or .44 Magnum, using .30/40 Krag rifle cases as the basic material; and also for .45 Colt. For the latter, use the .444 Marlin case which is a closer fit in the rear of the chamber than the slightly smaller .30/40. The same forming and loading instructions apply, as does the loading data. Unfortunately, there are no cases suitable for extending the capacity of smaller calibers such as the .38 Special and .357 Magnum in the same fashion. However, in .44/40 or .45 Colt caliber, one may use the slightly longer Remington "Five-In-One" blank cartridge case by opening up its neck.

The foregoing types of loads do not in any way affect continued use of the same gun with standard factory or handloads. Most people choose them for that reason since they don't particularly care to modify the gun so it doesn't perform well with standard ammunition. However, if one wishes to go all-out on the job and sacrifice some performance with ball ammunition, here is what to do. First, have the chambers reamed out so that they are continuous full-length cylindrical holes in the cylinder. In other words, eliminate the throat portion. Once this is done, prepare cylinder-length cases as before, and load them. Each caliber will then hold somewhat more shot.

Standard cartridges may still be fired in a revolver so altered, but accuracy will be considerably reduced because of the oversized portion of the chamber through which the bullet must pass unguided before it enters the barrel. Pressures will also rise considerably since the bullet upsets, then is swaged down as it enters the barrel. A better solution when bulleted ammunition is required is to simply handload the bullet and powder charge you choose into the full-length case, seating the bullet flush or below the mouth of the case. In this fashion, the bullet is guided by the case until it enters the barrel and accuracy is generally up to the standards of the same gun before alteration. In the .45 caliber for several years I used a two-ball load, consisting simply of two .45 caliber round balls seated down on top of the same powder charge and wad used for shot. Bullet lubricant was smeared in the case mouth over the top ball, and seemed to do a good job

　　　　　　　　　　　　　　　　　　　　　Modern Handloading

of preventing leading. A couple light stab crimps held the balls in place. Generally, both balls would strike within the kill zone of a standard man silhouette target at ten yards.

The .38 Special and .357 Magnum cases simply do not hold enough shot to really be worthwhile. However, loading them presents no problem. In both, I have long used 3.0 grains of Bullseye Powder over which a .357″ gas check is tightly seated skirt down. The case is then filled with Number 12 shot, leaving room for a second .357″ gas check seated skirt up and crimped heavily in place. No filler wad is used simply because of the reduction it would make necessary in the shot charge. While by no means nearly as effective as the large caliber, full-length case loads already described, these concoctions are adequate for polishing off a copperhead or timber rattler who bars your path at a range of four or five feet. And, that's really the type of shooting most people have in mind for handgun shot cartridges.

Assembling a useful shot load for use in even the large caliber autoloaders is quite a bit of a chore. Of course, if you can locate a few of the long .45 ACP shotgun cases, it is a simple matter to load them as already described above. Alternatively, for the same gun, use .308 or .30/06 brass trimmed to a length of 1.130″ and necked down so it will chamber freely. Then, either turn the case neck externally or ream it internally to a wall thickness of .010″ about ½″ deep. Anneal thoroughly all the reamed portion. After this, the case may be fireformed and loaded as already described, and will be suitable for single-loading in the big automatics. If you *must* have magazine feeding, then form cases in the same manner, but reduce the shot charge sufficiently that the mouth of the case can be tapered by forcing it

Remco shotcaps are loaded into charged and primed cases just like bullets. Speer shot capsules get the same treatment.

.38 Special case and shot charge on left shows relatively small amounts of shot that can be gotten into the standard case.

into a 45° conical recess cut in a die body or a block of steel. The relatively small opening left after the case is tapered inward is then plugged with cardboard wad cemented in place. Even when annealed dead-soft, such cases will often split at the tapered portion on the first firing. In my own personal opinion, this makes it hardly worth all the effort required to make the cases, but that is up to the individual.

Autoloaders in other calibers can be given essentially the same treatment. For example, cases can be made in the same way for the .38 super and various 9mm's from .222 Remington or 5.56mm military brass. Any caliber smaller than .38 or 9mm will give you nothing but headaches and inadequate results, so why bother?

In the past couple of years, several firms have attempted to simplify the production of handgun shotshells and also to improve results. The most spectacular results have been obtained by Thompson-Center Arms Company in its "Contender" single-shot pistol fitted with a rifled choke tube. This gun is chambered for the .44 Magnum bulleted cartridge and the barrel is rifled accordingly. However, the muzzle is threaded for the attachment of a tapered choke tube which contains very deep *straight* rifling lands and grooves. This rifling serves three purposes: first, it satisfies the requirement of the law that the barrel be rifled and not smooth-bored; and the straight grooves tend to halt the rotation given the capsule by regular rifling; the deep, sharp rifling serves to break up the heavy-walled plastic shot container which is loaded into the standard .44 Magnum case. Thompson-Center calls this special shot load, which is a production-loaded item supplied by that firm in Super Vel cases, the "Hotshot." It contains five-eighths ounce of shot and actually duplicates the performance of the three-inch .410-gauge shotshell. The loaded Hotshot cartridge is too long to be used in .44 Magnum revolvers, so is restricted to the Contender or similar single-shot pistols. In addition to supplying loaded ammunition, Thompson-Center also offers the shot capsules ready for handloading.

In addition to this, the Remco Corporation offers its "Shot Caps" in several calibers. These are clear plastic shot containers supplied with the shot charge sealed in place. They are available in .38/357, .44, and .45 calibers. In the latter caliber, the Shot Cap is offered in two lengths — one for the stubby .45 ACP, the other for the longer .45 Colt. In practice, these Shot Caps are simply seated in the case in the same manner as a conventional bullet. Unfortunately, in this writer's experience, it is impossible to obtain a sufficiently-tight crimp upon the springy body of the Shot Cap to prevent its shifting during recoil

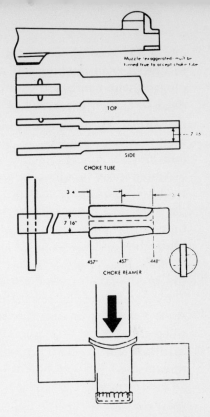

Rough choke tube is shown second and third from top, while top figure illustrates modification to revolver barrel to allow the tube to fit in place. After internal diameters of choke are drilled, the reamer shown is inserted from the rear and used to produce the choke. Shown at bottom is a simple punch and die set for making cupped over-powder wads in nonstandard diameters for handgun shotshell loads.

and tying up the gun after the first two or three shots. Perhaps if one used very light powder charges, producing less recoil, this would not occur. In addition, I have experienced considerable difficulty with the plastic Shot Cap *not* rupturing as it is intended to as it leaves the muzzle. This has resulted in the shot charge striking the target still completely encased in its plastic container. Though accurate records weren't kept, memory tells me that roughly 25 per cent of the .44 Shot-Caps fired failed to break up as claimed and struck the target as a single projectile.

As of this writing, Speer Products, makers of Speer Bullets, has introduced a satisfactory plastic shot capsule for loading .38/357 cartridges. Intended primarily for use in factory-produced Speer "Snake Loads," the capsules are also available to handloaders.

Occasionally I have received inquiries for shot loading data to be used in some of the straight-case rifle caliber, specifically, the .45/70 and .444 Marlin. Those individuals were advised to simply use standard .410-gauge shotshell loading data after cutting special-diameter wads with a home-made punch to fit their cases. Reported results have been as satisfactory as could be expected from the rifled barrels in question.

Special-Purpose Loads

From time to time, people get the idea that multiple-bullet loads would be more advantageous for defense or police use than the usual single-bullet cartridge. There is a good bit to be said for this approach. Under careful loading conditions, a two- or three-bullet load can be assembled which will place all of its projectiles well within the vital area of a man-sized target at normal combat ranges. One significant factor to be considered about this type of load is that the combined effect of three wounds is generally considered to be much greater than merely triple the effect of *one* of those wounds. Effect increases more geometrically than arithmetically. However, the idea has never really caught on well, probably because the presence of multiple-bullet ammunition in one's gun greatly reduces its point-target capability, and, it is often just as important in such situations to be able to make a single pinpoint hit.

In any event, such loads are generally assembled with two or three very short wadcutter-type bullets seated in the case on top of one another. A number of years ago the C-H Die Company made swaging dies (for use in its Swag-O-Matic Press) which were easily adjusted to produce .357″ diameter bullets in wadcutter form weighing only 60 grains. The bullet profile was such that a very short semi-wadcutter nose on one would fit into a similar-shaped cavity on the base of another. Thus, two or three bullets could be easily stacked and seated inside the case mouth in a single operation. However, when such bullets were made of lead only, the heat and pressure of firing tended to cause the individual parts to fuse together and function as a single projectile. I was able to correct this condition simply by placing a gas check on the base of each bullet. When this was done, the individual segments did not fuse together and took slightly divergent paths upon exiting the muzzle. Because of their very short length, actually less than their diameter, the individual bullets were highly unstable, but not so much so that they would not produce the results already mentioned. At the present time, I do not believe any manufacturer is currently offering dies for this type of bullet. However, I am certain that any of them would be glad to produce it on special order.

Actually, much more practical double-bullet loads can be produced without any special tools. Simply use a moderate powder charge for the *total* weight of bullet(s) involved and seat a stiff card wad over it. Seat a round ball of the proper diameter slightly below the mouth of the case, then seat a second ball on top of that, pressing it down only until the case mouth can be crimped upon its perimeter. Fill the space over the first ball with a fairly stiff bullet lubricant (smeared in by hand) before seating the second ball, and you have ample lubrication for both balls. This type of load produces acceptable accuracy, the spherical ball being ballistically much more stable than the disc-like segmented bullets just described.

It isn't unusual, really, for both balls of this load to print quite close together, practically cutting one another's holes. In .45 caliber, loaded in .45 Colt cases, I find this load particularly deadly. When a charge of ten grains of Hercules Unique is used, the two balls move out at well over 1,000 fps from a six-inch barrel.

Hardly a handloader alive hasn't been at one time or another asked by some acquaintance to make up some blank cartridges. Smokeless powders suitable for blanks are not available to the handloader. It is partly because the demand is too small for their sale to be profitable, and, likely, also because such powders are extremely dangerous should they inadvertently be loaded behind a bullet of any sort. Factory-loaded blanks are assembled with special blank powders which are designed and manufactured to burn *without* confinement. One will occasionally encounter such powders recovered from surplus military ammunition, and if their identity is known, it is possible to begin with *very* small charges and develop satisfactory blanks. Such powders should *never* be confined by more than a single card wad seated within the case mouth. Neither should any attempt be made to obtain automatic functioning by plugging or covering the gun muzzle. Even the special devices used for that purpose by military establishments sometimes cause blown-up guns.

By far the best bet is to rely upon commercial and military blanks. If you can't tolerate that, then use black powder only. Simply fill the case about two-thirds full of FFFG powder and top with a single card wad. Such loads may be messy and inconvenient, but they work well, and they are safe.

Shotshell blanks suitable for some purposes can be loaded with available smokeless powders. However, such blanks are restricted to uses where relatively heavy expelled wads are not hazardous, as in dog training.

Hi-Skor 700x powder is used: 60 grains in 10 gauge, 30 in 12, 25 in 16, 20 in 20 gauge, 15 in 28, and 10 grains in .410. In each gauge, the charge is placed in the case, then a wad column is built of a plastic over-powder wad and sufficient thick, hard, card (*not* felt or composition) to make the entire column 1¼″ to 1⅜″ long. All wads must fit tightly in the case and be seated solidly, though without excess pressure. The case is finished off with its standard crimp, normally a folded crimp except in 10 gauge. The finished crimp may require a spot of glue or heavy waxing to hold it securely in place, since it is not backed by any shot charge.

When fired, such loads project the wad column with considerable force, sufficient to penetrate half-inch pulp board at close range. They *could* kill and will certainly produce serious injury at close range. The light individual wads lose velocity very rapidly and become rather harmless beyond ten yards range, but sometimes the entire column stays together like a single slug.

Sooner or later many handloaders get the idea they'd like to load a few tracers, not for any really useful purpose, but just to do it. The loading presents no problem when proper tracer bullets can be found. However, the manufacturers *won't* sell tracer bullets except as loaded ammunition and only then to *bona fide* governmental agencies.

One often encounters small quantities of salvaged military tracer bullets. Aside from questionable reliability as tracers, they can be loaded with powder charges normally acceptable for the same weight and diameter non-tracer bullet. In the chapter devoted to military surplus ammunition, tracers were discussed in detail, and the same objections apply here. They serve no useful sporting purpose and can sometimes be injurious to the gun barrel. So, my advice is, "Don't bother."

One more caution: We've seen a couple of attempts at homemade tracers. One involved drilling a hole in the bullet base and packing same with wet, crushed match heads. The other was more sophisticated and used an incendiary chemical mix prepared by the loader. Don't — you'll only be asking for trouble.

Many correspondents have asked about loading shotshell tracers. Only W-W produces a satisfactory tracer element for shotshells and it is definitely not for sale to handloaders. Forget it. Much the same may be said for armor-piercing as for tracer. With one exception, the only effective AP bullets available are military salvage, often of questionable quality. That exception is the special bullets used by KTW in its armor-piercing pistol and carbine ammunition. These loads use bullets much heavier than normal, a state obtained by making the cores of a safe deteriorated radioactive metal much denser than lead. Such bullets, even if the maker can be persuaded to sell them (sales are normally restricted to police), are very costly, like $1.00 each. Like tracers, armor-piercing ammunition serves no useful sporting purpose, so is best left alone.

This by no means covers all the special-purpose loads that might come to mind, but perhaps it will help to avoid some of the problems certain to be encountered.

Modern Handloading

CHAPTER 22

Loading for Auto Rifles

SEMI-AUTOMATIC high-power rifles are common today, much more so than a mere ten or fifteen years ago. For example, a decade back only one domestic sporting model of the type, the Remington M-740/742, was in production. Since then we have seen Winchester, H&R, and Browning enter the field with popular and well-received models. In addition to all that, hundreds of thousands of self-loading military arms are in circulation, as well as several tens of thousands of obsolete Remington M-8/81 and Winchester M1905/07/10 rifles. In short, today, semi-autos probably rank *second* in overall sales, right behind the venerable bolt-action.

For many years most handloaders ignored autoloaders, while others simply felt it wasn't practical to reload ammunition for them. This was generally due to the fact that such designs are integrated with the cartridge to a high degree. *All* autoloaders depend upon energy generated by the cartridge to power the moving parts which unlock, extract, eject, cock, feed, chamber, and lock the mechanism to ready it for the next shot. Some older guns are rather critical in that they function correctly only with ammunition loaded to full power within rather narrow limits of bullet weight, velocity, chamber pressure, and time-pressure curve. Many attempts at handloading for such guns have been unsuccessful because proper attention was not paid to those factors.

Whether the rifle is recoil-operated like the Johnson military, the Remington M8/81, and Winchester 1905/07/10 or gas-operated like all the designs since WWII, handloaded ammunition that will make them function reliably can be produced.

So long as we stick with conventional jacketed bullets, it is simply a matter of loading to standard velocity and pressure with a powder of the same type and general burning rate as used in factory or arsenal ammunition for the same gun.

Since all common modern semi-autos are designed around existing standard cartridges, this presents no problem. About the only exceptions would be certain Soviet Bloc military arms such as the Czech SHE (7.62 x 45mm) and Russian SKS (7.62 x 39mm M43) whose cartridges have not been standardized here.

Back to jacketed bullets though. Simply select a standard weight bullet for the cartridge in question (ultra-light or extra-heavy, such as 90- or 250-grain in .308 or .30/06 won't do) and then select from reliable load tables a powder charge that will duplicate factory or arsenal velocity. Powder selection is somewhat critical in that those types burning unusually slowly or unusually fast *for that caliber* will produce problems. The best powder will generally be the one falling in the middle of those shown in the load tables for that bullet weight.

For example, published data for the .308 Winchester (7.62mm NATO) often includes powders as fast as IMR 4198 on through those as slow as H 4831. Right in the middle we will find IMR 3031 and IMR 4895 listed. Either of the latter two will be satisfactory. The others will either not produce reliable functioning or will "overdrive" the recoiling gun parts and possibly damage them. IMR 4198, being fast, will reach its pressure peak too soon, then drop off rapidly resulting in pressure at the gas port being too low to drive the recoiling parts fully rearward. Conversely, H 4831 will reach its pressure peak much farther down the bore, producing *too much* pressure at the gas port and overstressing the recoiling parts. All this while producing safe chamber pressures and identical velocities with the same bullet.

As already mentioned, the solution is simple enough: Use middle-range powders for the specific cartridge; use standard-weight bullets; load to standard velocities. If you *insist* on varying velocities and bullet weights widely, then problems will be encountered with gun functioning and with excessive wear and tear. However, the comments following shortly on lead-bullet loads will show how reliable functioning *can* be obtained with non-standard bullet weights and velocities. Obviously, so long as proper port pressures are maintained, any weight bullet *can* be used, but selecting a powder to do that becomes difficult.

First though, let's look at a particular problem gun and see how workable, reliable-functioning handloads were developed for it.

A near-new Soviet SKS was obtained from a Vietnam returnee and a quantity of Finnish military ammunition was procured to test it. After ascertaining the gun functioned correctly, the distance and angle the empty cases were ejected were measured and filed for reference. Disassembling cartridges showed us a 125-grain flat-base bullet and about 26 grains of unidentified powder. Military references indicated the military load produced 2410 fps at pressures just under 50,000 psi.

All of these factors indicated a rather fast-burning powder, on the order of IMR 4198 or 4227. New cases were formed from 6.5 x 54mm M-S brass; Speer 125-gr. bullets of suitable diameter were procured.

Loading began with enough 4198 to fill two thirds of the powder space and the charge was cautiously increased until reasonably good semi-automatic functioning was obtained. Velocity tests were then conducted and it was found that we were well below Soviet specifications. Further charge increases of 4198 resulted in the case being filled completely without reaching the desired velocity, and case ejection became more violent than with military ammunition, an indication the port pressure was too high.

The process was then repeated with faster-burning 4227. Proper velocity and positive functioning were obtained with a charge that just filled the case. At the same time, there was no indication of excessive chamber pressures, nor of excessive port pressures. After rather extensive shooting to insure performance was consistent, that load was standardized. That the load is reasonably well balanced for the SKS is indicated by it performing correctly in several other identical guns since, of both Soviet and Chinese manufacture.

This same sort of empirical load development can be applied to any bastard gun and cartridge and was used quite some time ago with equal success when the first 7.5 x 54mm French MAS 1949 semi-auto military rifle came to hand.

But jacketed bullets cost a pretty penny, usually over $5 per hundred in all but the smallest calibers. That's still too much money if you're given to a fair amount of rapid-fire practice. The solution lies in the use of cast lead bullets.

It has been said many times that such bullets don't work well in auto-loading rifles. Many reasons are given: lead clogs gas ports, sufficient velocity and energy can't be developed to function the actions, etc., etc. Don't believe it. Since the late 1940's, I have obtained perfect functioning and good accuracy from virtually every self-loader I could lay my hands on, including such odd ones as the French MAS 49, Soviet SKS, and even, during my Army days, automatic weapons such as the MP 44 and various machine guns and automatic rifles.

Probably by far the most "handloaded-for" auto rifle is the diminutive little "Carbine, U.S., Cal. 30, M1" popularized during a pair of substantial wars and countless revolutions since its debut in 1942. Personally, I think this gun owes a great deal of its popularity to the fact that it was the first U.S. service rifle small enough to steal easily!

We'll give this little gun its treatment first, though that places it a bit out of sequence in this dissertation. Properly, it should be grouped along with other gas-operated designs, but because of its special problems we'll cover it separately now.

Reams and reams of copy have been written on handloading for this little gun. Unfortunately, many of them overlook the basic problem: this particular design possesses a very narrow port pressure tolerance. As a practical matter, the minimum pressure at which the gun will function reliably is only very slightly below the maximum average pressure to which the cartridge is loaded. In addition to this, the gun has relatively light recoiling parts and short recoil travel; both conditions contribute to marginal reliability. In short, we have here a mechanism which does not permit the latitude in ammunition that can be accepted easily by more massive designs. Heavier recoiling parts traveling greater distances produce greater reliability.

This simply means that successful handloads must very closely approach the time-pressure curve of arsenal ammunition. We are fortunate here in that the performance range of the cartridge is well within the capabilities of lead bullets. From that point, it is simply a question of using enough of the right powder to produce the correct pressure and velocity. Actually, due to the fact that the lead bullets offers less resistance than the jacketed type of the same shape and weight, absolute matching both of pressure and velocity isn't possible; but, we can come close enough.

H110 powder was developed specifically for this cartridge and follows very closely the characteristics of the Winchester-Western ball powder developed for the original military load. It is the logical powder to use. Lyman bullet No. 311359GC, nominally 115 grains, was designed for the carbine cartridge and is of the proper weight and shape. A charge of 13.5 to 14.0 grains of H110 will produce correct reliable functioning of the average M1 Carbine when it is clean and in proper mechanical condition.

More often than not, carbine malfunctions are due to the gun rather than the ammunition. Failures to feed with properly shaped bullets are normally caused by the gun, while failures to extract or eject are more likely the fault of the ammunition. So, before condemning your ammunition, make certain you are placing the blame in the right place. Ammunition loaded as described *will* produce correct functioning if the gun is right.

Casual shooting — that is, only a few hundred rounds per year — with lead bullets won't usually produce any significant clogging of gas ports. However, we've examined a few M1 Carbines whose gas ports were badly clogged after firing as few as 2,000 to 3,000 rounds. They contained a very hard deposit that appeared to be a mixture of powder fouling, lead, and carbonized bullet lubricant. It had built up to the point that only a tiny hole remained. Soft bullet metal, lack of gas checks, and poor lubrication seem to combine to cause this. Using hard gas-check bullets and a proper lubricant will keep it to a minimum, but it still seems to occur occasionally. Any condition that normally contributes to bore leading will cause port clogging.

If you encounter gas port clogging, simple, *occasional* cleaning will solve the problem. Remove the gas piston retaining nut and piston to expose the port. Select a twist drill (high speed steel) or straight pin reamer that will just barely enter the port where it exits on the bottom of the gas cylinder. Grip the drill or reamer in a pin vise, then rotate it in the hole to scrape out the fouling. Take care that the point does not contact the far side of the bore and cause damage. It is best to fit a stop of some sort to prevent this.

Once the port is clean, periodic checking will tell you when it needs more work. Any time it reaches a point where the drill cannot be pushed through easily, it needs cleaning. The M1 rifle and any others with removable gas cylinders can be treated in the same way. However, the G-43 and some others do not permit cylinder removal. For them, simply drill a ⅛-inch hole in the cylinder directly over the port, then plug with a short screw which can be turned out for cleaning. Make certain the screw doesn't protrude inside to damage the piston.

Working up loads that use lead bullets, yet will function reliably in most high-power automatic mechanisms, requires mainly that we keep in mind just what makes those guns work. Recoil-operated designs will function reliably only when a certain minimum amount of recoil energy is generated. The Johnson rifle, which is the only mobile-barrel design around in any quantity, was designed to function with the standard U.S. .30 M2 load with its 150-grain bullet at about 2,800 fps.

The recoil energy required to actuate the mechanism under ideal conditions was established well below the average produced by that load in order to provide a power reserve to overcome combat conditions, dirt and lack of lubrication, deformed ammunition, low environmental temperatures, and a host of other variables. Since most other standard .30/06 factory and military loads produce essentially the same recoil energy, they, too, normally will function the mechanism realiably.

When it comes to making this gun digest lead-bullet loads, we immediately encounter the impossibility of obtaining 2,800 fps with a 150-grain bullet and still maintaining adequate accuracy. Two approaches can be taken to this problem. It is probably best to begin with the rifle. Remember that it is designed to function with somewhat less than full recoil energy under *ideal* conditions. By making certain it is always scrupulously clean and properly lubricated, we eliminate the need for part of that reserve power. By going a step farther and reducing friction (polishing out tool marks, removing burrs, etc.), we can eliminate the need for a bit more of that reserve.

We do not want to make any mechanical changes such as shortening springs, removing metal, etc., for to do so would cause undue stress on the gun when fired with standard ammunition.

Soft lead bullets do not produce the accuracy we desire at high velocities. We reduce this effect by going to the hardest lead alloy we can obtain and handle, pure linotype metal. Past experience has shown that acceptable accuracy can be obtained with properly cast linotype bullets at over 2,200 fps. This range can be extended a bit further by use of a gas-check design with check attached, and by careful selection of bullet lubricant. Alox-blend lubricants seem to produce the best results.

With that much accomplished, we are getting close to a solution to the problem. Since velocity is our limiting factor with lead bullets, and recoil energy is dependent upon both velocity and *weight,* we can increase energy for a given velocity by increasing weight. Simple, no? By using a 180-grain rather than a 150-grain bullet, we increase recoil energy. Should this not prove enough, there are still 200- and 220-grain bullets we can use. The heavier the bullet, the slower we can drive it and still produce the needed recoil energy.

In just this fashion, I developed a .30/06 load back around 1960 which provided perfect functioning in the half-dozen Johnson rifles in which it was tested. It consisted of standard military brass and Large Rifle primers, Lyman bullet No. 311334GC (170-grains) sized .309", and 42.0 grains of H4895 powder for a velocity approximately 2,400 fps. If you don't achieve the level of accuracy desired with this load, and feel the gun isn't at fault, work on up the line with heavier bullets at lower velocities.

This same approach will work with the old Winchester and Remington autoloaders already mentioned.

Gas-operated guns such as the M1 Garand, FN M1949, G-43, Tokarev, etc., require entirely different treatment. Such guns are unlocked, opened, and cocked by propellant gas diverted through a port to expand against a piston. The amount of gas diverted against the piston, and the pressure it applies to the piston are the critical factors. Generally, the pressure within the barrel at the time gas enters the port is regarded as the determining factor. It is expressed as "port pressure" in terms of psi.

Each gun is designed to function within fairly broad but clearly defined upper and lower port pressure limits. However, port pressure for one gun is not necessarily suitable for another of the same caliber. For example, the M1 has its port located very near the muzzle, where pressure within the barrel has dropped to less than 10,000 psi. The FN M1949 in the same caliber has its port about midway between chamber and muzzle, where pressure is much higher. The two designs compensate for this difference by utilizing less of the high-pressure gas; more of the low. This is accomplished by using different size ports and by allowing the gas to expand against the piston for different periods of time. In addition, piston sizes are slightly different.

What we must do to produce usable lead-bullet ammunition for gas-operated guns is to develop loads which generate port pressure above the minimum required by each gun. We can reduce that minimum exactly as we did for the recoil operated Johnson already described — clean gun, proper lubrication, reduced friction. In some guns, notably the FN M1949, we find a means to further reduce it. The FN is fitted with a gas regulator which can be adjusted to increase or decrease the amount of gas that acts against the piston; also the time during which it acts. By increasing both gas and time, we get more work out of lower pressure.

Guns not having a regulator of some sort can be given essentially the same treatment by carefully increasing the size of the gas port, but this ruins the gun (or at least that particular barrel) for use with standard ammunition. Too much pressure on the piston can damage the gun. Because of its gas regulator, I recommend the FN M1949 above all others for reduced loads.

So, with all those problems hanging fire, how do we develop loads that will function reliably with lead bullets in self-loaders, and still produce good accuracy?

Bullet selection remains the same. By using a very hard gascheck bullet and proper lubricant, we raise the velocity level at which the bullet will deliver the desired accuracy. In addition, we must use powder charges that produce relatively high pressures and that move the pressure curves as far toward the gas port as possible. Fast-burning powders will meet the first requirement, but not the second. By the time the bullet uncovers the gas port, pressure within the barrel may have dropped to a far too low value to provide the energy required. So, we go the other way, slow-burning powders whose pressure curve remains relatively high farther down the barrel. But, in order to secure adequate burning of such powders, we must use relatively heavy bullets and large charges. Even then, we may get unburned powder blown out the muzzle. However, I've found such loads can still be quite accurate.

As mentioned before, the location of the gas port is sometimes critical in load development. The closer the port to the chamber, the less trouble will be encountered in maintaining high enough pressure to actuate the mechanism. Conversely, a port near the muzzle or a long barrel will cause the most difficulty. However, a load that performs correctly in the muzzle-ported design will normally do so with the rear-ported gun, but vice versa does not necessarily follow. For this reason, all the loads that have satisfied me in the M1 Garand have worked well in .30 caliber FN rifles, while some worked up for the FN functioned only marginally or intermittently in the M1.

The four commercial full-bore self-loaders (Browning BAR, H&R Ultra, Remington M742, Winchester M100) are comparatively easy to load with cast bullets for all use a typical gas cylinder/piston setup installed well back on the barrel. It is actually located somewhat rearward of the barrel's middle. The primary purpose of placing the gas port closer to the chamber than on many military arms, is to permit the entire gas system to be housed within a sporting-style fore-end of conventional length. This location requires the system be designed to operate at a fairly high port pressure.

Such designs work best with fairly fast-burning powders such as 3031, 4895, and Norma 203. Faster powders move the pressure peak too far rearward resulting in below-minimum port pressures, and marginal functioning. Our work with these four guns has been limited almost entirely to .30/06 and .308 and the loads shown in the accompanying tables work well in all. Though it undoubtedly can be done, we've never attempted to work up cast bullet loads in the .243 and similar smaller calibers for which these guns are also chambered. Generally, the smaller the caliber, the more difficult it is to produce cast loads that will function a semi-automatic action reliably.

Loads with light bullets and faster-burning powders that may function perfectly in a commercial .30/06 often will give marginal performance, or simply won't work at all in an M1 rifle, particularly in cold weather or as the gun heats in rapid fire. For my own use, I insist on 100 per cent reliability, so choose a fairly heavy load. Over the years, many cast loads have worked reasonably well in the M1, but I have settled on Lyman bullet No. 311365GC (190 grains) sized .309, ahead of 46.0 grains of 4831. It produces about 2,100-2,200 fps. In over-size barrels it works better if sized to .310 or slightly greater.

This load will produce about two to three MOA accuracy in a good M1, and has produced many 100-yard groups hovering near an inch in a heavy barrel bolt gun. This load has also worked well in the FN M1949 and one converted G-43 in which it has been tried.

The FN M1949 and most .30/06 commercial self-loaders with their rear-positioned gas ports will give complete reliability with Lyman bullet No. 311375GC (170 grains) driven by 42.0 grains of H4895. Velocity is approximately 2,400 fps. This load will often function the M1 rifle, but its reliability is marginal there, as opposed to 100 per cent in the FN, with its adjustable gas port.

All other semi-autos except the M1 Carbine will generally function with the same class of loads that work in the FN. The commercial self-loaders, the G-43, Tokarev, FN FAL, M14, MAS 49, etc., all have rear-positioned gas ports where the pressure curve remains relatively high with any powder suitable to the cartridge.

Thus far, we haven't mentioned the other factors that contribute to reliable functioning. First, nearly all self-loading actions are designed to feed spitzer-type pointed bullets. To insure feeding with cast bullets, they must be of the same shape and be seated to approximately the same cartridge length as factory loads. In addition, the bullet must be held tightly enough in the case neck to resist the severe forces applied to it during its movement from magazine to chamber. Nearly all feed systems depend upon deflecting the cartridge into the chamber by causing the bullet to strike sharply against a sloping ramp or guide. If the bullet is soft, of improper shape, incorrectly held in the case, or if over-all cartridge length is too short, the conditions for correct feeding are upset — and it doesn't happen!

In addition, all bottleneck cases should be crimped rather heavily on the bullet to aid in keeping the bullet in place during feeding. Many seat-crimp dies will do the job fairly well in one operation. However, unless cases are of truly uniform length, crimp force (and, consequently, bullet pull which has a good deal to do with velocity uniformity) will vary a great deal. I prefer to trim cases uniformly, then seat and crimp in separate operations. A little more time is consumed in doing it that way, but the results are worth it.

Another feeding problem that occasionally occurs is easily avoided. Since feed ramps and bullet guides are designed for jacketed bullets, they are often rather rough; they may even have sharp ledges where one part transitions into another. When cast bullets strike and skid over these surfaces, pieces of lead sometimes are torn or scraped off. They can pile up and interfere with bolt movement and free passage of the cartridge. A stiff bristle brush can be used to clean lead scrapings out now and again, thus avoiding that problem. Polishing guide surfaces will cure the problem permanently and will not harm the gun.

One little gimmick I've never known anyone else to try, but it's almost as old as the metallic cartridge itself, is to lubricate cartridge cases to ease extraction. Here, where we want to use as light a load as possible and still get free functioning, at low pressures, lubrication is acceptable and safe. Not so, however, with high-pressure loads. The most effective and simple method I've found for accomplishing this is to spray a mist of wax over the loaded cartridges. Good quality furniture *wax, not polish,* in a spray can works best. If you prefer, though, heavy paste wax may be wiped on with a cloth. Don't overdo it, just a very thin coat that will dry thoroughly is what you want. Waxed cases will ease extraction just enough to provide reliable functioning with some loads that would otherwise be marginal.

There is one other type of self-loading design we haven't touched upon, the delayed or retarded blowback. The CETME and SIG rifles currently sold are of this type. They use fluted chamber walls to ease extraction by floating the case on a layer of gas during peak pressures. We've not gone very far in working with such guns, but we do know that relatively fast-burning powders are required in order to have pressures as low as possible as the bullet exits the muzzle. Loads approximating full power must be used or the action will not fully open. A load that has functioned correctly in one 7.62MM (.308 Win.) CETME Sport rifle consists of Lyman bullet 311332 (180 grains) and 37.0 grains of IMR 3031 powder. Functioning was marginal with dry cases but reliable with cases lightly lubricated. RCBS Resizing Lube was applied sparingly to the entire case with a cloth. A stamp pad would work as well. More work is necessary in this area.

CHAPTER 23

Handgun Loading

THE AVERAGE serious handgun competitor (if there is any such thing as an average) simply *must* shoot handloads for at least a good part of his match preparation. Many shoot nothing but handloads except in those matches where ammunition issued on the range is required. Some practice with handloads, but feel a little more confident in a match with factory fodder. Sometimes I think that last group just wants to be able to lay the responsibility elsewhere if the final scores aren't as good as expected.

The necessity for handloads is primarily economic, since cartridges costing 12¢ to 13¢ over the counter can be duplicated for as little as 2¢. Even so, some quite affluent pistoleros shoot handloads purely by choice.

We do have to accept one fact: A large percentage of home-brewed handloads are *inferior* in performance to the industry-produced item. Many reasons for this disparity exist, and every one of them must be eliminated if target grade loads are to be produced. These defects *can* be avoided, but it takes care and attention to detail.

I've assembled something like one-million-plus .38 Special wad-cutter loads and about half that many in .45 ACP caliber. Some were pretty sloppy, I must admit, but some were damn good. We've found that match-winning loads can be produced at very little increase in cost over those "plinking specials." How? Simply by refining standard loading procedures to produce maximum uniformity. Uniformity is the key, as in all other handloading.

The .38 Special wadcutter and .45 ACP semi-wadcutter are used almost exclusively in all serious competition, except those matches restricted to the .22 rimfire. There are still a few individuals using .32 and .44 or other calibers, but they are getting mighty hard to find. Also, the auto-pistol reigns supreme in both calibers. Few revolvers remain in use in high level competition. Good target loads therefore revolve around the .38 Spcl. and .45 Auto.

The criteria of a good target handload are few but explicit: Produce *perfect* functioning of the gun; produce maximum accuracy (groups at least as small as the X-Ring); be relatively simple and rapid to assemble; produce minimum recoil. Both the .38 Special and .45 ACP can be loaded to meet these requirements without difficulty.

.38 Special

For use in Colt and S&W autoloaders (Gold Cup and .38 Master respectively) and various custom-built guns built on the Colt frame. First, you must have top quality equipment. This means one of the better presses with close die/shell-holder alignment and a well-aligned, adjustable primer seating mechanism. Dies must be the best you can afford, preferably including a "carbide" (cemented tungsten carbide insert type) resizing unit, and seat-crimp and expander units carefully matched to the case and bullet being used. The powder measure must throw extremely uniform charges and, of course, this means you need a very accurate scale to check the measure. And, if you're casting your own bullets, a sizer-lubricator that insures concentricity and roundness of bullet.

Cases must be the best and most uniform you can get. All one make and lot is best; assorted makes and range pickups are worst. Uniform length is most important in securing a uniform crimp which affects uniformity of powder combustion. Any wide variations in ignition and, consequently, vertical stringing. Case volume (capacity) is important, but less so. Weighing and discarding cases varying more than plus or minus 3 per cent from the norm will help close up groups, but it's doubtful you can see the effect with anything less than a machine rest.

Other items are important, but the bullet is the heart of the cartridge. Of the cast variety, Hensley & Gibbs No. 50 and Lyman No. 35863 seated flush with the case mouth have achieved excellent reputations. Bullets must be perfectly cast, uniformly round, without fins or

Modern Handloading

voids. Weight must be uniform and until you *know* what your casting techniques produce in this area, weigh and segregate bullets into lots varying no more than 0.3 grain. Exact weight isn't nearly as important as weight uniformity in a given batch of loads. Over or under-weight bullets can be set aside until you accumulate enough to load and use them separately. Heavy or light bullets will shoot with equal accuracy, but to different points of impact.

Hard bullets sized to exact groove diameter or no more than .0005" over are usually most accurate. They must be round and concentric. Fresh linotype metal works very well, but other similarly hard alloys do too. Sizing dies must make a gradual transition from entry to final diameter. Any definite shoulder connecting the two will distort the bullet, reducing accuracy. Bullet bases should be wiped clean to remove any lubricant picked up during sizing.

For the individual who prefers not to cast his bullets, or who hasn't faith in his ability, factory-swaged wadcutters are the answer. They are furnished, sized and lubricated, ready to load, at quite reasonable prices. Being cold-formed under tremendous pressure from fresh lead wire, they are more uniform than any but the most perfect cast bullets.

I doubt that there is a domestically-produced primer that will not do its part in producing top-notch accuracy. While changing primers may sometimes shift center of impact, group size seems to hold up with

Factory-swaged bullets will generally run more uniform than those cast at home.

Handgun Loading

all the present makes. The important thing is to not mix primers in any one batch of handloads.

There are lots of powders that can be used, but nothing seems to match small charges of Hercules Bullseye. Nearly all .38 Special target autos function well with 2.7 grains and the 148-grain wadcutter bullet. Some will accept as little as 2.5 grains, a few require 2.8. Nothing is to be gained by trying to use different charges for different ranges. Use for everything the one that produces *perfect* functioning and best fifty-yard accuracy. That load will do right by you at twenty-five yards, albeit with some sight change.

Assembling components into target rounds is no more difficult than for plinking ammunition. It just requires greater attention to detail and uniformity. Perform each operation separately on convenient-size batches of cases, unless you're rich enough to buy a big progressive tool, such as the Star.

Check cases for length with a simple snap gauge unless you've already trimmed them to uniform length. Wipe clean and if to be resized in a steel die, lubricate lightly. If in a carbide die, lubrication is not required. Size full length and seat primer carefully, "feeling" it to the bottom of the pocket without applying crushing force. If your press doesn't allow this feel, use a separate priming tool such as the Bonanza or Lachmiller.

Expand the case necks and bell the mouths just enough to allow the bullet to be started with the fingers. Here is where you'll notice the difference if cases aren't of uniform length. The expanded case should grip the bullet tightly enough to keep it from being easily rotated with the fingers after seating but before crimping. The expander must therefore be matched to the after-sizing bullet diameter. If the case neck is too tight it will deform the bullet during seating, which may produce an occasional wild shot. This is one reason hard bullets are best. Don't expand the case any deeper than necessary, just to the base of the seated bullet. This may require shortening the expander plug of some makes, in order to also secure the proper amount of mouth flare.

Throw powder charges as uniformly as possible. If the measure feels different for a particular charge, throw that one out — it's probably light or heavy. Having the measure solidly mounted contributes to uniformity of charge weight. Practice with the measure until you've achieved maximum uniformity. Double-check to be sure that each case gets one and *only* one charge.

Start bullets in charged cases by hand. Most dies will straighten a bullet started crooked, but just might deform it in the process. Eliminate that potential problem by starting bullets as straight as possible. One

Modern Handloading

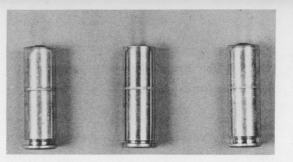

Left: belling or flaring should be held to the minimum that allows smooth seating of bullets without shaving lead. *Middle:* left and right cases are roll-crimped moderately; middle is very lightly taper-crimped. *Right:* primers must be seated to rest on bottom of pocket, not protrude above case head as here.

fellow I know uses a sleeve and plunger affair as shown to start bullets straight.

Adjust the die to seat the bullet to the correct depth *without* crimping the case. This is accomplished by backing the die out so the crimping shoulder does *not* touch the mouth of an expanded case. Take care to get all cases fully into the shell holder so the bullet doesn't bang into the die mouth. Seat bullets with a single, smooth stroke of the press handle. If any lead is shaved from the sides of the bullet, the case mouth is not flared enough. For autos the bullet must be seated flush with the case mouth for proper feeding through the magazine. For revolvers, the bullet is best seated so the case mouth is flush with the top of the crimping groove. Crimp case on bullet as a separate operation. Some makes, mainly RCBS, offer separate dies for this purpose. In the standard seat/crimp die, back out the seating stem, then adjust to crimp *without* the stem touching the bullet. The crimp must be uniform in depth completely around the bullet and need be only slight. Mashing the case mouth heavily into the bullet serves no useful purpose in target loads. Uniformity is more important than degree of crimp. Either roll or taper-type crimp may be used, but better uni-

Properly uniform case length is essential to crimp in revolver ammunition; essential to headspacing and ignition in autos.

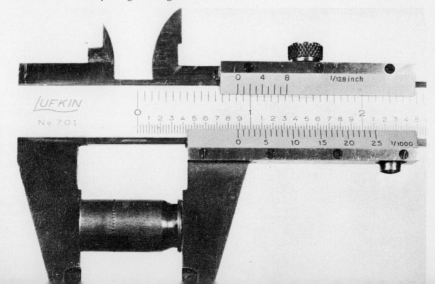

formity can be obtained, in my experience, with the latter. Gil Hebard Guns supplies a special "Taper-Lok" die for this purpose.

Each operation should have been followed by careful inspection, but now carefully check each completed round. Any that aren't exactly right should be set aside for non-serious use. Wipe off any lubricant with a cloth moistened with solvent and box the cartridges. Now that they are loaded perfectly they should be kept in that condition, not tossed loosely in a coffee can and banged around. Plastic compartmented boxes are excellent protection and cost little.

Loaded in this manner with components as described, your .38 Special wadcutters will shoot right with the best factory load, but get careless just once and you'll wind up with a gun malfunction or a wild shot that can cost you match.

Everything said thus for applies equally to the .45 ACP, but it poses a few special problems that need particular attention. Bullets *must* be hard to insure reliable feeding. In addition, they must have a properly-shaped point. In normal feeding, the nose of the bullet strikes the gun's feed ramp and is deflected upward into the chamber. Soft bullets will be deformed in varying degrees and improper shapes simply won't feed. H&G No. 68 and Lyman No. 452460, 210 and 200 grains, work very well, as do the 185-grain H&G No. 130 and Lyman No. 452488.

Swaged (not cast) .45 bullets in both 185- and 200-grain weights are readily available. The 200-grain seems best for fifty yards, and as an all-around bullet, while the 185 is rather a twenty-five-yard item.

This case headspaces on its mouth against the forward end of the chamber. Short cases will cause weak ignition by virtue of the primer receiving a lighter firing-pin blow. An excessively long case can prevent the slide from going fully into battery. Entire batches of cases should be sorted carefully. Discard short cases and trim long ones. It isn't practical to simply trim all cases to match the shortest since this makes them *all* too short. Correct length for use in a given gun can be established by using the chamber as a gauge. A resized case dropped into the chamber should seat with its head flush with the rear surface of the barrel tang, or no more than .006" below it. If a large quantity is available to select from, enough "gilt-edge" cases may be selected in this manner to reserve for match use only.

Case mouths should be chamfered only very lightly since it weakens the crimp and increases the rate of neck splits. After neck expansion, cases should be a very tight fit on bullets. Expansion should be carried only deep enough to allow easy straight starting of the bullet, about halfway to the base of the seated bullet. When properly hard, the bullet will expand the case the rest of the way without being deformed.

Left: auto chamber may be used as case length gauge. Cases entering .006" or more are too short, as this one is. *Right:* simplest chamfering tool is this unique design by Lee.

Left: Opposite is result of case not being tight enough on bullet. Impact with feed ramp has driven bullet deep into case. *Right:* in severe cases, a heavy cannelure may be rolled into case to form internal shoulder against which bullets seat.

Case mouths must be crimped tightly on the bullet because of the tendency of the impact against the feed ramp to shove it deeper in the case, resulting in gun malfunctions and vertical stringing. Yet, case mouth diameter must remain large enough to provide positive headspacing. The long taper crimp is best for this purpose and should be applied only heavy enough to produce correct functioning in your gun.

Heavier powder charges are, of course, required for the .45 than the .38 and 3.0 grains of Bullseye will reliably operate most target-grade guns with both 185 and 200-grain bullets. Many shooters use 3.5 grains for fifty-yard slow-fire work, and 3.0 grains for timed and rapid-fire at twenty-five yards. The .45's greater recoil makes this practicable, while not so with the .38 Special. Personally, I prefer to use a single powder charge for both ranges.

So much for automatic-intended handloads. For any caliber to be used in revolvers, all comments regarding the .38 Special apply fully. Lighter powder charges may be used in revolvers to reduce recoil since positive functioning is furnished by the shooter.

For use in the .45 caliber revolver, loads should be assembled in .45 Auto Rim cases selected as outlined for the .38 Special. The .45

Handgun Loading

ACP case can be used, but requires three-round, half-moon clips which are not only an inconvenience, but cushion firing-pin blow which interferes with uniform ignition. It is not necessary to select cases so carefully for length, nor to have so tight a grip on the bullet as for the auto. Expanding, bullet seating, and crimping should be carried out as outlined for the .38 Special. And, of course, much lighter powder charges may be used if desired.

Since .45 Auto Rim cases are usually more costly than military ACP brass, many revolver buffs use the latter in clips for practice, the former in matches. Next time you start to blame the gun, the weather, the tax assessor, or your mistress for a "wide 8" in an important match, just ask yourself how much of the preceding dope went into the loading of your ammunition. Might be you just didn't load it good enough to stay in the running.

A few comments are in order on high-performance handgun loads, though all the foregoing applies as well.

Of particular importance in hot revolver loads is the necessity for tight assembly of bullet to case. This means a very tight fit of bullet in case and a heavy *roll* crimp into the bullet cannelure and crimping groove. If the bullet is not held very tightly, recoil will "walk" it out of the case to protrude from the cylinder and tie up the gun. Not a very healthy situation.

Just the opposite condition exists in autoloaders where poorly-secured bullets will be driven *deeper* into cases by recoil and feeding loads. This is easily avoided by a combination of tight assembly, cannelured case, and crimp as outlined under bullet seating.

Many of the top handgun loads recommended today produce pressures we wouldn't even have considered a decade or two back. Pressures in the 20,000-plus range for .38 Special and 40,000-psi range for the magnums are not unusual. At those levels there is little margin for error and powder charges must be checked meticulously. A few tenths of a grain over can put you in real trouble.

Additionally, more attention should be paid to case and bullet variations. A bullet running .0015" oversize won't cause any trouble at 15,000 psi, but at double or triple that level it can easily push pressures into the danger zone. No more than groove diameter is best for bullets in such hotshot loads. Some factories load their high-velocity bullets about .0005" *under* groove diameter for the same reason.

Many heavy handgun loads were developed in balloon-head cases. Today's solid-head cases produce excessive pressures with many such loads, especially in the .44 Special which was so thoroughly magnum-ized by handloaders in the 40's and 50's. The lesser volume of solid-head cases raises pressures with such loads alarmingly. The NRA *Illustrated Reloading Handbook* treats this subject in detail, with specific pressure tests. If in doubt, reduce powder charge one grain or more. Just remember that the older loading data in the larger calibers *often* refer to balloon-head cases.

Handloading for shortguns is no more difficult than for rifles, but for best results, these different techniques should be used.

CHAPTER 24

Loading for Black Powder Guns

A GUN has the unique facility of hardly ever *wearing* out. It may be lost, destroyed, or ruined by abuse, but to be simply worn out is highly unusual. People think nothing of junking a $4,000 automobile, but they will hang on to a $40 rifle or pistol forever, no matter how obsolete and/or inefficient it might be. Great-gran'pappy's old rifle or six-shooter will be kept in the closet or attic generations after he's gone, and long after all of his other property has been used up, worn out, sold, given away, thrown away, etc. I suppose this is so because any gun is a highly personal item and sometimes seems to almost absorb the spirit of its owner. It becomes as much him as himself. Maybe, too, it's because of the aura of heroism and romanticism that surrounds *old* guns. What other object carried by a revered (or otherwise) ancestor is so likely to have taken part in some forgotten or unproclaimed epic of courage and perseverance. Sure, there's no evidence he ever fought off Indians, bank robbers, or marauding raiders with it — but he *might* have, and somehow that's a comforting thought.

Anyway, old guns come to light from dusty storage every day, and often if they are cartridge guns — breech-loaders, if you will — there is someone who wants to shoot them.

But the ammunition for those guns of the black powder era (with a few notable exceptions, such as the .44/40, .45/70, etc.) hasn't been

Top: Winchester M1886 is typical of old black powder guns that can be put to use by handloaders. (Photo by B. T. Ramspott) *Bottom:* Winchester M1876 .

made for thirty to fifty years. A few rounds in any given caliber can usually be dug up from dealers in collector's cartridges, but shooting them at today's prices is like lighting a cigar with a $20 bill. Even when found, many such cartridges won't fire because of age and deterioration, and many that will are unsafe for the same reasons.

But all is not lost. Very nearly *all* of the centerfire cartridges of the late 1860's onward can be reproduced in one way or another by the astute handloader.

Proper cartridge cases to take modern primers are the first necessity. Generally speaking, most can be made up by re-forming some case that is currently manufactured. We've covered that rather extensively in the chapter devoted to case forming.

There are other sources. People hoard ammunition and components and it is still often possible to obtain at least a small supply of cases by running a want ad in those two journalistic classics, *Shotgun News* and *Gun Week.* Not too long ago I acquired two sealed boxes of sixty-year-old unfired Rem-UMC .44/77 Sharps cases in just that fashion. Gun collectors often wind up with small supplies of cases or ammunition they can be persuaded to sell, like the Winchester collector who acquired in trade with a Model 1886 a full case of .50/110 Express ammunition. If eyes and ears are kept active, you'll eventually find at least a few original cases. Beware, though, of old *fired* black powder cases. More often than not they were fired decades ago and set aside without cleaning. The result is usually internal corrosion so extensive the case is ruined, even though it looks solid enough outside. The inside will be deeply pitted and badly weakened. Often such cases are reloaded and emerge from the first firing looking like a scorched lace doily; full of holes and fractures.

If cases were washed before being laid away, the interior will be smooth, not pitted, and should be okay to use. If mouths seem unnaturally hard and brittle, beware. They may have been damaged by firing with mercuric primers. A case badly weakened by mercury will crack or break instead of bending under heavy thumb/finger pressure at the mouth.

One more thing, primers. Old primers are likely to be mercuric at worst, and deteriorated at best. Primed new or fired cases should be decapped (and clean them while you're at it) and reprimed with fresh modern primers. Obviously, you don't want to ruin the cases with old mercuric primers, nor do you want misfires after going to all the trouble to load them. Save time and frustration by repriming.

The same can be said of old loaded black powder ammunition you might obtain. Best results will be obtained if it is broken down, salvaging only the case and bullets, then reloading with fresh powder and primers, for the reasons given above.

Of course you'll need bullets. Even though later black powder style loads often used jacketed expanding bullets, you can get along nicely without them. At safe and practical velocities cast lead bullets will perform just as well on either targets or game. Fortunately, bullet moulds in almost any imaginable or useful caliber and weight can be ordered special from Lyman. In addition, many original moulds are in circulation and can be obtained, as with cases. If you insist on jacketed bullets, some custom makers can supply them, or you can have dies made up and swage your own. Personally, I don't think it's worth the cost in dollars and effort.

If you really want to be original and don't mind the work of extensive gun and case cleaning, black powder loads are the simplest. They are also in many ways the most satisfying. It seems to me to be cheating at least a little when we stoke gran'pappy's Sharps or Winchester with smokeless powder that was unknown when the gun was made. There are more kicks in that flame-shot cloud of white smoke that billows out when the hammer falls than in the sterile crack of a smokeless load.

Black powder loading is by far the simplest, and also the safest insofar as the old guns are concerned. You can't put an unsafe amount of powder in the case. In short, there is no way with black powder

Modern Handloading

that you can produce pressures high enough to damage the gun or the shooter, so long as the gun is in good mechanical condition.

By good mechanical condition we mean that the breech locks up snugly and securely; that the locking surfaces aren't peened and battered; that the firing-pin is a snug fit in its hole in the face of breech block or bolt; that firing pin retractors and retainers are functional; that the firing-pin nose is smooth and properly shaped and protrudes the correct distance; and that the chamber is not bulged or badly pitted. Condition of rifling controls accuracy, of course, but has little if anything to do with functional safety.

To get back to black powder loading, use a cast, plain base, *soft* lead bullet (about 1/20 tin/lead alloy) properly lubricated and sized to groove diameter or up to .0015" more. Establish the powder charge as that amount of FFG black that fills the case to about 1/16" *above* the point where the base of the fully seated bullet will be. Once that is established, set your measure accordingly. No need for concern about actual weight in grains. This produces slight compression of the powder charge when the bullet is seated, the conditions under which black behaves best and most uniformly. The only useful modification to this procedure is to substitute FFFG granulation powder in the very small cases such as .32/20, .38 Long, and the like. The smaller granulation burns better in the small cases.

If for use in a single-shot arm or to be single-loaded, the cases need not be resized at all so long as they will chamber freely and will hold the bullet snugly enough that it does not shift in handling. In many instances this will allow loading without tools other than a decapping punch made from a nail or ice pick, and a charge cup.

The procedure is simple: press out old primer with the punch; lay a new primer anvil-up on a smooth hard surface and tap the case down over it with a block of wood; dump in powder charge; press bullet into case mouth solidly against powder with fingers or by tapping down with a stick of wood. No crimp is needed.

Loading in this manner (without resizing) subjects the case to the absolute minimum wear and tear and insures maximum life. A good case will stand hundreds of loading like this.

If case necks become too loose to hold the bullet snugly, resize necks only ⅟₁₆"-⅛" at the mouth to give a better grip. Alternatively, increasing bullet diameter slightly (no more than .005") will tighten up the assembly. Often such a slight increase will not effect accuracy, so is a preferable solution when maximum case life is important.

The old-timers often reloaded the same small batch of cases scores, even hundreds of times in this way. Buffalo hunters reloaded each day's crop of empty cases daily and used them season after season. To insure continuing good case condition, they carried a water bottle and brush afield and scrubbed fouling out of empties during lulls during the day.

Where ammunition is to be handled roughly or used in repeating rifles, cases must at least be neck-sized to grip the bullet tightly. If for a tubular-magazine rifle, then the case must be crimped tightly on the bullet as mentioned in the chapter on bullet seating. Generally speaking, the older repeating rifles have rather "springy" action making frequent full-length resizing also necessary in tapered or bottle-neck cases.

One thing is certain, your cases loaded with black powder will have a short life unless washed completely free of fouling soon after each firing. Corrosion will set in very quickly, especially if humidity is high. The simplest and most effective method is to decap cases first, then boil them vigorously in soap or detergent and water. Done the same day as firing, while the fouling is still relatively soft, this will clean the cases well and no scrubbing is required. Otherwise, a stiff bristle brush and a prolonged soaking period will be required to get the job done.

But many people who want to shoot the old-timers don't care for all that black-powder mess. Then, too, the laws of today make black powder increasingly difficult to obtain, even sometimes requiring a special license for purchase and possession.

Nothing less than smokeless powder loads will do under those circumstances.

The problem with smokeless isn't in duplicating original black-powder ballistics; there are plenty of modern powders capable of that. It is a bit of a problem, though, to do so while staying within the pressure limitations of those 75- to 100-year-old guns and while obtaining uniform ignition.

Cases intended for black powder are big. The cavernous .45/70 used 70 grains of powder to produce its 1300-odd fps. A mere 20-25 grains of some of today's smokeless powders is required to produce the same performance at safe pressures. This means a relatively tiny powder charge rambling around in a huge case. If it is all piled up at the front of the case at the time of firing, ignition will be relatively poor and velocity reduced over that produced if it's back at the case head. In straight cases this is easily solved by seating a tight card or composition wad snugly on the powder before seating the bullet. Such wads are easily cut with a length of sharpened tubing tapped through a stack of cardboard laid on an endgrain hardwood block. Held back in this fashion, the powder will ignite and burn uniformly.

Bottle neck cases require different treatment. In the older days we poured cornmeal or Cream of Wheat over the powder charge to fill the case, then compressed it in seating the bullet. So long as the "filler" material was compressed, it didn't mix with the powder and everything worked well. However, to protect against mixing that *might* be induced by rough handling, as large a card wad as could be pushed through the case neck was placed between powder and filler. Some shooters report equally good results with a loosely-crumpled, elongated wad of toilet tissue replacing the filler. Once through the case neck, it expands to fill the remaining powder space and keeps the powder where it belongs.

In recent years the use of Dacron or Kapok fluff has largely replaced cornmeal and such. The fibre, obtained by shredding upholstering pads or worn-out cushions, is rolled into a loose ball that can be pushed through the case neck. It must be large enough so that once inside it will expand to grip case walls and hold back the powder. The same method is discussed in more detail under reduced loads.

Simply dumping the correct powder charge in the case will produce functional ammunition. But, if you want top accuracy and uniformity, one of the foregoing methods of keeping the charge localized near the primer must be used. So much for ignition.

Chamber pressure is the other major problem. It is really a problem of the guns involved; a problem of understanding that guns of varying degrees of strength will be found chambered for the same cartridge. Some makes and models will handle twice as much pressure as others. This makes it aboslutely essential that cartridges and loads be matched carefully to specific guns.

For example, the .45/70 may be loaded with smokeless powder and a 350-grain bullet to nearly 2,000 fps around 45,000 psi for use in Winchester Single-Shot and M-1886 rifles. This load will wreck a U.S. M-1873/88 Springfield trap-door single-shot chambered for the same cartridge. The Springfield limit is about 20,000 psi, and there are other .45/70's that will handle pressures about half-way between the two just listed.

Because of this, there are at least two, sometimes three, load levels one must adhere to, depending upon the gun in which the ammunition will be fired; assuming, of course, that you do want maximum power from the gun/load combination. Frankly, since most guns are shot simply for fun, there is no real need to load them to the hilt. The probability of error will be reduced and gun life will be increased if you stick to the lighter loads for *all* black powder guns.

To clarify this area, those old guns can be divided roughly into three groups or classes. Group I contains the weakest rifles and consists of the following and similarly-designed and constructed models:

GROUP I

Repeating	Single-Shot
Winchester M1873	Wesson Tip-Up
Winchester M1876	Springfield M1866/88
Whitney-Kennedy	Hopkins & Allen
Stevens Tip-Up and M44	Remington No. 2 RB
Ballard	
Maynard M1882	

Group II could be called, "medium-strength" guns, limited to 30-35,000 psi, and includes the following:

GROUP II

Repeating	Single-Shot
Marlin M1881/88/89	Sharps Side-Hammer
Whitney-Burgess	Remington No. 1 RB
Colt-Burgess	Winchester Low-Wall
Colt Lightning	Wesson Falling Block
Bullard	Whitney
	Wurfflein

Group III represents the top of the line, guns which will generally stand heavy loads up to 40-45,000 psi, some even more:

GROUP III

Repeating	Single-Shot
Winchester M1886/92/94/95	Sharps Borchardt
Marlin M1893/94/95	Winchester High-Wall
Remington-Keen	Stevens M44½
Remington-Lee	Remington-Hepburn
Winchester-Hotchkiss	Peabody-Martini
	Martini-Henry
	Farrow
	Remington No. 5 RB

These gun variations need not present any problems providing you identify each correctly instead of lumping them all as "just another .45/70 (etc.)" and then match the loads correctly. Separate load data for each of the three different groups is given here. Load data group I goes with gun group I, but

Modern Handloading

is also okay for groups II and III. Load data group II goes with gun group II, but must *not* be used in gun group I. It's okay for gun group III. Load data group III may be used only with gun group III.

Aside from all this, loading for obsolete black-powder calibers is done in accordance with the usual procedures described throughout this book.

LOAD DATA GROUP I

Caliber	Bullet Wt.	Bullet No.	Powder
.22-15-60 ST.	48	225415	3.4/Unique
.22-15-60 ST.	60	22536	3.4/Unique
.22-10 May.	45	228151	3.0/Unique
.22-13-45 Win.	45	228151	3.0/Unique
.25-21 ST.	86	25720	4.5/Unique
.25-25 ST.	86	25720	5.0/Unique
.25-20 SS., ST.	86	25720	4.0/Unique
.28-30 ST.	120	287221	11.5/4198
.32 L. Bal.	90	299153	4.0/Unique
.32 X-Long	115	299154	7.0/4227
.32-20	115	3118	10.0/4227
.32 Ideal	150	32360	5.0/Unique
.32-35 ST.	165	3117	11.0/4227
.32-40	165	319247	13.0/2400
.33-40 Pope	218	338301	16.0/4227
.35-30 May.	250	39875	6.0/Unique
.35-40 May.	250	35875	6.0/Unique
.38 L. Colt	150	358160	4.0/Unique
.38 X-Long Bal.	150	358160	6.0/Unique
.38 X-Long Wsn.	145	358161	6.0/Unique
.38-40 Win.	170	40143	17.0/2400
.38-50 May.	250	375165	12.0/R7
.38-55	252	375248	16.0/R7
.40-40 May.	270	415179	12.5/R7
.40-60 Win.	210	403168	18.5/2400
.40-60 May.	270	415179	16.0/R7
.40-63 Bal.	330	403149	17.5/R7
.40-65 Bal.	330	403149	30.0/3031
.40-70 May.	270	415179	17.0/R7
.40-70 Bal.	330	403149	19.0/2400
.40-85 Bal.	370	403171	32.0/3031
.40-90 Bal.	370	403171	32.0/3031
.44 Long Bal.	200	419180	16.0/2400
.44 X-Long Bal.	240	419182	17.0/2400
.44 X-Long Wsn.	240	419182	17.0/2400
.44-40 Win.	205	42798	17.0/2400

LOAD DATA GROUP II

Caliber	Bullet Wt.	Bullet No.	Powder
.44-77	400	446110	22.0/2400
.44-100 May.	430	. . . . . .	36.00/3031
.44-100 Bal.	500	446187	40.0/3031
.45-60 Win.	300	457191	22.0/2400
.45-70 U.S.	405	457124	23.0/2400
.45-70 U.S.	500	457125	22.0/2400
.45-75 Win.	350	457192	22.0/2400
.45-100 Bal.	550	457132	40.0/3031
.50-70 U.S.	450	515141	20.0/R7
.50-95 Win.	300	515139	23.0/2400
.22-13-45	45	228151	3.4/Unique
.25-20 Win.	85	257238	7.5/2400
.25-20 ST.	90	257312	5.0/Unique
.32-20 Win.	115	3118	10.0/4227
.32 Ideal	150	32360	5.5/Unique
.32-30 Rem.	125	31356	8.0/Unique
.32-40 Rem.	150	3113	13.0/2400
.32-40 Bal./Win.	165	319247	13.0/2400
.32-40 Bul.	155	311157	13.0/2400
.38-40 Rem.	250	375165	9.2/Unique
.38-40 Win.	170	40143	17.0/2400
.38-45 Bal.	190	375164	9.2/Unique
.38-50 Rem.	250	375165	16.0/R7
.38-55	252	375248	17.0/R7
.38-56 Win.	252	375248	30.0/3031
.40-45 Rem.	260	403169	18.0/2400
.40-50 Sharps	260	403169	18.0/2400
.40-70 Sharps	330	403149	22.0/2400
.40-75 Bul.	260	412174	21.0/2400
.40-90 Sharps	370	403171	36.0/3031
.40-90 Bul.	300	415175	35.0/3031
.42-Russ.	385	427103	22.0/2400
.43 Mauser	400	446110	22.0/2400
.44-40	205	42798	20.0/2400
.44-60 Sharps	400	446110	22.0/2400
.44-77	400	446110	23.0/2400
44-90 Sharps	500	446187	45.0/3031
44-105 Sharps	500	446187	45.0/3031
.45-70 U.S.	405	457124	23.0/2400
.45-2 1/10 Sharps	405	457124	23.0/2400
.45-2 4/10 Sharps	550	457132	24.0/2400
.45-85 Bul.	300	457191	23.0/2400

LOAD DATA GROUP III

Caliber	Bullet Wt.	Bullet No.	Powder
.45-2 6/10 Sharps	500	457125	45.0/3031
.45-2⅞ Sharps	500	457125	46.0/3031
.45-120-3¼ Sharps	550	457132	48.0/3031
.45-90 Win.	300	457191	24.0/2400
.50-70 U.S.	450	515141	23.0/R7
.50-90-2½ Sharps	450	515141	50.0/3031
.50-115 Bul.	300	515139	45.0/3031
.25-20 Win.	85	257312	9.0/2400
.25-35 Win.	110	257325	13.0/2400
.25-36 Marlin	110	257325	20.5/3031
.30-30 Win.	170	311291	28.0/4895
.30-40 U.S.	170	311413	14.5/4227
.303 Br.	170	311413	14.5/4227
.32-20	115	311316	12.5/4227
.32-40 Win.	174	319295	14.0/2400
.32 Win. Spd.	180	321297	16.0/2400
.33 Win.	200	338320	32.0/3031
.35 Win.	250	358318	34.0/3031
.38-40 Win.	170	40143	19.0/2400
.38-55 Win.	265	375296	22.0/3031
.38-56 Win.	265	375296	32.0/3031
.38-70 Win.	265	375296	23.0/2400
.38-72 Win.	275	375167	34.0/3031
.38-90 Win.	217	37582	38.0/3031
.40-50 Sharps	260	403169	20.0/2400
.40-60 Win.	210	403168	18.0/R7
.40-65 Win.	260	403169	24.0/2400
.40-70 Sharps	330	403149	35.0/3031
.40-72 Win.	330	406150	37.0/3031
.40-82 Win.	260	403169	36/3031
.40-70 Bal.	330	403149	23.0/3031
.40-90 Bal.	320	403171	37.0/3031
.40-110 Win.	260	403169	41.0/3031
.405 Win.	290	412263	40.0/3031
.44-40 Win.	205	42798	22.0/2400
.45-60 Win.	300	457191	22.0/2400
.45-70 Win.	350	457192	45.0/3031
.45-90 Win.	350	457192	48.0/3031
.45-2 4/10 Sharps	405	457124	24.0/2400
.45-2 4/10 Sharps	550	457132	24.0/2400
.45-2⅞ Sharps	500	457125	46.0/3031
.50-70 U.S.	300	509134	30.0/R7

LOAD DATA GROUP III (Cont.)

Caliber	Bullet Wt.	Bullet No.	Powder
.50-70 U.S.	450	515141	25.0/R7
.50-95 Win.	300	515139	24.0/2400
.50-110 Win.	300	515139	52.0/3031
.50-110 Win.	450	512138	50.0/3031

Bal: Ballard Br: British Bul: Bullard May: Maynard Rem: Remington
Russ: Russian SS: Single Shot St: Stevens Win: Winchester Wsn: Wesson

CHAPTER 25

Ammunition for Early Breech-loaders

WE GENERALLY think of handloading only in reference to the truly self-contained cartridge. However, there exists a series of transition gun/cartridge combinations that quite a few people like to shoot. Generally these guns fall in the "Civil War Carbine" classification and came into being during the early part of that conflict in various attempts to give horse-soldiers a weapon more easily and rapidly reloaded than the muzzle-loaders of the day.

They all share one common characteristic; some form of cartridge is loaded in the barrel at the breech, but is ignited by a conventional percussion cap installed on the breech block. Such ammunition is classified as "separate-primed" since the cartridge proper does *not* contain any ignition device.

The cartridges fall into two further classes; "combustible-case" and "metal-case." The former is typified by the Sharps linen cartridge, the latter by the early Maynards. They are found usually in large calibers and many variations will be encountered.

The Sharps type consisted of a cylindrical powder container made of linen, of a diameter to enter the chamber freely, with one end closed. The linen envelope was filled with powder, then a bullet inserted in the open end and glued in place.

The Sharps falling-block action was opened and the linen cartridge inserted bullet-first in the chamber. The breech was closed and a sharp lip on the upper edge of the breech block sheared off the protruding rear of the linen envelope, exposing raw powder to a flash hole extending through the breech block from a conventional percussion nipple. A cap was placed on the nipple and when struck by the hammer, its flame flashed through to ignite the powder charge. Obturation was supplied rather haphazardly by a sealing ring in the breech block face. The linen envelope was consumed by the powder flame, leaving the chamber unobstructed for subsequent loading. It worked reasonably well with guns in new condition, but they wore rapidly, and then things got a bit sticky.

Though not as satisfactory as later metal-case cartridges, the Sharps saw a good bit of use during the Civil War. It formed an important step forward.

The Maynard is quite different, consisting of a conventional-looking brass cartridge case made by soldering a wide rim to a tubular body. The center of the head (formed by the rim) is pierced by a small flash hole. Powder and bullet are loaded in the case just as in modern cartridges. The cartridge is chambered and then fired like the Sharps by a percussion cap whose flame ignites the powder through the hole in the case head. The empty case is then extracted manually or automatically, depending on gun model.

Both types of cartridges can be made up easily if you've a yen to shoot those early breech loaders. The Sharps is best made of nitrated paper rather than the original linen. The steps are shown in the accompanying drawing. First take good bond paper and soak sheets of it in a saturated solution of potassium nitrate for an hour or so, then hang them up to drain and dry. Keep away from high heat or open flame.

Cut the paper in strips and roll around a waxed and polished cartridge stick to provide about ⅛" overlap. Glue the overlap and slide off the stick. When dry, press a tissue-paper plug in one end as shown and secure with thinned glue. Place powder charge in tube, then glue bullet into open end, and finish by dipping bullet in melted lubricant.

The tube should be of a diameter that will enter the chamber freely, but not excessively undersize. In length make it so that the tissue-paper end is just flush or slightly below the chamber mouth when the cartridge is pressed into the chamber as far as it will go. The tube can be made longer so as to be sheared off by the breech block like the original linen Sharps, but this is not necessary. The cap flash will punch right through the tissue and ignite the powder adequately. The shearing action scatters loose powder through the action which is some-

322   **Modern Handloading**

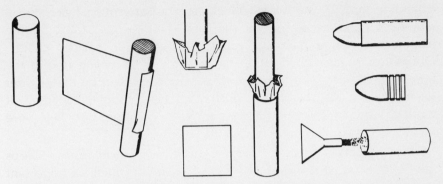

Method used in making paper combustible cartridges for the Sharps and similar breech-loading rifles requiring them. Paper used for body must first be nitrated, while the tissue used to close the rear need not be. Tube mouth may be glued or tied to base of bullet, or if a tight fit, the lubricant will probably hold things together well enough.

times ignited upon firing and the flash can be quite disconcerting, one of the major objections to the original system.

The Maynard type is a bit more difficult to produce, but once a case is made, it will last for many firings. The case may be built up as were the originals, or machined from solid brass.

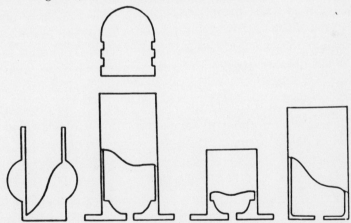

These partially sectioned views show the general construction of brass, aluminum or other metal cases and gas checks for use in shooting separate-ignition breech loaders requiring an obturating case. Left is a Burnside-type gas check which must be dimensioned to fit the gun closely and is used only to permit muzzle-loading of the gun. Third is one of the Maynard type for the same purpose. Second and last are full-length cases of Maynard and Smith type respectively and may be used to load conventional cartridges. Best material is dead-soft brass, with aluminum second; construction may be solid or built-up of separate head and body soldered together.

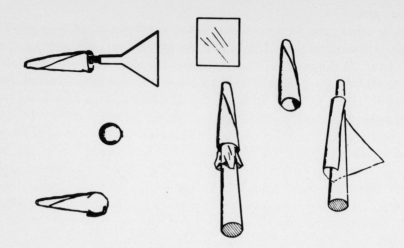

Method of making combustible paper cartridges for use in percussion revolvers. Wood form must produce a tapered paper tube that will just accept the ball in its mouth and which will fit in the revolver chamber. The ball may be held in place by turning over excess paper as shown or by trimming off excess, then gluing ball in mouth. Body paper must be nitrated, and better accuracy and less leading will result if bullet end of complete round is dipped in melted lubricant. Combustible cartridges may be made in this same way for use in rifled muskets with minie bullets, but will not work well in regular patched-ball rifles with round ball.

For the former, select thin-wall brass or copper tubing as large in diameter as will freely enter the chamber, then cut lengths about ¼″-⅜″ longer than chamber depth. Have someone with a lathe turn up a mandrel, then *spin* one end of each tube closed, forming as square an outer edge as possible. Ideally, have him leave about a ⅛″ hole at the center of the closure. Unless you find about .010″ wall thickness, the open end will require thinning before a bullet can be seated.

Select brass sheet of a thickness that will just allow the breech to close on it. Cut out discs of the proper size and drill a 1/32″-3/64″ hole in the center of each. Sweat-solder these discs to the closed end of the tubes, trim to length, and there is your Maynard case. Load with powder and bullet in the usual manner. If powder has a tendency to dribble out the flash hole, put a single layer of tissue or nitrated paper inside before filling with powder.

Solid cases are produced simply by turning dead-soft brass rod or bar stock down to essentially the same form. However, the head (web) may be left much thicker than in the built-up case when this is done. Note no specific dimensions are given because of the many variations

Modern Handloading

found in original guns. Take measurements from the gun and work from them.

Many other guns used cases similar to the Maynard and they may be made in the same way. The Smith used a rubber case without rim, and it can be duplicated in brass, copper, or aluminum. The Williamson resembled the Maynard but had a percussion nipple screwed into its head. Simply turn a solid brass case and substitute a nipple for the flash hole.

One deviate from the above is the Burside with its tapered brass or copper case inserted from the front of the chamber. This, too, can be turned from brass and re-used, but far simpler is to purchase new plastic Burnside cases from Dixie Gun Works and use them: Incidentally, DWG offers several varieties of ready-made cases for other early breech-loaders and can thus save you the trouble of making your own.

Nearly all separate-primed breech loaders can be provided with ammunition by the above methods. In loading, follow basic black powder rules: proper diameter and weight soft lead bullet; powder to fill the case. *Don't,* however, *ever* attempt to use smokeless powder. It is pure poison to those weak old actions. Remember, black-powder only.

CHAPTER 26

Handloading Mathematics

THE HANDLOADER really need not get deeply or often involved in math. In fact, the average hull-stuffer probably will spend his life at the game and never have a greater problem than figuring out how many of a given charge he'll be able to get from a 11-oz. or 1-lb. powder cannister.

Far more valuable than the ability to calculate various values is to have at hand a few tables for simple things like bullet energy. Sure, you can calculate that value when velocity and bullet weight are known, but unless you are a practicing slipstick manipulator, the answer is gotten quicker and with less probability of error by simply flipping to the appropriate table in the back of this book. And, if you're a slipstick man, you already know more about how to get the right numbers than I do.

There are several useful formulas we'll list here in a bit; but for a moment let's talk about formulas that *don't* exist. I often get queries that run something like this: "My .30/06 load with 4350 and 180-grain bullet produces 2700 fps. Tell me how to figure the right powder charge to produce 2800 (2900) fps, also with other powders."

Many people seem to think there are formulas and methods whereby they can whip up a pat answer in a case like that. There *aren't*. Under laboratory conditions a ballistic engineer can calculate a rough approximate answer, but even then his final answer is produced by test shooting. For a handloader in the field it's impossible.

Sometimes the query is the opposite: "Tell me how to figure what the pressure will be if I cut powder charges by 10 per cent." No way!

Another query: "I obtained a large quantity of 4064 powder, but all my loads have been developed with 4350. Tell me how to figure how much 4064 to duplicate all my 4350 loads." There's no formula for this either.

I was once approached by a well-educated engineer just getting into ammunition work. He wanted assistance in calculating *accurate* load data, and was amazed when told it couldn't be done; that top accuracy could only be developed empirically. Small arms ballistics simply is not an *exact* science as found in many fields. Where electrical and hydraulic engineers may calculate cold facts and performance precisely, ammunition makers may not. They calculate roughly for pressure and velocity, then hope, juggle variables, and pray, until the ammunition either does what is desired or proves that it can't.

For these and similar reasons, handloading use of mathematics is rather limited to measurement of performance after the fact, instead of in *producing* performance. Too many variables are involved, and you must have known constant factors to calculate.

All the same, there are a few basic procedures you may find useful.

Bullet Energy

1. Square the velocity of the bullet in fps.
2. Divide 1 above by 7,000 (to convert to pounds).
3. Divide result in 2 above by 64.32 (twice the acceleration of gravity).
4. Multiply result in 3 above by weight of bullet in grains to produce the striking energy in foot-pounds.

Velocity Correction For Temperature

1. Determine velocity at existing temperature.
2. Determine temperature deviation from 70° F.
3. Multiply 2 above by 1.7.
4. Add or subtract result in 3 above to or from measured velocity to correct to 70° F. at which all published data is correct.

To correct published velocity to existing temperature, multiply temperature difference from 70° by 1.7 and add or subtract result to or from published value.

To Determine Case Capacity

1. Determine weight in grains of water case will hold.

2. Divide 1 above by 252.78 to obtain capacity in cubic inches.
3. Divide 1 above by 15.43 to obtain capacity in cubic centimeters.

Correct Measured Instrumental Velocity To Muzzle Velocity

1. Measure distance in feet from muzzle to mid-point between the two chronograph screens.
2. Multiply measurement above by 0.64.
3. Add result in 2 above to instrumental velocity to obtain muzzle velocity, or

If working from published data listing instrumental velocity at X feet, multiply X by 0.64 and add result to velocity shown.

Both above methods ignore the effect of SD, BC, and FF, whose effect is insignificant with small arms projectiles at the distances normally involved.

Compute Recoil Velocity

1. Add 1.75x powder charge weight to bullet weight.
2. Multiply result in 1 above by muzzle velocity.
3. Divide result in 2 above by gun weight in lbs.
4. Divide result in 3 above by 7000 to obtain recoil velocity of gun in fp.

Compute Recoil Energy

1. Square recoil velocity.
2. Multiply result in 1 above by gun weight in lbs.
3. Divide result in 2 above by 64.4 to obtain recoil energy in fp.

To Obtain Circumference of a Circle

Multiply diameter by 3.1416.

To Obtain Area of a Circle

1. Square diameter.
2. Multiply 1 above by .7854.

To Obtain Sectional Density of a Bullet

1. Convert bullet weight to lbs.
2. Divide 1 above by bullet cross sectional area in inches.

Beyond those, just thumb over to our tables and you should be able to find what you need.

Modern Handloading

CHAPTER 27

Mass-Production Handloading

THE AVERAGE single-station loading press, when combined with a good powder measure and a few other accessories, will allow one to process about 200 metallic cartridges per hour. In short, if you have a well laid-out bench and organize your work efficiently, you'll have no trouble progressing from fired cases to loaded ammunition at that rate. No one person in a thousand needs to work any faster, though speeding up the processes does have some appeal. As far as individual use is concerned, probably only the devoted pistol shooter really consumes enough ammunition regularly to justify greater production capacity. Shotgunners fall in the same category, but that field is covered elsewhere in this volume.

The first step upward from a single-station tool is the so-called turret tool. This outfit generally has a rotating block or turret at the top which is drilled and threaded to hold several dies, normally at least four, sometimes as many as twelve. Dies are placed in the turret in sequence and properly adjusted, and, in addition, a powder measure can be set up at the station immediately prior to bullet seating. With the tool thus set up, a fired empty case is placed in the shellholder, run through the resizing die; the turret is turned one notch for the neck expanding and decapping die, and case is reprimed; then the turret is rotated to bring the powder measure into position and the charge is thrown; then further turret rotation brings the bullet seating die into position. When all these operations have been completed, that fired case emerges as a fully loaded round without having once been removed from the shell-holder or otherwise handled unnecessarily.

By virtue of having eliminated all the repeated handlings of the cartridge case and by concentrating all operations into a small area, the average turret tool will very nearly double one's production rate. Of course, in order to take advantage of this, it is necessary to have a very orderly and efficient loading bench layout where all components are immediately accessible and no time need be wasted away from the stool. Using either a Redding Hunter turret press or the economically priced Lyman Spar-T, I have been able to sustain production rates of 300 to 400 rounds per hour of pistol calibers such as .38 Special and .45 Automatic.

However, just casual operation of the tool won't produce these results. Several things must be done to avoid wasting time. Probably the most important is the substitution of a tungsten-carbide resizing die to eliminate the need for lubricating cases prior to loading. This is practical, I might add, only with straight-side, pistol-caliber cases. As described elsewhere, this T-C die permits cases to be resized dry without ruining either them or the die. However, the cases must be cleaned and this requires tumbling or washing in advance of the loading session. In addition, you must provide yourself with several shallow wood or cardboard boxes for components and a clean, neat, orderly working area layout. The accompanying example shows a layout that has worked very well for me and should, with minor variations, perform as well for anyone else who is right-handed. Southpaws may well have to reverse the setup.

The sequence begins as follows: Pick up a case from the fired case box and insert it into the shell-holder, guiding it with the thumb and forefinger into the die mouth as the working hand starts the press handle downward; as the case enters the die, a primer is picked up from the primer tray and placed in the collar of the priming arm; after the ejected primer falls free from the fired case and before the tool handle is pressed downward to pull the case out of the die, the priming arm is swung forward into position under the case; the working hand pulls the case down over the primer while the left hand reaches up to the turret head; the tool handle is raised slightly to allow the primer arm to spring free, and the turret is rotated into the next position; the tool handle is worked through a full cycle to expand the case neck and bell the mouth; as the case is drawn from the expanding die, the left hand rotates the turret into the third position wherein sits the powder measure; the tool handle is moved to bring the case mouth up into gentle contact with the bottom of the drop tube and the left hand actuates the powder measure; the tool handle is moved downward to withdraw the case, while the left hand rotates the turret into the fourth position; the left hand picks up a bullet from the box and starts it into the case mouth and guides the case/bullet assembly into the seating die as the working hand thrusts the tool lever   upward; the working hand lifts the tool handle upward to withdraw the complete cartridge from the die, and if a separate crimping die is being used, the left hand rotates the turret into its fifth position; the working hand thrusts the tool lever downward to carry the cartridge into the crimp die and then upward to withdraw it; at which point the right hand is removed from the tool handle and plucks the completed cartridge from the shell-holder while the left hand picks up a fresh fired case and places it in the shell-holder, then rotates the turret around to the starting or number one position to prepare for the next sequence.

With practice and proper convenient layout, the foregoing operations take only a few seconds. Considering that with a half-hour of preparation and two hours actual loading work you can assemble 500 rounds of pistol ammunition with little effort at all, it would seem that the conventional turret-type tool is more than adequate for even the most over-zealous handgun shooter. However, such is not always the case. There are much more sophisticated tools on the market, which, even though being fully manually operated, will still turn out 700 to 1000 rounds per hour with two operators. The same tools will easily produce from 500 to 600 rounds per hour with a single operator. The appeal of such tools is considerable to those who want to load their ammunition as quickly as possible and is often sufficient to justify the $300 to $400 a tool of this type will cost.

The Star Progressive Reloader is the most typical of this type, though over the years a more or less exact copy was marketed under the name Phelps, and there have been several earlier designs rather dissimilar but capable of the same type quantity production. All of these tools are generally termed "progressive" reloaders and achieve their high rate of production by performing one operation on each of several cases simultaneously; that is, one case is decapped, another is reprimed and neck-expanded, another is charged with powder, and a bullet is seated in another all simultaneously with a single stroke of the tool handle. Thus, once such a tool has had all stations filled with cases progressed to that point, it kicks out a completely loaded cartridge with each stroke of the handle. Once you have developed the necessary arm and shoulder muscles, the production rate is determined only by the speed at which you can operate the handle and by your facility in keeping the various powder, case, and primer reservoirs refilled. As already mentioned, a two-man operating team can easily produce nearly 1000 rounds per hour with a Star tool all day long.

Of all the tools of this type mentioned, only the Star, manufactured and marketed by Star Machine Works, is currently in production. Older tools capable of essentially the same performance will be found bearing the label PESCO, Buchanan, Dirks, and others.

Perhaps a brief explanation of the functioning of the Star tool is in order. As can be seen by the photograph, the dies are arranged in a circular plate and are thrust down over the cases which are held stationary by another circular plate whose edges are notched to form shell-holders. After each downward stroke of the tool die head, it is then raised, and the "shell plate" is rotated one die space counterclockwise and a fresh fired case is inserted in the first station. Once all stations are filled, each station performs its own function and after each indexing movement of the shell plate the completely loaded

Star Progressive tool with .38 Spcl. (installed) and .45 ACP die sets.

Modern Handloading

cartridge either falls through a hole in the base or, in the event of a rimless type, is plucked out with the fingers.

The first station contains a resizing die, usually of tungsten-carbide and only resizes the case. Station two contains a decapping stem and pin which removes the fired primer to fall through a hole in the base. Station three combines mouth and neck expansion with priming, in that the expander enters the case to bear upon the inside of the case head at the same time a cam-operated primer-seating punch forces a fresh primer into the primer pocket. That primer is carried into position by a horizontal slide actuated by a cam inside the base of the tool. A brass tube primer magazine supplies the slide. Station four places the case under a fixed-charge, pistol-type powder measure attached to the die head. A triangular cam actuates the powder slide and the next station is for bullet seating. In addition, the Star tool is equipped with sufficient stations to roll-crimp or taper-crimp as a separate operation after bullet seating.

Manually-operated progressive tools have one disadvantage in that they do not permit visual checking of the powder charge to insure that it did in fact get into the case. However, the Star especially is fitted with a safety interlock device which prevents a second throw of the handle until the shellplate has been indexed properly, and this eliminates the probability of a double powder charge. Of course, the interlock can be removed or deactivated, but that is simply asking for trouble.

By way of speeding up Star tool operation even more, various accessories are offered. Most prominent among them is the Hulme case feeder which is available separately or as a factory option. From time to time, other makers have offered automatic indexing devices for the shellplate, automatic bullet feeders, and devices to feed extra primer magazines in sequence as needed. In addition, some custom loaders have designed and installed electric, air, or hydraulic power systems. These power packages generally do not greatly speed up tool operations, but they eliminate the fatigue factor and thus allow more uniform sustained operation.

Because of the nature of their operation, especially with large powder charges, it is essential that progressive tools be operated smoothly and carefully to prevent powder spillage. When the shellplate is indexed, anything less than a very smooth movement will cause powder to slosh out of the case and eventually work its way into the innards of the tool and gum it up. If copious amounts of oil or other fluid lubricants are present, this condition is made even worse. For this reason, I have always preferred molybdenum disulphide or a similar dry lubricant in all of those areas which might be reached by spilled powder. So long as they are kept scrupulously clean, Star and similar tools work easily and rapidly. However, if spilled powder or other debris is allowed to accumulate, malfunctions will be encountered frequently. For this reason, it is most desirable that at the end of each working day or after each period of more than two or three hours' use that the shell plate be removed and all working parts in the base be thoroughly cleaned and re-lubricated with a dry compound. When a dry lubricant is used, cleaning involves nothing more than opening up the parts and using compressed air or a syringe to blow out powder granules and brass shavings and primer residue.

In recent years truly large-quantity producers of handload ammunition have more or less dropped the manually-operated tools in favor of those electrically-driven production-type loading *machines*. As this is written, only one such rig is in production, the "Ammo-load," (formerly known as Auto-load) which appears to be a development of the old Dirks power-driven loader of a couple decades back. The Ammo-load is a straight-line progressive machine which accepts clean fired cases at one end and spits out loaded ammunition at the other. In between, all essential operations are conducted in much the usual fashion by a vertically-moving die head containing the necessary tools. In addition, it is fitted with sensing devices which automatically stop the machine if the powder charge varies beyond specifications or if a case or cartridge becomes damaged. Sensing and automatic shut-off devices are absolutely essential with a tool of this sort which can produce upward to 3000 rounds of ammunition per hour. Anyone obtaining such equipment should be warned to *never* disconnect or otherwise make inoperative any of the sensing and shut-off systems.

Prices of such power-driven automatic loading machines are far beyond the reach of the average handloader. They do, in fact, run around $5,000 and upwards for the basic tool, and it is easy to spend several thousand dollars more in adapting them to a particular use.

Where a club or a group of individuals wishes to band together in order to save time and money assembling large quantities of hand-loaded ammunition, the Star tool mentioned earlier is probably the best choice. It is not only relatively inexpensive for the result obtained, but requires only two people for even maximum-speed operation. However, if your group doesn't feel like shelling out some $400 for a fully-equipped rig, lots of ammunition can be produced quickly with conventional single-station tools on a production-line basis. Ammuni-

Modern Handloading

tion can be produced just as rapidly as with a two-man crew on the Star, but more people are required.

This type of operation is based upon a separate press and die set up for resizing, another next to it for mouth expanding and decapping, a priming tool at the third station, a powder measure at the fourth, and a fifth loading press and die for seating bullets. The first three stations can all be operated at about the same speed, but the powder measure man can work much faster, which leaves him some free time to be employed in starting bullets in cases in loading blocks, thus helping to pick up the time lag for the bullet seater who has the slowest job of all. With a plentiful supply of already-cleaned and tumbled cases and components, five people can crank out several hundred rounds of ammunition per hour with very little effort. And, if they can't all get together at the same time, it is a simple enough matter to isolate the first three operations so that, say, on Monday night a couple of fellows come in and do the sizing and decapping, and on Tuesday someone comes in and does the priming. However, the throwing of powder charges and seating of bullets must not be separated. They must be accomplished at the same time. To let cases stand which have already been charged with powder is asking for trouble in several ways.

Where large quantities of ammunition are being loaded rapidly, it is no longer practical to individually wipe each cartridge free of bullet lubricant and/or other debris it might collect in passage through the various operations. By far the simplest method of producing clean, shiny, dry handloads is to tumble them after loading. The same tumbling equipment, methods, and material described earlier in this volume does very well for removing bullet and sizing lubricant, fingerprints, etc., from finished ammunition. However, cost can be reduced by using common sawdust for the tumbling medium. Since the cases were thoroughly cleaned before loading, they have no tight-clinging corrosion or other material to be removed. Only fingerprints and lubricant are present, and they come off easily. Common lumber-yard sawdust will do the job very neatly and quickly if lightly moistened with a non-flammable solvent. Simply charge the tumbler with sawdust and pour in a few ounces of solvent, run the tumbler a few revolutions to spread it throughout the sawdust, then dump in the cartridges. Don't overdo the solvent —use only enough to just barely moisten the sawdust—an excess might well penetrate around the primer or bullet to weaken either primer or powder charge. Tumble loaded ammunition for only a few minutes or until it can be seen that the oil, grease, and fingerprints have been removed. Excessive tumbling will result in the bullets looking somewhat the worse for wear if they are of soft lead. There is no need for this inasmuch as a very short period of tumbling will remove all undesirable material and leave cases and bullets clean and dry.

In reality, individuals and clubs are not the major users of large quantities of handloads. Commercial ranges and law enforcement organizations use them by the tens of millions annually. While law enforcement agencies need not be concerned about the various laws applying to handloading that exist, the custom loader does. Many individuals who occasionally load up a few hundred rounds for sale to friends and acquaintances are not even aware that they are violating federal law. Anyone who reloads fired cartridge cases for sale (that is, not for his own use) is considered to be engaged in "Manufacturing Ammunition." As such, he comes under the provisions of Federal Law which requires that a "Ammunition Manufacturers License" must be purchased and that federal excise tax must be collected on the product and must be paid to the Federal Treasury in accordance with the pertinent rules and regulations. Naturally, this involves the regular quarterly preparation of the proper forms and of keeping exceedingly accurate records to show clearly how many cartridges have been loaded, the price, and the amount of the tax. While federal agents have not been particularly obnoxious in enforcing these rules in the past, insofar as the casual loader is concerned, the present climate of firearms and ammunition legislation would seem to indicate that anyone violating this law might well find himself in serious trouble. Agents do check custom loaders regularly in this respect.

The simplest way to avoid this problem is to insure that in loading for friends and acquaintances, you use only fired cartridge cases supplied by them and that each and every one of those self-same cases are returned in loaded condition. The law differentiates between ammunition loaded in cases which are your property and which are the property of the individual for whom the ammunition is being loaded. The requirements for both the Ammunition Manufacturers License and the payment of federal excise tax do not apply when you simply provide the *service* of loading cases belonging to the customer. If, however, the cases are your property, procured from no matter where or how, then you are supplying *all* the components and are actually *manufacturing* ammunition and all the rules apply.

This is mentioned and emphasized because it seems such a simple and logical step for the competent (sometimes not so competent) handloader to gradually develop his hobby into the business of supplying custom-loaded ammunition to friends first, then acquaintances,

Modern Handloading

and then on a regular retail or wholesale basis without giving thought to requirements of the law. It cannot be emphasized too strongly that to do so without an Ammunition Manufacturers License and without complying with the federal excise tax regulation can get one into serious trouble.

And, then, there is the not so small matter of product liability. If an individual buys a box of factory-loaded ammunition and one of those cartridges damages his gun or injures him in any way, he has the right to sue the company not only for actual damages (such as medical bills, lost wages, gun repair, etc.), but for punitive damages as well. If you sell a box of handloads to an individual or if you simply reload his fired cases, and a similar incident occurs, he may take exactly the same action. In short, if you sell a fellow a handload and it hurts him or his gun in any way, he has the right to drive you into court and may well obtain a judgment that will make you a pauper the rest of your life. In light of today's trend towards ever-increasing manufacturers' liability, one must be very cautious about selling any handloaded ammunition. The only sure protection against damage suits is adequate product liability insurance.

It is extremely difficult to obtain such insurance on products (ammunition) assembled in a home workshop by an individual in small quantities. Generally speaking, no company will write such coverage without first inspecting the tools and premises and making certain that the maximum protection from error in the processes exists. Frankly, I don't know of a single individual's loading shop and setup which could pass such an examination. Consequently, if you start selling handloads, you are very likely letting yourself in for a big chunk of trouble. The guy you thought was such a good friend may turn out to be something less if he thinks one of your handloads cost him a finger, a good gun, or put a few scars on his face.

Most of today's successful custom loaders managed to squeak by for as much as several years without being called into court, until their business had expanded to the point that they could acquire tools and equipment which would satisfy insurance companies. Virtually all of the larger custom loaders exist primarily upon law enforcement business. The amount of revolver ammunition, mostly .38 Special wadcutter loads, which is consumed in police training annually, almost beggars description. Some metropolitan departments and states consume many millions of rounds per year. While some of them set up their own custom-loading operations to meet this need, it is generally more practical in the long run for them to ship their fired cases to a custom loader periodically and have them returned in the form of loaded ammunition. I know of several custom loaders who supply from five to ten million rounds of .38 Special handloads to police departments every year. The individual who is really serious about developing his hobby into a business should certainly explore this particular field. It is far more lucrative and safer in the long run than attempting to peddle small quantities of indifferent loads to individuals or to gunshops.

In the end, the major disadvantage of fast reloading is the tendency toward reduced quality and uniformity; thus, performance. Any time a single operation is speeded up, it is most important to insure that none of the vital inspection and control operations are downgraded in the process. Speed isn't everything.

CHAPTER 28

Home-Brewed Tools and Equipment

IT SURE is nice to have all that shiny new equipment the manufacturers keep developing for us. It is all convenient, and, oh, so very useful, but it *does* cost money. Maybe the individual items don't bulk so high, but after a year or three of piddling around, you may suddenly discover you've sunk several hundred dollars into items you really didn't absolutely need. One-shot or occasional jobs don't really require that. Back in the old days — please note I didn't say *good* old days — lots of handloaders made many of their tools and accessories. Or, if it was less trouble, they modified existing items to fit new handloading jobs that came up.

One of my earliest handloading memories is of a grizzled old character sizing lubricated cast bullets by pushing them through a hole in a piece of angle-iron held in his bench vise. He hadn't even drilled the hole himself. It was a rivet or bolt hole in the piece of scrap when it was picked up. He had polished the hole to size, and mirror-smooth with emery paper wrapped around a dowel. Slow and laborious, perhaps, but cheaper than buying a sizing die. I've made many a bastard-size die since by drilling, reaming, and polishing a hole through the center of a piece of ⅞x14 threaded steel rod, then screwing it into a loading press and using a rod in the shell holder to push already-lubed bullets through to size. For occasional use, such dies don't even need to be hardened. Not only does this work for bullet sizing, it can be used for neck-sizing cases. In fact, once a number of different sizes have been accumulated, you'll find many will serve both purposes. For example, a die made to size .338 or .340 bullets is just about right for resizing .30 caliber case necks. A die made for sizing .45/70 bullets is just about right for sizing some .44 caliber necks; and so on, and on. A polished hole will do lots of things.

Such a die is even more useful if you do a bit of case forming. Just cut a 45° or less (including angle) chamfer around the entry end of the hole and it may be used to push back case shoulders, as in making .308's from .30/06. The chamfer is easily cut with a common hardware-store countersink in a hand brace, then polished smooth. Used with care, such a die will set a shoulder back or reduce neck diameter just as well as a $35.00 special-order forming die.

This same type of sizing die can be used to reduce jacketed bullets slightly in diameter. Maybe you need .318" bullets for an old 8 x 57 Mauser and only .323" size is available. That .005" reduction can be had by pushing bullets carefully through a .319"-.317" hole. Likewise, .308" bullets can be reduced to .304" for pre-1911 Swiss service rifles, and so forth. I once squeezed nearly 100,000 .315" 7.65mm Mauser bullets down to .3085" just that way, but must admit to using a hydraulic press for power, rather than doing it by hand. Lazy, I guess.

Much has been said about jacketed bullets being inaccurate after being reduced as above. Theoretically they should be, and by bench-rest standards, they may be. Yet, I know quite a few 7mm shooters who use Italian 7.35mm military bullets swaged down by George Spence from .298" to .284", and they think they shoot fine. George says he's never had an accuracy complaint yet, even on those same bullets squeezed to 6.5mm (.264").

Along the same lines, keep a few spare expander plugs handy and you'll always be able to polish one down to get a size you need but don't have. Plugs are cheap enough, but it may take a couple weeks to get what you need if the local shop doesn't have it in stock.

Expander plugs are file-hard, but by screwing them on a rod and chucking it in your electric drill, abrasive cloth or stones can be used to reduce them without too much trouble. Always remember to polish as smooth as possible, and make sure it's a *steel*, not carbide, plug you're working on.

If you've a regular bench lathe handy, there is no end to the tools and dies you can make. Unfortunately, few of us can afford such niceties unless already in the machine-shop business. However, I've recently acquired a sort of miniature metal-working lathe that costs little enough for almost anyone, yet can be used to make all manner of items. This is the briefcase-size UNIMAT distributed by American Edelstahl.

On the UNIMAT you can make decapping stems, lock rings, any of the dies already mentioned, expander plugs and rods, bullet-seating punches, and all manner of other goodies. One thing I find it useful for is making up full-length or neck-sizing dies in odd calibers for which I don't want to spend the $35 or so a die maker might charge. Such a die is made in two pieces. The neck and shoulder are first cut and polished in one piece as the drawing shows. The neck portion is drilled and reamed, then polished to size. The shoulder cone is then cut to the proper angle with the tool bit, and polished smooth.

The die body is drilled undersize, then brought to proper taper and diameter with standard taper reamers which can be had quite reasonably in many sizes from the mail-order tool houses. The two parts are then carefully aligned and silver-soldered or tack-welded together, then finish-polished inside to remove any scale or solder overflow. Unhardened, such dies won't last forever, but they will suffice for a few hundred rounds, and that's usually more than enough to satisfy your wildcatting whim or urge to try some long-obsolete caliber that has cropped up.

If a bit more energetic, you can make standard one-piece dies. It just involves first making on the UNIMAT a half-reamer with which to cut the die cavity. Half-reamers are relatively easy to make, but we won't go into them here. Check Howe's *The Modern Gunsmith* if you want to know more.

In resizing there is another seldom-used gimmick that will often pull a job out of the fire. CERROSAFE or other low-melting alloys can be used to *cast* resizing dies. It requires supporting an oiled unfired or resized empty case in a small metal container (cut-down juice can) or cardboard tube, then pouring molten metal around it, clear up to the head. The case is then removed and the "die" cleaned up with files, lightly chamfered at the mouth, and the neck end cleared (of can metal) so a knockout rod can be inserted.

Though relatively soft, this die may be used in an arbor press or vise to resize 100 or so cases before becoming excessively worn, and then it can be melted down and recast. One important thing: Case mouths must be turned over slightly as shown before entering the die. If this isn't done, the sharp mouth edge will dig into or scrape the die surface and ruin it. Accomplish this by pressing the case mouth into a conical hole or countersink in a block of metal or hardwood. Naturally, cases must be properly lubricated for resizing in this type of die.

Another project useful in these times when deliveries on some bullet moulds take so long is making them — well, almost — yourself. It is almost always possible to find a bullet mould just a few thousandths of an inch *under* any special diameter you might need. That being the case, it isn't too difficult to clamp the blocks together, set them up in the UNIMAT 4-jaw chuck, then grind a tool bit to bore out the mould to the size needed. Open up the grooves that form the driving bands first, then clean up the minor diameter so grease grooves are about the same depth as before alteration. Unless diameter is increased a great deal, the point cavity won't need opening up unless you are real fussy.

A set of blank mold blocks can be drilled and bored out in much the same way to produce virtually any bullet you might contrive.

Of course, having a UNIMAT or similar outfit for handloading jobs will enable you to do quite a bit of small gun work if you like, but that is another field entirely.

You'll probably eventually wish for an arbor press or a hydraulic press of some sort if you get into extensive case forming and bullet swaging. Both are expensive, but the hydraulic rig will serve for both, and you can make one easily as the accompanying drawing illustrates. All it really amounts to is a thick steel base plate, three or four uprights, a top plate containing a die or punch hole, and a hydraulic jack. Dimension the whole thing to fit around the jack to be used. Welding is okay, for assembly, but you'll find it simpler if standard hardware-store threaded rod and nuts are used. That way, once the two plates are cut and drilled, you can do the rest of the work with a wrench and hacksaw. Securing the jack to the base plate with a couple machine screws will keep it aligned under the die hole and save a bit of trouble during use.

Ever have a sudden need to pull bullets in some caliber your regular puller won't handle? Pick up a pair of cheap discount-store pliers and grind or file grooves in the jaws to fit the bullet. Leave the groove surface a little rough so they can get a good grip on the bullet. Put a loaded round in the shell holder of your press and run it up until the mouth of the case is just below the top edge of the die hole. Clamp the altered pliers on the bullet — tight — and jerk the press handle down. The case will pull off the bullet, as the pliers are supported by the top of the press. Handle extensions on the pliers will help.

Have you had trouble in getting a bullet lubricator-sizer die the *exact* diameter you think is necessary? There are two solutions. The first is to lap an *undersize* die large. Turn a brass rod that will just barely pass through the die, oil it, and roll it in fine flour abrasive.

A home-made hydraulic press can be as big and as fancy as you want. Joe McPhillips made this one for case forming, bullet swaging, drawing bullet jackets, etc.

A simple hydraulic press can be made of rod, plate, and an old jack.

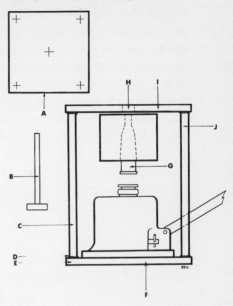

Work it through the die, then push and pull, all the while rotating the rod. Replenish abrasive as it wears away, and check die diameter frequently. It may take some time, but with care you'll eventually get the size you want. The other way is to have the inside of an over-size die hard-chromed, then lap it to size as just described. The lapping is easier if done in a drill press with the power *off*. Chuck the rod, then clamp the die in alignment with it. Use the press handle to move the lap through the hole, rotating the chuck about ¼ trun by hand at each stroke. This method avoids the "funneling" of the hole often produced by pure hand work. Incidentally, the same procedure can be used to open up the neck of a resizing die.

You might, in a pinch, need a means of seating primers in some odd-size case, or, maybe it's something standard and there just isn't time to get a store-bought tool for the job. Get a steel strap hinge and, if it's

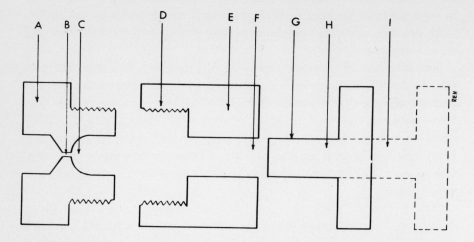

(A) nose punch; (B) bleed hole; (C) nose cavity; (D) assembly thread; (E) die body; (F) body cavity (G) punch lap to fit body; (H) base punch; (I) vary length to suit bullet.

a rimmed case, drill a single hole in one arm as shown, that will accept the body but will stop the rim. Drill and tap a 3/16" hole centered over the first in the other arm. Seat a short screw in the second hole and file it to form a priming punch. Drop the case in place, thumb a primer into the pocket, then close the hinge arms to seat the primer.

If a rimless case, drill a second smaller hole overlapping the first. Then file the edges so when the case is moved from the larger to the smaller, they will enter the extractor groove and support the case. Center the primer seating punch over the smaller hole and you're in business.

Maybe you've got some hard-to-find cases whose primer pockets are a bit oversize, but are otherwise in good condition. Make an anvil rod as large as will enter the case mouth and support it solid vertically. Then, make a hollow punch as shown that will just encircle the primer pocket. Drop case over the rod, center the punch over the pocket, and rap it with a hammer. This will move brass inward to reduce pocket diameter. Most uniform results will be obtained if brass is moved in a hair too far, then the pocket is sized with crimp-removing swage of the type made by RCBS.

Most case trimmers work well enough when only a small amount of brass is to be removed, but they aren't of much use if ¼"-½" must be chopped off. Unless you've an ungodly number to shorten, a modified tubing cutter works well. Simply grind a bevel as shown on the guide rollers so that when the case shoulder is pressed against the rollers, the cutter wheel falls in the right spot. If the neck is too long to allow this, bend a guide from sheet metal and screw it to the frame of the cutter. With practice, cases can be shortened by this rig much faster than you would imagine.

Several types of cartridge case length gauges are available, but you usually don't have the exact length at hand when needed. Simply cut and file a wide "U" from sheet metal or plastic with the proper distance between the arms of the U. Alternatively, keep a long ¼" or ⅜" machine bolt handy with a large-diameter washer clamped against the head by one nut, and another out front between two more. Shift the front washer to provide the proper distance between the two to gauge your cases.

Having trouble keeping spare decapping pins handy? Pick up a couple feet of steel music wire the same diameter as the pins in your dies. Cut off a length anytime you need a new pin, polish the ends, and the problem is solved. A length of wire doesn't stray as easily as those short pins.

Cases do occasionally get stuck in resizing dies. Pullers to get them out are avilable, but you can do as well without them. Simply turn the head of a ¼ x 28 or similar machine screw to fit into the press ram head or shell holder head. Then, when a case gets stuck, remove the die from the press and drill out the primer pocket. Put the die back in the press, with the case puller in the ram. Turn the puller into the hole in the case, exerting pressure through the ram so the threads dig in. The offending case can then be pulled out.

Home-built bullet swages are another solution to some problems that crop up. Our drawings show their construction quite clearly. All you need is a cylindrical body of the right diameter, a flat punch or plug for the bullet base, and a profiled punch to form the bullet nose. Such dies may be used to bump-up lead or jacketed bullets to larger diameter; to change bullet shape; or to form new bullets from lead slugs or cores and jackets. You can get by without hardening the die body, but if the punches aren't hardened, they will upset under pressure and jam tight. Case hardening with KASENIT compound and a hand torch is all that is needed, so this presents no problem.

As you make a few tools and accessories, more and more ideas for such work will occur to you. Some will merely save money, others will enable you to do jobs for which tools aren't commercially available. Either way, it's a worthwhile effort. Building some of your tools at home will increase your enjoyment of handloading. Many other ideas will occur to you. We've only scratched the surface here.

CHAPTER 29

The Shotshell

THE SHOTSHELL as we know it today was probably the first truly successful, completely self-contained cartridge. Those early shells (they really should be called "cartridges," you know, since the word "shell" has meant an explosive artillery projectile since very early in the muzzle-loading era) didn't look much like today's product. Even so, they differed hardly any in design from the compound, paper/brass case in universal use until displaced in this country by plastics during the last few years. In fact, the majority of today's plastic cases (the so-called built-up type) differ not at all in mechanical design from those of over 125 years ago. The materials of the case and load are different and that is where the real improvement lies.

The typical shotshell consists of a case containing a battery cup primer (described earlier under primers) in its head. Into the case is placed first a powder charge; then an over-powder wad to seal gases; a column of filler wads; a shot charge; and finally a crimp or crimp-plus-wad closure at the mouth to tie all together.

Upon firing, the primer flash ignites the powder which generates the gases to drive the shot charge down the barrel. Just as in metallic cartridges, but from that point forward there are differences.

First, as the shot and wads are driven out of the case, the crimp is forced open and the case mouth unfolds into a portion of the chamber

346

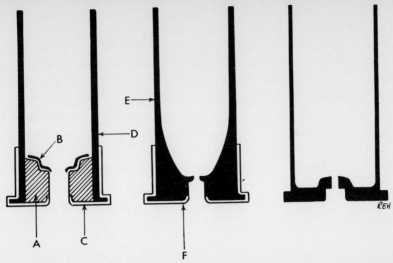

Left is traditional built-up shotshell case of tubular body, base wad, and metal head. Originally made of paper, now more commonly of plastic. *Right* is old-style drawn-brass shotshell case, made exactly like metallic cartridge cases. *Middle* is the latest thing, with head and body formed in one piece from plastic.

intended for it. Shot and wads then pass through a short reducing taper (called the "forcing cone") and are reduced slightly in diameter in the process. Simultaneously the filler or cushion wads are compressed longitudinally between the inertia of the shot charge and the expanding propellant gases; and the over-powder wad prevents those gases from escaping forward to mangle the shot charge. The cushioning effect of the collapsing filler wads reduces the tendency of the shot to be deformed and smashed together into multi-pellet clumps, thus insuring better patterns.

During all this, the case must first expand under gas pressure, tight against chamber walls, to seal the gas in until the shot and wads have left the barrel; then, as in metallics, it must contract sufficiently to permit easy extraction.

As the shot leaves the muzzle, the wads must drop behind immediately, and not be blown into the shot cluster to disrupt it. To do this they must have light weight and high drag.

And, after all this, as far as the handloader is concerned, the fired case must remain serviceable for several times re-use. Virtually all U.S. shotshell cases made during the last few years will meet this requirement, except for a few private brand names made to sell for the lowest possible price. The more recent cases are the best, for only during the past decade have manufacturers actually been concerned with designing "reloadability" in the case. Where a case life of three or four loadings was considered acceptable fifteen years ago, four or five times that is expected today.

The Shotshell

First we have the built-up paper case of antiquity, produced in only relatively small quantities now, but still available in abundance. The body is rolled and glued up of many layers of wax-impregnated paper. The wax gives the paper resiliency and moisture resistance and also increases its resistance to abrasion. Equally, it gives the assembled tube enough springback to contract after firing and permit easy extraction.

The body is inserted into the metal head, then a thick pulp, composition, or rolled-paper base wad is forced down inside, against the head. It expands the end of the body into the rim fold and tightly against the interior of the head. Some makers add light knurling of the head, which is likely as cosmetic as it is useful in binding the whole together.

It is this base wad that often gives trouble in subsequent loadings. Heat, pressure, gas turbulence, and extraction stress and strain combine to loosen it in its seat; and this reduces the head's grip on the body. The worst result of this can be a case body still in the chamber after the extractor yanks the head out. Plastic overlays and sturdier

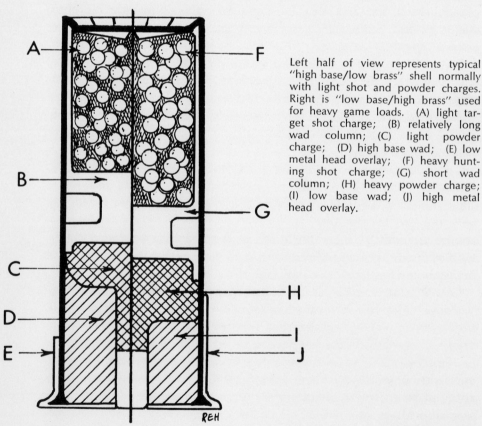

Left half of view represents typical "high base/low brass" shell normally with light shot and powder charges. Right is "low base/high brass" used for heavy game loads. (A) light target shot charge; (B) relatively long wad column; (C) light powder charge; (D) high base wad; (E) low metal head overlay; (F) heavy hunting shot charge; (G) short wad column; (H) heavy powder charge; (I) low base wad; (J) high metal head overlay.

wads have reduced this problem, but it still exists with paper cases, and to a very minor degree with some built-up plastic varieties.

With use and with age, especially when exposed to low humidity and high temperature, wax evaporates from the paper body and it loses its resiliency. In severe instances, it may split on the first firing, or after only one or two loadings.

Being not 100 per cent waterproof, paper cases can also absorb moisture when wet or exposed to high humidity for long periods. If loaded while this happens, cases swell so they won't chamber, even after being dried out. If empty, though, they can be restored as described later on.

Far better, both for factory loads and reloading, are plastic cases made in exactly the same way. The same head is fitted with an extruded plastic tube body held in place by a plastic base wad. The plastic is inert, won't absorb moisture or dry out, is dimensionally stable, and is far stronger and more resistant to abrasion than the very best paper. Remington developed and introduced this type of built-up plastic case in the early '60's and has since been followed by all other major makers. Today, this type case is the most common in all the popular gauges.

So, we have this one basic design, but of two different materials, that consitutes the bulk of today's shotshell ammunition, and consequently, is the most often reloaded.

A second type of plastic case is becoming more and more common. This is the one-piece CF (Compression Formed) case introduced by Winchester in the mid-60's. In it, case head, body, and primer pocket are formed under pressure and heat from a single slug of plastic possessing the resiliency and strength required. By this process, a single-unit case with none of the faults of the built-up type is produced very quickly. In addition, case walls may be given any reasonable profile, permitting greater thickness at areas of greatest stress. This is typified by the Winchester AA case with its sharply tapered walls and hemisphere-like powder space.

CF and similar cases are said to possess enough rim strength to withstand normal extraction loads unaided. Even so, all such cases produced by major manufacturers here are fitted with a conventional drawn sheet metal head. It not only adds strength, especially for handloading use, but preserves the traditional "bright brass" appearance without which, it is feared, consumers might not be so enthusiatic about plastics. Even in ammunition the cosmetic aspect is important to sales appeal.

There are, however, a few foreign and one domestic one-piece case whose heads are not reinforced. Most are injection-moulded and made of material different from CF cases. None have proven entirely satisfactory, including the WANDA made in Houston, Tex., and most do not permit the folded crimp preferred by U.S. shooters. None have achieved any significant market acceptance. When encountered by handloaders, they require special handling to varying degrees.

Generally speaking, plastic cases offer much longer reloading life, in addition to providing smooth and more trouble-free gun functioning, especially under adverse conditions. The handloader who has access to a good supply of the older paper cases need not throw them away, but he'll be much happier loading plastics. Of the lot, CF cases have the greatest reloading life, built-up plastics next, and paper last. The other one-piece plastics vary so much it is impossible to generalize about them, some are good, some are very poor.

The less-popular and obsolete gauges will usually be encountered only in paper cases. This means that if you have a yen to load 8 or larger bores, or the odd intermediates such as 14, 24, 32, etc., there won't be any plastics to smooth the way. Supplies of cases and wads for such odd sizes are imported by a very few shops and are available only sporadically.

Other types of cases are sometimes encountered. Until the success of plastics, drawn, solid-brass cases were produced for (primarily) military and police use. Their major advantages were moisture and abrasion resistance, and freedom from being bulged by magazine spring pressure in riot and service guns kept fully loaded for long periods of time. Paper cases simply couldn't stand the gaff and be relied upon to produce absolute reliability under those conditions.

U.S.-made brass cases accept standard large pistol primers (Boxer type). Foreign production usually requires large Berdan-type primers, though a few were made for the battery cup type.

Limited quantities of cases have been made from aluminum for the same reasons. Both aluminum and brass cases are drawn exactly as are metallic cartridge cases. Properly loaded, such cases last virtually forever unless damaged or abused. I know of brass military 12-gauge cases made by Remington that have been loaded over 100 times and are still thorouhly serviceable.

At the beginning of the 1960's Alcan developed and marketed a built-up metal case. It consisted of a plated steel head, a soft zinc-alloy body tube and a composition base wad covered by a zinc overlay. This case was constructed exactly like contemporary paper types and today's built-up plastic, but could not be used with a folded crimp.

This case produced phenomenal reloading life for the period, twenty-five to fifty loadings being reported. However, it was followed very shortly by Remington's built-up plastic case which was much easier to load, did not require special dies and wads, and which permitted a folded crimp. The availability of good plastic cases killed Alcan's metal design in a very short time. Today they are rare, and were made only in 12 gauge.

It would appear that the shotshell has reached the epitome of its development in the compression-formed case. No doubt other improvements will come along as in the past decade, but if we wait as long for them as before, they won't occur in my lifetime.

Even so, plastic cases aren't yet perfect, and they are deficient in one respect which is of little interest to the average temperate-clime handloader-shooter. During a visit to Finland, I discussed shotshell design and production at length with manufacturers there. I was advised that in the bitter cold of the arctic, even the best (and they had tested them all) plastic cases often split upon firing, while top-quality paper cases did not. Virtually all plastics otherwise suitable for cases lose their resiliency and become brittle as glass under prolonged exposure to arctic temperatures. This isn't surprising, since other materials and even metals behave the same and often fail under normal loads and stresses. Ask anyone who has lived there or participated in arctic tests of military weapons. Since few American shooters are ever likely to hunt at -40°F or below, this is a rather insignificant shortcoming. The Finns, though, (and shooters in other arctic areas) remain wedded to paper cases for this reason. Personally, I'll stay indoors when it's that cold, and can settle for plastic cases.

CHAPTER 30

Primers, Powder, Wads, and Shot

THE CASE isn't much good alone. Primers have been covered earlier in this volume, so need little attention here. Battery cup designs are used almost universally except in solid metal cases. Design details and dimensions differ somewhat among various nations, but even those differences are lessening rapidly as more and more shooters travel and exchange information.

Since shotshell powders are generally fast-burning and easily-ignited, primers are not so critical a factor as in metallic ammunition. However, some shotgun firing mechanisms are notoriously weak and inefficient, so primer sensitivity becomes highly important. Over/under guns with sharply angled firing pins are particularly bad in this respect. Primers made sensitive enough for 100 per cent reliability in some older gun designs may approach handling safety limits. Fortunately, modern gun designs take this into consideration and will ignite all standard primers well.

Selection of primers revolves primarily about securing a gastight seat in the primer pocket. While all domestic makes are *nominally* the same size, slight differences in battery cup diameter, taper, and radius *may* result in primers of one make fitting poorly, too loose *or* too tight, in another make of case. Independent primer makers, such as CCI and Alcan, solve this problem by giving battery cups sufficient

352

taper that they will enter all standard pockets, yet are large enough at the rear to form a tight seal in the largest, all without requiring excessive seating pressure.

Some primers are "hotter" than others, in that they contain more compound and will therefore produce better ignition of the slower powders used in heavy magnum loads. These primers are identified in the chart contained in the chapter on primers you've already read.

As already mentioned, shotshell powders are among the fastest burning of all. They must be to suit shotgun conditions and to burn efficiently at the 8,000-10,000 psi range in which scatter-guns operate. They are comparable to pistol powders in this respect.

Generally speaking, the fastest powders must be used for light loads; medium-speed powders for so-called field loads; and the slowest for top magnum loads. Also, the larger bores require faster powders than the smaller. Fast Red Dot performs beautifully in 12 gauge, but not at all well (or even safely) in the 28 and .410.

Using the most-popular 12 gauge as an example, we find the entire load range is best served by powders falling into three different burning rates, though in a pinch two will do.

Light loads such as 3/1⅛ skeet, requires fast Hercules Red Dot or DuPont Hi-Skor 700x, producing 8-9,000 psi and 1150 fps. The 3¾/1¼ heavy field load is best driven by Hercules Unique or DuPont SR7625, producing a bit over 1300 fps and 10,000 psi. Any attempt to obtain this velocity from Red Dot, or similar fast powder, will produce excessive pressures.

Magnum loads (either 2¾" or 3" case) in the 4/1⅝" range require even slower DuPont 4756 or Hercules Herco to reach the required 1200-1300 fps velocity at safe pressures. The faster powders can't drive the heavier shot charge that fast without producing excessive pressures.

Dropping down to smaller gauges, by the time we reach the 20, we find that Red Dot and Hi-Skor 700X are too fast even for light Skeet loads, so the next slower powders, Unique and 7625, fill the bill, and heavier loads require even slower powders such as Herco.

Down at 28 and .410 level, even slower powders are required and Hercules 2400 and DuPont IMR 4227 *rifle* powders become necessary to hold pressures down to acceptable limits. Thus, the pattern is set.

This doesn't mean that there are only two powders suitable for each load level in each gauge. There is a large number of other powders similar to those mentioned, and a great deal of overlap exists among them all. This is best understood by referring to our shotshell loading data tables, and by examining the powder burning rate chart shown elsewhere. In reality, there is no reason whatever for reaching out to

Primers, Power, Wads, and Shot   353

Evolution of over-powder wads progresses from a cardboard disc, to a cardboard cup, and now a plastic cup.

pick some marginally-suitable powder. Simply select a standard load, then choose from the two or more powders listed for it in the loading data tables. To do otherwise is to risk poor results at best, or a blown-up gun at worst. The biggest, and most dangerous, mistake you can make is to try to get by with one fast-burning 12-gauge powder by using it occasionally to assemble heavier loads and loads for the smaller bores. Use the correct powder or forget about loading.

Remember — large bores and light loads need fast powders; small bores and heavy loads need slow powders. Use what is right, not what is convenient.

Wads are far more important to load performance than once believed. The over-powder wad seats tight against the powder; it must be stiff and strong enough to remain whole under 10,000 psi gas pressure; must fit the case tight enough to stay in place; and it must fit the bore tight enough to prevent gas from escaping around it to disrupt the shot charge. For over a century, a simple disc of stiff cardboard .135-.200" thick served this purpose. Or, at least we thought it did. We know now that it performed indifferently much of the time. Some improvement was gained by forming the cardboard or paper pulp into a shallow cup, open side down, so that gas pressure expanded it outward for a more efficient seal.

But, by the late 1950's and early 60's, moulded plastic over-powder wads of cup form became available. They proved to be vastly more efficient, though considerably more costly, than cardboard. Except in the cheapest loads, plastic cup-type wads are standard here today. They are the best we've had yet.

By sealing better, both in the case and in the barrel, the plastic O-P cup wad improved powder efficiency, so much so that it was soon found that charges developed with card wads produced excessive pressures with plastic wads. Extensive tests, to show this problem clearly, were conducted by the NRA and published by the *American*

Rifleman in the mid-60's. It was found that merely substituting a plastic cup wad for card could raise pressures to nearly proof-load levels. Consequently, new loading data was developed, resulting in a difference of several grains in the amounts of powder required to produce identical results with the two different wads. Depending upon the load and powder type, charges may differ from 10 per cent to 15 per cent — rather a significant amount.

Today, most loading data sources show different tables for card and plastic O-P wads. Glancing at a Hercules table before me, I see that plastic wads produce the same performance with one to two grains less of Green Dot in 12 gauge than required by card wads. With some other combinations, the difference is substantially greater. For this reason, it is always necessary to verify which type wad was used to develop the load data you are using. All data before 1960, and a good portion of that around 1965-66, which is not clearly identified, is for use *only* with card wads. Be guided accordingly when working from old references.

As for choosing O-P wads, by all means use the plastic cup type if you want best performance. However, money can be saved with card wads. If you are shooting cottontails and quail at ten to fifteen yards in heavy brush, you'll never be able to see the improved performance of plastic wads. Likewise for Sunday-afternoon fun shooting over a hand trap, so why not use the cheaper wads? I do. In addition, you'll be forced to use card wads in odd gauges and in metal cases for which plastics aren't made.

As for performance, all the current plastic cup O-P wads are so near alike, there is no reason for choosing one over the other. All depend upon their thin skirt being expanded tightly against the case and bore by powder gases to achieve a secure seal. Some are easier than others to load. Those fitting the case snugly and not provided with an air escape hole will sometimes be popped out of position after seating by compressed trapped air. Some have more flare to the skirt than others and are slightly more difficult to seat. In the end, though, all perform the same.

Filler wads serve two purposes. Originally, they were used to take up some of the space formerly filled by bulky black powder. Case and loaded shell dimensions were standardized during black powder's heyday. So, when more efficient smokeless came along, occupying far less space, the emptiness was filled with wads to keep external shell dimensions the same. In addition, filler wads cushion the shot against the violent acceleration produced as the powder begins to burn. In this respect, wads function just like a recoil pad, collapsing part of their length to cushion the blow. Without this cushioning effect, individual shot are badly deformed and stray from the pattern, as air acts on their irregular surfaces to push them unpredictably about.

If you doubt the value of this cushioning, load a half-dozen shells with soft, springy wads, then an equal quantity with a section of hard-wood dowel replacing the wads. The wads give; the dowel doesn't. Fire both at the pattern board and watch how much worse are the patterns from the dowel-loaded shells. You'll be convinced. You'll be even more convinced when you discover how much harder the dowels make recoil. A springy filler wad cushions *you* as well as the shot.

Over the years, factory loads have used filler wads cut or moulded from paper pulp, hair felt, cork, and various composition materials similar to insulating board; sometimes waxed for lubrication, sometimes not. All provided some degree of cushioning, but felt and cork gave the best results, and, incidentally, cost the most.

Prior to WWII, few good filler wads were available to the handloader. Then Homer Clark began making wads independently from insulating board down in Alton, Ill. They provided good cushioning, and their availability had much to do with the growth of shotshell handloading. These wads were lubricated by rolling them across a hot steel plate covered thinly with molten wax. The wads picked up just enough wax to do the job. Though less popular now than then because of other developments since, such wads perform extremely well; at least as good as those then used in most factory loads.

At the same time(and earlier), many a handloader cut his own wads from felt or insulating board with a special "wad cutter punch" or a piece of sharpened pipe. Some installed sharp hollow punches in drill presses to speed up the job.

Probably spurred by Alcan's efforts, the major ammo makers finally made their wads more available to shooters, and other independent makers got into the act as well. One or two importers brought in excellent cork wads from abroad for the hard-to-please handloaders.

Eventually, though, the obvious became apparent, and single-unit wad columns were moulded from flexible plastic, combining filler wads with the O-P wad. Though differing in detail, all functioned the same. Fingers or other open structures were extended upward from the O-P wad to the height needed, then capped with a floor or cup to support the shot charge. Upon firing, the structure above the O-P portion collapsed, folded, or compressed to provide cushioning action. As further developed, these wads provided better cushioning than even the best felt or cork types. Today they are supplied in all but the cheapest factory loads, and are widely available under nearly a dozen names and in as many different designs, to handloaders.

Modern Handloading

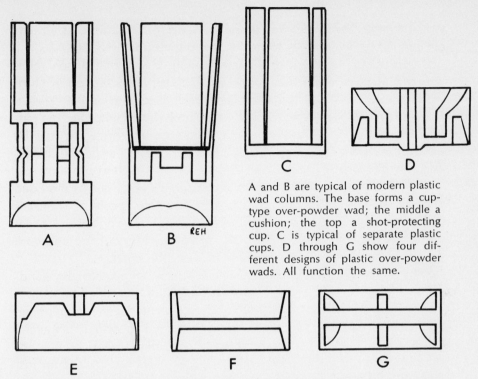

A and B are typical of modern plastic wad columns. The base forms a cup-type over-powder wad; the middle a cushion; the top a shot-protecting cup. C is typical of separate plastic cups. D through G show four different designs of plastic over-powder wads. All function the same.

Regardless of advertising claims, all function in the same manner and provide nearly the same degree of cushioning.

Single-unit plastic wads do have disadvantages. They cost more than old-style cut wads, and must be made in several lengths or heights to accommodate different cases and loads. Where one could split or stack cut wads to match any case and load, it is necessary to buy a different length single-unit plastic wad for each requirement. Even so, unless you're flat broke, the added loading convenience and performance more than made up for this and the extra cost.

As with O-P wads, the modern plastic single-unit wad columns do give the best results on target and are the most convenient to load. But they aren't necessary for many types of shooting and they do raise costs. By using old-style cut wads, you can save ½-¾ cent per shell. And, believe me, for fun shooting or short-range scattergunning, you won't be able to tell the difference. A squirrel or quail ten to twenty yards off, doesn't require the *ne-plus-ultra* plastics. After all, one hell of a lot of game was killed for over 125 years without them. Very impressive skeet and trap records were also set without them.

Selection of the proper plastic wad unit boils down to the length your particular case/load combination requires. They all function much the same and give about the same results, so close to it you'll never know the difference. Just check the data tables and pick the one that meets your requirements.

Unfortunately, many relatively new handloaders don't even realize that plastic wad units aren't always necessary, or, for that matter, even desirable. They do reduce apparent recoil ever so slightly, and they do produce slightly better patterns, and they do it all with less powder. But, they aren't actually *necessary*. Those who've never tried the old method with cut wads should give it a try, just so they'll know how it's done.

Shot protectors didn't originate with plastic wads, though today they are usually made as a single integral unit. Back in the 50's and early 60's Alcan and other companies introduced soft plastic cups to contain the shot charge during its passage through the bore, and Winchester introduced its polyethelene shot wrapper in the now-famous "Mark V" shotshell. Both served the same purpose, that of preventing naked lead shot from rubbing against the bore. Shot pellets contacting the bore are scrubbed flat on one side and air pressure acting on that flat causes them to skitter about in flight, usually away from the main part of the charge. This reduces pattern density and uniformity. Placing a plastic shield between shot and bore greatly reduced this and improved patterns significantly; so much so that barrels bored modified for old-style loads often produce full-choke patterns with shot cups.

Alcan finally combined its plastic O-P wad into a deep shot cup and placed cut filler wads in the bottom of the cup. This worked well and is still available under the name Flite-Max. Other makers simply formed the shot cup on top of a plastic O-P/filler wad unit. This has since become the standard, a single, one-piece moulded unit combining O-P and filler wads and a shot protector. In all designs

Left: one early form of shot cup and wad column combined is this Alcan Flite-Max, which contained a conventional fibre filler wad. *Right:* typical of independent brand wad columns available today is this Verelite, by Pacific Gun Sight Co., offered in three color-coded lengths to suit diffferent loads.

Modern Handloading

the cup is split longitudinally in three or four places, so that air pressure opens up the cup and causes it to drop off the shot charge during the first few feet from the gun muzzle.

Nearly all of these shot protector wads are formed in one piece, and must be made in several lengths to suit different cases and loads. A notable exception is the Sullivan Vari-Wad which is in two pieces, the O-P wad or base, and a shot cup, which can be easily assembled to produce three different lengths. Posts extending from the base support the cup and provide the collapsing/cushioning action in firing. The base may be assembled to the cup by inserting the posts in any of three different sets of holes to provide different lengths. This is less convenient than one-piece wads, but allows one to get by with one rather than three or more wads on hand, and that keeps the investment down.

Naturally, it is impossible for manufacturers to provide single-unit shot cups to suit every case and load, or even for every gauge. At most each maker supplies three lengths in the popular gauges. These suit nearly all recent domestic cases and standard target and light field loads. Thus, they lack the flexibility that can be had from cut filler wads, which are so easily varied to meet any real or imagined need. Details of modifications and improvisations to meet special loading problems will be found farther along in this book.

Shot is what does the job. It does the killing and all the other components serve only to deliver it. There is little to be said about shot, for it is extremely uniform among all makers, and the handloader merely dumps it into his prepared cases.

"Standard" shot is made of antimonial lead alloy and is what was once called "chilled" shot, meaning it is considerably harder than pure lead. Since the harder shot is, the less it is deformed in the barrel, and the better patterns it will produce, some shooters want it harder still. For them, several makers supply "extra hard" plain shot and copper or nickel-plated shot. Staunch supporters of these claim substantially improved long range performance, and often use them for waterfowling and competitive live pigeon shooting. All the same, few handloaders can see any real difference on target. Standard shot seems fully as good a choice and is much cheaper than plated types.

All makes of shot appear equally good, but don't get caught with a batch of reclaimed shot, unless the price is really right. It has already been fired once, then mined from the range and washed and screened. Consequently, it contains many imperfections that will reduce pattern density and uniformity. It's okay for short ranges and fun shooting, nothing more.

Shot choice, then, boils down to suiting the size and charge weight to the game. Far be it from me to tell a man what size shot to use. One fellow I know always uses No. 2 on ducks; another uses 7-½ or 8, and sometimes even Skeet loads. Both kill ducks, but they shoot differently. As an objective observation, I'd say most people tend to use shot about one size larger than is best for game. Remember the old adage, "If there ain't enough shot in the pattern to hit, it don't matter how big they are." More smaller shot will often pick you up out of a gunning slump. Hits count more than misses.

The last component is one we seldom see in popular gauges today, the over-shot wad. The advent of the folded crimp has done away with it, except for buckshot and some odd gauges.

The O-S wad lies on top of the shot charge and is held tightly in place by having the case mouth turned in 180° upon it. This is the old standard roll-crimp many youngsters have never seen. This wad and crimp is much more difficult to apply, and requires different tools, than the modern folded crimp. It is also less perfect in that the wad sometimes deflects enough pellets to produce a hole in the pattern, what is called a "blown" or "doughnut" pattern.

CHAPTER 31

Shotshell Loading
Tools and Equipment

SIMPLE BASIC loading tools developed early for shotshells, but loading was slow and laborious until only recently. With few exceptions
all operations were performed separately by hand. About the closest
thing to a machine involved was a hand-cranked, belt-driven crimping
head which had to be turned fast enough to soften waxed shell
mouths by friction heat. The very idea of loading a case of shells
between dinner and the ten o'clock news as we do today would have
boggled the imagination. Anyway, who *needed* to load ammunition
all that fast?

As near as I've been able to determine, only Lyman (Ideal) made
a fast-operating progressive loader way back when. It featured several
different die stations like modern tools, and even had its own built-on
powder and shot measure operated by a foot chain. It was the elite
tool of its day, but not many saw use. Most shooters were content
with the much cheaper single-stage hand tools with which diligent
effort could produce at most one or two boxes (twenty-five rounds)
in an hour. And with the existing quality of paper cases and the lack
of complete resizing as we know it, those reloads didn't always do
too well.

Since WWII, though, the more or less standard modern type of
integrated shotshell loader has developed.

This Texan Model FW is typical of modern low-to-medium price shotshell loaders combining all operations in a single, permanent setup.

Metallic tools developed around a basic frame to which individual dies and other tools were assembled in turn for the various operations, and accessory items like powder measures were built separately. Shotshell tools took a different route whereby all dies and holders, as well as shot and powder measures, were integrated into a single unit which allowed processing each case through all operations without any adjustment or change. This came about because of so few shotshell gauges and the few loads needed for them. Versatility for dozens of calibers and hundreds of loads wasn't needed as for metallics.

Usually such loaders consist of a solid base to which is attached one or two vertical posts. Upon the posts rides a die head carrying five (usually) die or work stations, powder and shot reservoirs, and a charge bar containing sized cavities for dropping powder and shot charges. The die head is moved vertically by a handle and linkage, forcing dies down over each case as it is moved successfully by hand under the stations.

Using a typical all-in-one press of the type marketed by MEC, Pacific, etc., functioning and operation is generally as follows for the low-medium cost models.

Modern Handloading

1. The empty case is placed on the base under the decapping station, and the handle is brought down. This punches out the fired primer and the decapping rod straightens the case mouth. The case may also be resized, either fully or partially, at this station.
2. The case is next moved to the priming station and as the handle is pulled, a rod enters and presses the case over a primer placed in the appropriate hole in the base. The rod also presses hard against the base wad (if present) and reseats it if necessary. Pressure here may be applied through the rod to straighten dished case heads.
3. The case is then moved to the charging station and the handle is moved to lower the drop tube inside the case, and the charge bar is moved to drop a powder charge into the case.
4. The charge tube is raised and the wad column is placed in the wad guide. The drop tube is lowered (by the handle) to seat the wad column, then the charge bar is moved in the opposite direction to dump shot in the case.
5. The fully charged case is moved under the crimp-start die which is then lowered gently to turn the case mouth in at about a 45° angle and re-establish the crimp folds properly.
6. The prepared case is moved to the crimp station and the handle is brought all the way down firmly. This forces the resizing die fully over the case, bring it back to size, and completely forms the crimp. The handle is moved smartly to full up position and an ejector plunger forces the loaded round from the crimp-size die, ready for use. A final sizing operation is incoporated with crimping, even if the case has already been fully resized, because crimping pressure might otherwise bulge case walls and create chambering difficulties.

Not all tools function in that same way, though all perform the same operations on the case. A few carry the case up into the dies instead of the dies down over the cases; some place the dies in different sequence; some use two separate charge bars instead of one. Some place the work stations in a U, some in a circle, and some in a straight line, but they all function essentially the same way.

More costly tools use the same dies and many interchangeable parts, and movements, but include sturdier construction, auto primer feeds, automatic cycling of the charge bar, and other refinements.

Next step up are the more robust true progressive tools, still with the same dies, which incorporate a rotary shell holder or plate which when filled holds a case under every work station. Thus all operations are performed simultaneously upon separate cases. Then, with

Fast production, more sophisticated shotshell tools are represented by this big, circular-progressive Pacific DL 366, which produces a complete shell for each stroke of the handle.

the plate indexed one space after each operation, a completed shell is turned out with each handle stroke. Obviously, such tools can be operated more rapidly than the others and several will turn out more than 500 rounds per hour.

Generally speaking, the true progressive tools are of much heavier and sturdier design and may be expected to load tens of thousands of shells with relatively little care and maintenance, other than being kept clean and lubricated.

There are a few other types of loading tools intended for special purposes. For example, the Lyman EASY LOADER may be had on special order with a roll-crimp station. In it the folded-crimp die is replaced by a rotating roll-crimp head which is spun rapidly by helical cam grooves on its shaft as the tool handle is moved.

Another design variation is found in the basic Ponsness-Warren system where the case is resized full length as a first operation and then remains in that die for all operations. This means a die that moves successively under each work station. The progressive P-W Du-O-Matic contains an indexing shell plate carrying eight permanently installed full-length sizing dies for this purpose. This system eliminates flexing of the case under loading and handling pressures, insuring a more uniform loaded shell.

The height of sophistication is found in the better true progressive

Modern Handloading

tools fitted with hydraulic actuating systems controlled by a foot pedal or thumb switch. Only one such model, the MEC HYDRAMEC, is commercially available, but numerous custom conversions have been made. Hydraulic actuation reduces operator fatigue and increases *sustained* output, but does not increase (in fact usually decreases) short-period maximum production rate. Needless to say, such tools are a bit costly, around $350 and up. The average handloader has no need for them and can hardly justify their cost.

Then, there is the basic hand-type loading kit typified by the LEE LOADER. The only real justification for such tools is their very low cost. Operation is slow and inconvenient, but scores of thousands of handloaders get their start that way.

The Lee Loader consists of simple ring dies and punches placed on a base, and operating power is provided by a hammer or leaning heavily on them. All operations are performed mechanically the same as fancier tools and perfectly acceptable ammunition is produced.

During the period immediately after WWII several companies introduced separate shotshell loading dies which could be used in con-

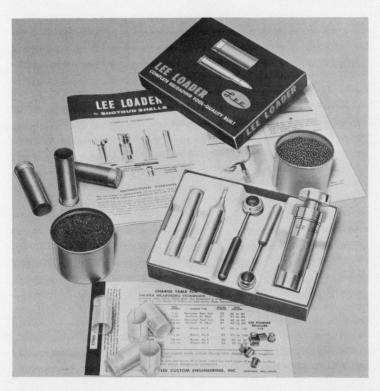

Lee Loader is typical of hand or pocket-type tools that can be adapted at least in part to several calibers.

ventional metallic presses. They produced fine ammunition, but required dies to be changed for every operation, as well as the use of separate shot and powder measures. They have virtually disappeared from the market since the integrated tool appeared, but are still available from Hollywood and Herters. At one time Hollywood could supply such dies in the odd gauges for which current conventional loaders are not made. In fact, their greatest value is in that area.

All the foregoing tools were originally made primarily for paper cases. The advent of plastic cases raised new problems in that crimping and sizing operations could not be performed properly on both types of cases in the older dies. Initially some makers offered two types of dies, but combination dies were quickly developed to satisfactorily handle plastic and paper indiscriminately. Today all makers supply only the combination dies on their tools. Older tools you may encounter *might* be fitted with paper-only dies, a point to remember when buying used gear.

It is interesting to note that while metallic cartridges require adjustable powder measures to accommodate a myriad of powders and loads, shotshells really do not. As mentioned, there are only five popular gauges, and three more or less standard powder charges to meet all needs within each gauge; one for skeet, one for heavy field loads, and one for magnum loads. Consequently, all loading tools are supplied with a simple fixed-cavity charge bar into which fit interchangeable inserts. These inserts are machined to contain the several standard powder charges and shot charges. Thus, changing loads is merely a matter of replacing inserts. Because of load standardization and limited numbers of loads, this is entirely practical and is a far less costly system than that used in loading metallics. Most manufacturers supply tools already set up and adjusted for skeet or trap loads, but will meet any reasonable load requirement on special order. By far the largest percentage of shotshell loading is for skeet and trap use, thus this procedure.

The experimenter or the person addicted to non-standard loads is forced to obtain a separate adjustable powder measure, and also a special adjustable shot measure. Generally a large-capacity, large metering chamber powder measure may be used to measure shot, but some companies offer modified designs especially for shot. Best of the lot is the patented Hollywood which has provisions to prevent the shot-jamming often encountered in powder measures.

While nearly all loading tools are sold set up for one gauge and case length, most have provisions for conversion to all popular gauges. Conversion kits are sold for the purpose and are easily installed, though the job is often quite time consuming. Generally, each different gauge requires new shell holders, decapping stem, wad guide, drop tube, primer seating rod, crimp-start die, crimp-resize die, and powder and shot charge bushings. Exchanging and adjusting all those parts is a chore, so many handloaders simply purchase a complete tool for each gauge. Changing case lengths is less involved and usually requires only a new crimp-resize die, plus minor adjustment of other work stations.

Because shotshell loading tools are generally integrated units, there is no need for a wide range of accessories such as is found for metallics. A good powder scale is needed to check accuracy of powder and shot charges, and a shell reconditioner is essential to obtaining maximum life from built up cases. It uses a die and punch to reform distorted metal heads (common in cases fired in well-worn guns) and to re-seat loose base wads. A shell ironer or waxer is also useful in restoring worn or damaged cases. It contains a heated recess into which a deformed case mouth is pressed, and the heat softens the plastic and re-forms it so a more secure crimp can be produced. Wax is added in the recess for paper shells where it serves the same purpose by penetrating the worn paper case mouth and strengthening it.

The fellow who must economize to the hilt can also use a case trimmer. With it worn case mouths can be cut off, then the case continued in use with modified dies and shortened wad column. Not all repeating or autoloading guns will function well with shortened cases, but doubles and singles will.

One other accessory is valuable in extending loading life, especially with paper cases. Primer pockets sometimes become too loose to hold primers securely while still in otherwise good condition. Williams makes a Primer Pocket Peener which tightens pockets for several more loadings. A hollow punch and base set simply squeezes the metal head and base wad inward to reduce pocket diameter, at the same time also re-seating the base wad. Of course, this gadget cannot be used on solid C-F plastic cases which do not have a compressible base wad.

There are, of course, accessories offered with loading tools. They take the form of auto-primer feeds, primer-tube fillers, wad guides, oversize shot and powder hoppers, measure baffles, Lyman's micrometer-adjustable charge bar, special dies, crimpers and crimp starters, etc. None is really essential to good loads, but they do offer convenience and speed. Most of them constitute more or less gilding of the lily.

CHAPTER 32

Loading Operations

IN DESCRIBING shotshell loading tools we've touched (but only touched) upon what must be done to that bushel or so of fired cases you want to convert to good, safe, shootable ammunition. Let's take it all step by step and explore the pitfalls along the way.

Sorting cases comes first. Naturally, they must be segregated by length. Those long 3" magnums require a different setup. At the same time, sort cases by type and make. Usually you'll just want to pick out the one in the majority and set the rest aside. Sort also by height of base wad. Use high base (wad) cases for light target loads; low base for those heavy field loads. The factories do it that way when they make them, regulating case volume to suit the load by varying base wad height. For a light powder/shot charge, high base; for lots of powder and shot, low base. Brass height doesn't mean much as far as base wad height is concerned, except that *generally* high brass means low base. The only way to be sure of base wad height is to look closely or use a length of dowel as a plug gauge. Striped with colored tape, such a dowel makes sorting easy.

All the while, inspect cases for condition. If the head is dished or the rim is bent, set aside for possible rejuvenation later. If the head is loose on the body, better ashcan it, though a very *slight* amount can sometimes be corrected in a shell restorer. Bulged heads get the

368

heave-ho if any cracks are in evidence, otherwise they can probably be salvaged later. Look inside to make certain the base wad is present (yes, they do occasionally blow out in firing) and down where it belongs. Again, *slight* looseness can be corrected.

Especially on paper cases, look where the body meets the head for tiny pinholes burned through, and for cracks or splits. Ashcan them, for surely such cases will soon separate, leaving the front of the body in the gun, or even halfway up to the bore to create a burst barrel on a subsequent shot. Plastic cases seldom show these defects, but check anyway.

Look for crushed or flattened case mouths, usually the result of being stepped on. If not too bad, they can be restored in a shell saver, but sharp creases or cracks mean the ashcan. Pay particular attention to the mouth. If it is torn or cracked, or eroded to paper-thinness, it won't hold a decent crimp and must be discarded or shortened. If the mouth is simply soft and mushy, it needs restoring, waxing if paper, heat-forming if plastic.

Top, left: burn-through at edge of metal head spoil case; *top, right:* torn mouths are cause for rejection; *left:* split cases obviously must be junked.

Once the good cases are sorted from the bad, you can check the loading data tables and select the load and components you'll need. No point in starting unless you have them all ready to hand. Also, before starting, recheck the powder and shot charge bushings or measures. The tool might be set up for another load you used last time. Give the rest of the tool a good look too. Is it clean and lubricated? Are all the pins and springs in place? Nuts and bolts tight? Parts move freely? Bolted down solidly? These things are a lot easier to check *before* you get a case jammed in the tool because of them.

Wipe the cases free of grit and dirt. If they happen to be the C-F type, washing is simpler, since they won't absorb moisture. Just make sure they are completely dry before loading. Water trapped in primer pockets will give you fits later with misfires.

Contrary to metallic processes, shotshell cases don't require lubrication before resizing. Paper cases are wax-impregnated, and plastics are naturally slick, and sometimes even waxed at the factory to make them even more slippery for easy extraction. The only exceptions would be old, dried-out paper cases. If they show signs of sticking in the dies, rub on a dab of candle wax or paraffin, or maybe give them a shot of spray-wax, a genuine hard wax, not a polish. Usually, paper cases that badly off will already have been rewaxed at the mouth, and that wax will provide all the lubrication needed.

Decapping is first, and some tools combine this with partial or full-length sizing. According to tool instructions, either place the case on the base under the decapping pin or slip it up *over* the decapping

Left: decapping; right: repriming.

rod. If the latter, correct alignment is assured; but if the former, take care that the decapping pin contacts the primer battery cup centered. If it does not, it will tear the base wad or enlarge the primer pocket, ruining the case.

Pull the tool handle down smoothly, but with enough force to complete the stroke without hesitation. If resizing is involved, especially full-length to include the metal head, a good deal of effort will be required. Make certain the stroke is completed, clear down to the stop, or, in the case of tools without a stop, until the punch and die bottom solidly on the case. In some tools, a shoulder on the decapping rod re-seats the base wad during the last fraction of the stroke.

Check the head and mouth of the decapped case for visible defects, Place a fresh primer, flange-down, in the priming bushing, then set the case over it, making certain it is centered. Pull down the handle and watch to insure the priming rod enters without snagging the case mouth. Continue the stroke and *feel* the primer into the case. The rod pushes the case over the primer, but in some older tools it is possible to bulge the case head inward by too much primer seating pressure. If this occurs, it will cause misfires in some guns. During seating, note if the primer fits tightly. If it enters the case with hardly any pressure, the pocket is oversize and that case should be discarded or reworked.

Examine the primer to make certain it is fully seated, then move the case to the charging station, making certain the wad guide spring fingers enter the case mouth without snagging. If the mouth retains too much crimp to allow this, expand and flare it slightly with a waxed, tapered, hardwood dowel.

Lower the drop tube (powder and shot) into the case, then push the charge bar in the proper direction (as indicated by tool instructions) to drop the *powder* into place. Raise the drop tube above the wad guide.

If using a single-unit plastic wad column or shot cup, place it in the wad guide, then lower the drop tube to force it down fully against the powder. Here follow the load table recommendation for wad-seating pressure, checking that pressure on the indicator scale of the tool. Some load combinations require so-called "zero" pressure, others need up to eighty pounds or so. Zero isn't actually zero. Enough pressure must be used to put the wad into firm contact with the powder charge, and that may be as much as ten to twenty pounds. Zero actually means the minimum amount of pressure needed to seat the wad fully; and that means in full contact with the powder.

If you are using nitro card wad over the powder and composition filler wads, they may all be seated as a unit, stacking the entire column in the wad guide. However, if very high wad-seating pressure is required, the filler wads will be pre-compressed or crushed, destroying at least part of their cushioning effect. So, at high seating pressures, it is best to seat the card wad alone first, then follow with the filler wads and just enough pressure to put them in place. The same procedures apply if a separate plastic cup O-P wad is being loaded. Paper cases should have 9/16" of mouth above the shot for a proper crimp; plastics 7/16".

As the entire wad column is seated, leave the drop tube down and move the charge bar (again, as the instructions indicated) to dump shot into the case. Then raise the drop tube and move the fully-charged shell under the crimp-start die. Some older tools will have a fixed starting die with an index mark facing you. With them it is necessary to turn the case so a crimp fold falls under the index. Most modern tools use a self-indexing start die which senses the location of the crimp folds and rotates to accommodate them.

Fired plastic cases do not always require a separate crimp-starting operation, but even so, its use will insure more uniform crimping. Fired paper cases and *new* cases of any type always require use of a crimp-start die to establish or re-establish the proper fold locations. It should be noted that both 6- and 8-point crimp folds will be encountered. Paper cases use 6 folds, plastics use 8 and the crimp-start die *must* match the crimp to be formed. Incidentally, paper cases originally fired with a roll crimp may have the folded crimp started as outlined above if in good condition.

In any event, the crimp-start die is pressed down over the case just enough to establish the folds and turn the case mouth in about 30°-45°. Too much pressure will bulge the case or spoil the ultimate crimp. Generally, use the least amount of pressure that will insure a proper final crimp. Experiment.

Transfer the case to the final crimp-resize station. As mentioned earlier, the case must be strongly supported during crimping or the pressure will bulge its walls. Crimping requires more effort than the other operations and should be done smoothly in one stroke. In modern double-action crimping dies, the mouth of the case is first turned over flat by guiding surfaces as it is forced into the die; then during final travel, a punch advances into the center of the crimp area to give it the typical recessed form which locks the fold in place securely. All this requires power; power applied smoothly and rapidly. Hesitation can cause rumpled crimps.

When all is correct — powder, wads, and shot leaving the proper

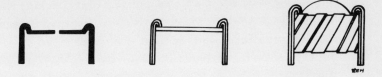

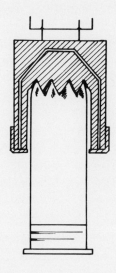
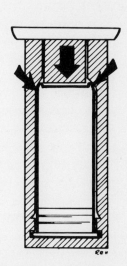

Left: Winchester A-A plastic shell crimped with Lachmiller Super Jet. Top, left to right: star crimp; roll crimp on shot; roll crimp on slug. Bottom, left to right: star crimp-start; finish crimp.

amount of case mouth — the crimp folds will meet precisely in the center. If too little mouth is available, a small hole will appear at the center, and is acceptable if not big enough to let shot escape or the crimp to bulge. Too much will cause a jammed-up wad at the center of the crimp. Such a crimp won't hold well.

In any event, a *forcible* upward handle stroke is required in most instances to eject the completed round from the crimp die. It is tight in there after all that pressure. Many handloaders habitually leave the shell to "set" a few seconds in the crimp die, or withdraw it part way and repeat the stroke. They feel this improves the crimp. It might, but I doubt it.

Once finished, crimp durability of paper cases can be improved by a drop or two of melted wax over the crimp center. This not only adds strength and water resistance, but wax absorbed by the paper increases case life. Wax will waterproof plastic shells, but adds nothing to case longevity. Some people water-proof plastics with a fat drop or two of clear lacquer or water-glass (sodium silicate), since it is easier than wax to apply. One or two shooters I know stick small circular pressure-sensitive patches over the completed crimp. These are the small, round, peel-off stickers made for price labels. They are cheap, and even available in several colors, so you could color-code handloads if desired.

Usually, cases loaded and crimped only a few times don't need any help as above, but, as mouths soften and erode thinner, a few more loadings can be eked out by reinforcing the crimp in one way or another.

Inspect the finished round closely. The crimp should be tight and firm, fully recessed, and slightly rounded or tapered on its outer edge. Some older dies leave virtually a square outside shoulder on the crimp and this will sometimes interfere with smooth feeding in some guns. Special dies were made to round off this edge and lightly taper the case afterward. Modern dies do it all at once.

The question of roll crimps arises, for sooner or later many hand-loaders will need to use them. They are necessary in loading: rifled slugs, large buckshot, short cases, non-standard gauges, etc. In fact, for any purpose where there isn't room for a folded crimp, or where a standard fold-crimp die won't work.

The roll crimp may look complicated to one who's never seen it applied. Actually it's very simple, and the basic tool for doing the job hasn't changed in over a century. The crimp-head is in effect a female die to match the finished crimp form, but, instead of simply being pushed against the case mouth, it is also spun rapidly, thus developing friction heat which softens the case crimp area (paper *or* plastic) and makes the forming easy. Hot and soft, the case mouth will easily follow the desired contour.

Old-time tools used hand or foot power to spin the crimp head. That's not only hard work, but the head doesn't spin really fast enough to do a good job. The best method now is a Lyman crimp head chucked in a drill press. For this, the job is simplified if you make a case holder as shown from any good hardwood.

But first, if cases are paper, make sure the mouths are well waxed. Then process them through your regular tool, up through dropping the shot charge. Usually about 5/16" to 3/8" of case is needed above the shot to produce a good crimp, but this will vary according to wall thickness. Make certain you are leaving the proper amount before charging a large quantity.

Place a "B" card wad from Alcan or Federal in the case mouth and push it down so it levels the shot charge. A short length of dowel is handy for this. Slip a fully charged case in the case holder, all the way up to its rim, and squeeze the handles tightly to clamp it there. Slide the case across the drill press table under the spinning crimp head; lower the crimp head smoothly, with just enough pressure that in three to four seconds it rolls the mouth over into a full crimp. Don't apply excess force or the case will be bulged, and don't keep the head on the case any longer than necessary or it will get too hot. Quick and easy does it. If you overdo it with plastics, the case mouth will melt from the heat.

In the finished crimp, the turned-over case mouth should be smooth and shiny and should butt up tightly against the O-S wad without any gaps or flared-out excess. The shot charge should be somewhat compressed by this pressure, but not enough to cause dimples or bulges in the outer surface of the case.

Incidentally, paper cases that are worn enough that they don't hold a folded crimp perfectly, can be helped a bit by being run through a roll-crimp operation after being fully loaded. Just drip in a bit of wax and hit them lightly with the spinning crimp head. The crimp will be firmed up considerably.

You'll note no final sizing die is involved in roll crimping. For this reason, any length shell may be crimped with the proper gauge die, from shortened cases to the longest magnums. It is possible, though, to bulge unsupported cases by too much crimping pressure, causing chambering problems. Should this happen, run the offending shell through a conventional sizing die afterward. If it's bulged too much to enter the die, well, just forget it.

Metal cases, particularly the old Remington military 12-bore variety, are often handloaded and require somewhat different treatment. They are not processed on conventional integrated tools.

First, if the cases are so much expanded that they won't chamber freely, they must be resized, as are other metallics. A special resizing die available from Lyman or Hollywood is needed, the latter used in a metallic loading press, the former in a large vise or arbor press. Decapping and repriming is then done by hand. A simple punch takes care of the former, and for the latter, the case may be tapped with a dowel down over a primer set on a metal plate.

The powder charge is then placed in the case and the *card* O-P wad seated with a length of dowel. Since metal cases have thinner walls than paper or plastic, large diameter wads are necessary to secure a tight fit. For example, 12-gauge cases require 11-gauge wads, and so forth. Proper wads are available from Alcan. Filler wads are next seated (also special size), followed by the shot charge.

For use in double or other break-open guns, the case need not be crimped, and crimping will shorten its useful life. Simply seat a "B" card wad over the shot, then secure it in place with a bead of white glue, Duco cement, or waterglass. Melted wax will also serve, but not as well. Since no crimp is used, height of wad/shot column is not critical.

However, to permit use in repeating guns, such cases must either be crimped, or else have the mouth turned over into a smooth radius after loading, as described above. To crimp, leave only 1/16"-1/8" of case protruding past shot charge and seat a B wad. Make certain the wad is level,, then run the charged case carefully into a conventional folded-crimp die *only* until the mouth is turned over roughly 90°, forming a very narrow lip to hold the wad in place. It may be necessary to polish out a spare crimp die until it will accept the metal case freely.

An alternate crimping method consists of merely cutting a 45° countersink in a block of steel which is then placed over the case mouth and tapped gently to bend in the lip. Use is simplified if it is drilled through for a length of dowel to bear on the O-S wad and keep it from bouncing about during the tapping. Completed metal-case loads may be waterproofed with lacquer, wax, or waterglass over the O-S wad.

With standard loads, brass cases will seldom require resizing so long as they are fired in the same gun. So treated, they will last for countless reloadings.

Virtually all shotshell loading revolves around just those operations described. It matters not whether you are using a $10.95 Lee Loader or a $400 power-driven Hydramec, it's all the same as far as conventional loads are concerned. Special loads are another matter, and are covered in another chapter.

CHAPTER 33

Special-Purpose Loads

ALL MANNER of special loads can be made up. One popular at small-town turkey shoots is a 2 dram/2 oz. equiv. 12 gauge with very small shot, even as fine as No. 12. There they shoot at "luck targets" where the largest number of shot in a one-inch square (or the *one* shot pellet closest to a mark; the variations are many) wins. The more shot, the better the odds, and those boys are out for blood, no holds barred.

Loads of this type are handled conventionally, except that separate adjustable powder and shot measures are needed, and one must be prepared to use non-standard wads.

Then there is the "scatter load," sometimes called "brush load," intended to spread widely at short range. The most common method of accomplishing this is with shot separators as shown. One vertical or two vertical arranged as an X are easiest to use, since the shot charge can be dropped in place normally after the separators are in place. The card-wad separators are more trouble since 1/2, 1/3, or 1/4 of the charge must be dropped alternately with wads being seated. Aside from the unorthodox handling of shot and separators, loading is conventional and all operations can be performed on standard tools.

Incidentally, Remington makes its plastic Post Wad which gives some "scatter" effect. It may well give all the spread you need and

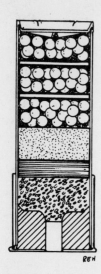

Left: scatter load using card wad separator; Right: same, with longitudinal card stock separators.

has the advantage of *not* requiring those special operations.

Buckshot loads present some problems. The first is that shot so large doesn't meter through a measure or charge bar. Individual pellets must be counted out for each charge and dropped into the case. In sizes of O and larger, even that isn't enough. In order to make the proper number fit into the case, they must be neatly stacked in precise layers. All of the larger sizes also require a roll crimp, there not being enough flexibility of the shot charge to allow the downward intrusion of crimp folds. As indicated in the data tables, buckshot charges in all but the smallest sizes are by pellet count, not weight. Aside from this special attention to shot charge and the roll crimp, loading is as usual.

Rifled slug loads are more popular than ever, due primarily to "shotgun slug only" deer seasons in many thickly-settled states. Standard hollow-base rifled slugs are available comercially, but they may also be cast of pure lead, then have the rifling swaged to form in Lyman tools and dies. Actually, the smooth, unrifled cast slugs appear to be as accurate as the rifled variety, so why go to all the trouble.

Accuracy with this type slug is greatly increased when a hard nitro card wad at least 1/8" thick is seated between it and the filler wad column. The best combination is a plastic cup O-P wad, composition filler wads, and one or two .135" or .200" nitro wads in that order. The nitro wad protects the weak slug base and drops cleanly from it in flight, while conventional filler wads do not.

Slugs require a roll crimp applied *without* any O-S card wad. The case mouth is simply rolled over tightly against the ogive of the slug itself. This is accomplished very easily with the standard roll-crimp head. Loading is otherwise conventional. Several other varieties of slug, some rifled, some not, are commercially available. All require some deviations in loading procedures and are supplied with full instructions. Several, notably the Vitt and Brenneke, have a complete wad column screwed to the rear of the slug where it remains throughout flight and impact. Then there is the Blondeau of spool shape (wasp-waisted) with a hole down its middle; air passing through the hole is claimed to stabilize the slug. Another is the MB1, which uses a 500-grain .50 caliber bullet enclosed in a plastic sabot which drops off at the muzzle. Of the entire lot, the Vitt and Brenneke seem the most popular and effective, while the MB1 is superior to all for retained velocity and energy and penetration.

Tracer loads are mentioned only because they don't exist. There are no suitable tracer elements available to handloaders, who are cautioned against trying to brew up some.

Satisfactory shotshell blanks can be loaded with smokeless powders. Loads are contained in the data tables and are self-explanatory. Just don't try to substitute powders.

Cases for all special purpose loads may be decapped, resized, primed, charged with powder, and have O-P wads seated on any conventional tool, with the exception that non-standard powder charges must be thrown separately from a good, well-regulated powder measure.

Filler wads may present a problem in that standard sizes available may not make up to the correct column length. In the old days, everyone was prepared with a sharp knife to split composition wads into any thicknesses needed. That is still the simplest way out, or, occasionally adding an extra card wad or plastic O-P wad will bring the column to correct length.

If you are using plastic shot cups, altering wad column length isn't so easy. Shortening it isn't practical, except by surgery on the Sullivan Vari-Wad, or by obtaining the Alcan Flite-Max cups with fibre insert, then slicing the insert back. Lengthening the column isn't too difficult. Just use a plastic O-P wad under the entire unit, or seat a thin wad in the bottom of the cup a *la* Winchester Win-Wad. Filler wads or plastic wads may be seated normally once brought to size, but many simply use a length of dowel to push them down by hand instead.

When unusually heavy or light shot charges are used, the depth of the plastic cup won't be right. If shot extends beyond the cup mouth, okay, patterns will still be better than if none of the shot were protected. Very light charges leave excess cup protruding to foul up the crimp. Of course, you can slice off the excess with a sharp knife or razor blade, but seating a thin wad in the bottom of the cup as a filler is easier.

Aside from separate adjustable powder and shot measures and a roll-crimp setup, no special tools are required. Even the very simple Lee Loader will do the job nicely, as will most integrated tools. However, if you really want to go first class, get an old Hollywood Senior Turret metallic press and a set of shotshell dies, shell holders, primer punch, etc. This setup is the most versatile of any I've used, and its turret operation makes it just as fast as many integrated loaders. The dies are adjustable in all respects and you'll be hard put to come up with a requirement they can't meet.

Nothing is much more frustrating when trying to inform one's self by reading a new book than to have to go to other references to find out what the words mean. To avoid that problem in this volume, we have compiled here a list of those words and terms most likely to appear strange to those of you who are not greatly experienced with the various aspects of handloading. We hope it makes your reading easier and more pleasant.

Action: Breech mechanism of a gun, by which it is loaded and unloaded, and which houses most of the moving parts.

Antimony: A silvery common metal melting at low temperatures and almost universally used to harden lead for bullets, bullet cores, and shot; often in conjunction with tin.

Anvil: In a primer or cartridge case, a fixed point against which the priming mixture is crushed, and thereby detonated by the action of the firing pin.

Back strap: The rear portion of that part of handgun frame to which the stocks are attached.

Barrel: The part(s) of a gun through which passes the bullet or shot, traveling from breech to muzzle.

Battery cup: In reference to a shotshell primer, the steel flanged cup (usually brass or copper plated) into which the primer cap and separate anvil are assembled.

Berdan: Col. Hiram Berdan credited with development of a type primer dependent upon an anvil integral with the case.

Black powder: A mixture of charcoal, sulphur, and saltpeter used as a propellant. Gives off much smoke when burned. See *Smokeless powder*.

Bolt (cylinder stop): A movable stud protruding through a revolver frame into a notch in the cylinder to hold said cylinder in alignment with barrel.

Bore: The inside of the barrel of a gun.

Bore diameter: In rifled arms, the diametrical measurement between tops of lands.

Bullet: The projectile *only*, not to be applied to the cartridge.

Bullet mould: Metallic device with a cavity(s) into which molten lead (or lead alloy) can be poured and allowed to harden to form the projectile.

Caliber: Bore or groove diameter expressed (in English) in decimals of an inch, otherwise in the metric system. Frequently compounded to indicate powder capacity of cartridge case; to show date of adoption; to show case length or to show proprietor, etc. E.g., .357 S&W Magnum, .22 Rem. Jet, 38-40 Winchester, etc.

Cannelure: Circumferential groove pressed or cast in a bullet, generally to allow crimping the case to the bullet. Also a groove in a case to provide a seat for the bullet. Sometimes used purely for identification or ornamentation.

Caplock: Used of a muzzle-loading gun whose ignition system employs a percussion cap, a small metal cup containing a detonating mixture. This cup, placed on a "nipple," transmits flame to the powder charge when struck by the gun's hammer.

Cap, primer: The cup in a shotshell battery cup primer that contains the priming compound.

Cartridge: A complete round of ammunition, made up simply of a cartridge case, primer, bullet (or shot) and powder.

Cartridge case: Commonly, the brass or copper envelope that contains powder, primer and projectile, but applicable to shotshells, too, whether of all metal, paper and metal or plastic and metal, or even all plastic.

Case forming: Action of converting by various means a cartridge case of one caliber to use as another caliber.

Case rim (flange): The thin circular portion of a cartridge case gripped by the extractor to remove it from the chamber.

CF: Centerfire (cartridges); those ignited by means of (generally) a separate and replaceable primer located centrally in the head of the case.

Chamber: That part of the bore, (or

Crimp: The bending inward of the case mouth perimeter, in order to grip and hold the bullet, or to keep the shot in a paper case intact.

Cylinder: In a revolver, a cartridge container that rotates (generally) around an axis parallel to and below the barrel.

Cylinder latch: A part, usually actuated by one's thumb, to disengage the cylinder of a revolver so the arm may be opened for loading.

Decapping: Punching out or otherwise removing a fired primer from a cartridge case.

Dent, oil: An inward depression in a cartridge case caused by lubricant trapped between it and the resizing die.

Density, loading: A percentage value indicating the portion of the cartridge case volume filled by the propellant charge.

Die, resizing: A cartridge-case-shaped cavity in a block of metal into which the case is pressed to be reduced to acceptable shape and dimensions.

Ejector: The device(s) at the barrel breech or within the action that forcibly expels the fired case from the gun. See *Extractor*.

Ejector rod: A rod protruding under the barrel of a revolver which is pressed rearward to extract and eject cases from the cylinder.

Energy: In bullets, the amount of work done, at given ranges, expressed in foot pounds.

Engraving: When a bullet is driven into the rifling and the lands cut or 'engrave' into the bullet surface.

Erosion: More or less gradual wearing away of rifling by combustion gas, heat and bullet friction.

Expander: A plug or ball forced into a cartridge case mouth to produce the correct diameter to permit seating and holding a bullet.

Extractor: Device that removes or par-cylinder, in a revolver) at the breech, formed to accept the cartridge.

Chamber cast: An impression of a gun chamber obtained by filling the chamber with a molten or liquid material which hardens without significant shrinkage.

Clip: See *Magazine*

Cordite: A British form of nitro-cellulose/nitroglycerine propellant extruded in strings or rods the length of the cartridge powder space. Considered highly erosive to barrels.

Crane (yoke): In a solid-frame, side-swing revolver, that part which is pivoted to the frame, (receiver) and carries the cylinder.

tially removes the fired cartridge case from the chamber. See *Ejector.*

Fire-form: To shape a cartridge case intimately to the chamber by firing it therein.

Firing pin: A part of the action, actuated by the trigger, that strikes the primer and fires the cartridge.

Flame temperature: The nominal temperature produced inside a cartridge case during combustion of the propellant.

Flash hole: The small hole connecting primer pocket to interior of case, through which primer flame passes.

Fluxing: Stirring and adding grease to molten lead to improve its casting qualities and float out impurities.

Flintlock: Used of a muzzleloading gun fired by means of a piece of flint, held in the hammer or "cock" jaws, striking against a steel "frizzen." Incandescent particles of steel scraped from the frizzen fall into a "pan" holding powder. This ignited powder flames through the "touch-hole," thus firing the main charge.

Foil: In reference to a primer, the thin disc of waterproof material pressed over the priming compound and lacquered in place to exclude moisture.

Follower: A metal platform in a clip or magazine that pushes the cartridges upward at the proper angle for feeding into the chamber.

Front strap: The front portion of the part of a handgun frame to which stocks are attached, sometimes including the trigger guard.

Gas check: A cup (usually copper) used on the base of a lead bullet to protect it from hot powder gases.

Gilding metal: A copper-zinc alloy used as bullet jacket material; usually 5% to 10% zinc.

Grain, powder: Commonly and confusingly used to mean individual kernels or particles of propellant. Particularly confusing since propellant is weighed in grains avoirdupois.

Grooves: Spiral cuts in a bore which cause the bullet to spin as it travels down the barrel.

Groove diameter: In rifled barrels, the diametrical measurement between bottoms of grooves.

Groove, lubricating: A groove around a bullet to contain lubricant.

Group: Number of shots fired into a target (number and range optional), usually with one sight setting.

Half jacket: A type of handgun bullet in which a thin, soft copper alloy jacket covers only the surface in contact with the bore.

Hammer: A part of the action (in some

guns) actuated by the trigger. The hammer drives the firing pin against the primer, thus igniting the cartridge powder charge.

Hand: That finger-like part attached to hammer or trigger that rotates the cylinder of a revolver when the arm is cocked.

Hang-fire: A cartridge which fires as long as several seconds after firing pin strikes primer.

Headspace: For rimmed cartridges, the distance from the face of the breechblock to the barrel seat for the forward surface of the case rim. For a rimless bottleneck cartridge, the distance from the face of the breech-block to a predetermined point on the shoulder of the chamber. For rimless straight cartridges, the distance from the face of the breechblock to the shoulder or ledge in the chamber.

Hollow point: A type of bullet containing a cavity in its point to promote expansion upon impact.

Lands: That portion of the bore remaining after the rifling or grooves have been cut.

Leade (Leed): The beginning of the rifling where the lands are tapered in thickness to provide clearance for the bullet.

Leading: Lead deposited on bore by bullets passing through.

Magazine: Device or reservoir to hold extra cartridges, of many types and names. "Clip," once reserved for the metal strip from which cartridges are fed into a magazine well, now refers to separate, detachable magazines also, as with those for self-loading pistols.

Matchlock: An early form of firearm in which the priming charge was ignited by a cord or "match" of slow-burning material.

Muzzle: End of barrel opposite to breech; point from which bullet or shot leaves barrel.

Muzzle pressure: Gas pressure in a gun barrel at the instant the bullet exits the muzzle.

M. E.: Muzzle Energy. See *Energy*.

M.V.: Muzzle Velocity. See *Velocity*.

Nipple: On muzzle-loading arms, the small metal cone at the rear of the barrel (or cylinder) through which the flame from the percussion cap passes to ignite the powder charge.

Pan: See *Flintlock*.

Percussion cap: Small metallic cup containing fulminating material that explodes when struck by gun's hammer. See *Nipple*.

Pistol: Reputedly derived from Pistoia, an early gunmaking center in Italy. Any small, concealable, short-barreled hand weapon, generally *not* a revolver.

Port pressure: Gas pressure in a gun bar-rel at the instant the bullet passes over a gas port therein.

Powder, ball: Smokeless nitrocellulose propellant formed into small balls in emulsion. Often balls are flattened to oblate spheroid forms by rolling.

Powder, flake: Nitrocellulose propellant rolled into thin sheets, then cut into square or diamond-shaped individual flakes; not perforated. Typical of European propellants.

Powder, stick (extruded; IMR-type; rod): Nitrocellulose propellant extruded in round or string form, usually containing a central perforation, then cut to length to form short individual cylindrical granules.

Pressure: The gas pressure generated in a cartridge on its being fired, usually expressed in (greatest) pounds per square inch (p.s.i.).

Primer: In a centerfire cartridge, the small cup containing a detonating mixture, which is seated in a recess in the base of the case.

Primer, corrosive: A primer containing any compound which produces hygroscopic residue which tends to promote rapid rusting of barrels.

Primer, extrusion: When the soft metal of the primer cup flows back under pressure into space between firing pin and bolt or breech face, but does not rupture.

Primer, mercuric: A primer containing fulminate of mercury, the combustion products of which amalgamate with and embrittle case brass.

Primer, pierced (perforated): When a disc of cup metal is blown out into the firing pin hole in the breech; usually caused by excessive firing pin clearance and protrusion combined with high chamber pressure.

Propellant: Powder. The chemical compound serving as the fuel in the cartridge. Generally classed as either black or smokeless.

Ratchet: A notched ring centered in the rear of a revolver cylinder and engaged by the hand to turn said cylinder.

Rebated-head: A form of cartridge case in which the flange provided for extraction is of smaller diameter than the adjacent base of the case.

Recoil: The backward thrust of a gun caused by the reaction to the powder gases pushing the bullet forward.

Recoil shield: That portion of a revolver frame which supports the heads of the cartridges to prevent their moving out of the cylinder.

RF: Rimfire cartridges. Those containing their primer mixture in the rim, which is where they are struck by the firing pin.

Rifling: Spiral grooving cut into the bore of rifles and handguns to impart spin to their bullets, thus assuring point-on flight and accuracy.

Rim: The projecting edge of the base or "head" of certain cartridges.

Rupture (separation), case: A failure wherein cartridge case separates in two parts around its perimeter just ahead of the rim or extraction groove.

Safety lever (block): In modern revolvers a part that moves to prevent the hammer from going fully forward unless the trigger is deliberately pulled fully to the rear.

Shoulder: The sloping portion of a bottleneck cartridge case that joins the body and neck.

Sizing: In handloading cartridges, sizing (or resizing) brings the fired cartridge case back to the (full or partial) dimensions of the new or unfired case. Bullets are also sized.

Smokeless powder: Gunpowder which gives off almost no smoke when burned. See *Black Powder*. Usually made by nitrating and otherwise chemically treating purified cotton waste.

S.P.: Soft Point: A term applied to bullets with partial metal jacketing, having some lead exposed at the point.

Sprue (sprew): The excess portion of a cast bullet filling the funnel-like area through which molten lead enters the hold. Cut off before bullet is used.

Sprue-cutter (plate): The pivoted plate on top of a bullet mold through which molten lead enters. Swung aside when bullet is hard to cut of excess lead sprue.

Swage: A die containing a shaped cavity into or through which a bullet is pressed and confined by one or more shaped punches to give it shape and dimensions desired.

Throat, barrel: That portion of a revolver barrel that is "funneled" to facilitate entry of the bullet as it passes from cylinder to barrel.

Throat, cylinder: That portion of a revolver chamber through which the bullet must pass before entering the barrel.

Top strap: The top portion of a revolver frame passing over the cylinder.

Trajectory: Curved path of bullet in flight, a parabola.

Twist: Pitch of the rifling, usually uniform, and expressed in turns or part-turns in so many inches. Less common, "progressive" or "gain" twist, usually starting at a rate at breech that becomes progressively faster.

Underlug: On S&W revolvers an integral protrusion of the barrel which houses the forward portion of the cylinder locking mechanism.

Upset (slug): Wherein a bullet expands radially and is foreshortened by violent acceleration given it by propellant gases. Also accomplished in swaging to increase bullet diameter.

Velocity: Projectile speed, usually measured in feet per second (f.p.s.) at the muzzle and other distances such as 100 yards, 200 yards, etc.

Vent: Orifice through the nipple.

Venting: Cutting air-escape grooves in bullet mould block meeting surfaces.

Wad, base: A fibre (usually) cylindrical plug in the head of a built-up shotshell to provide a seat for the primer, and to bind head and body together.

Wad column: The entire stack of wads in a shotshell between powder and shot.

Wheellock: Used of a muzzle-loading gun fired by means of a piece of flint or tyrites, held in the hammer jaws, which is held over a serrated steel wheel. This wheel, set in motion by a tensioned spring, protrudes through the bottom of the "pan" (wherein powder has been placed) and bears against the flint. Sparks are created, as in the flintlock, and the gun is fired by a flame passing through the touch-hole.

Following are Ballistic Data Charts for all standard calibers of rim and centerfire handgun and rifle ammunition and shotshells. Their major value is to serve as a basis for comparison of your handloads with what the factories are supplying.

Ballistics for Standard Centerfire Rifle Ammunition Produced by the Major U. S. Manufacturers

Cartridge	Wt. Grs.	Bullet Type(g)	Velocity (fps) Muzzle	100 yds.	200 yds.	300 yds.	Energy (ft. lbs.) Muzzle	100 yds.	200 yds.	300 yds.	Mid-Range Trajectory 100 yds.	200 yds.	300 yds.
218 Bee	46	HP	2860	2160	1610	1200	835	475	265	145	0.7	3.8	11.6
22 Hornet	45	SP	2690	2030	1510	1150	720	410	230	130	0.8	4.3	13.0
22 Hornet (c, d)	45	HP	2690	2030	1510	1150	720	410	230	130	0.8	4.3	13.0
22 Hornet	46	HP	2690	2030	1510	1150	740	420	235	135	0.8	4.3	13.0
222 Remington (e)	50	PSP, MC, PL	3200	2660	2170	1750	1140	785	520	340	0.5	2.5	7.0
222 Remington Magnum (c, d)	55	SP, PL	3300	2800	2340	1930	1330	955	670	455	0.5	2.3	6.1
223 Remington (c, d, e)	55	SP, PL	3300	2800	2340	1930	1330	955	670	455	0.5	2.1	5.4
22-250 Remington (a, c, d)	55	PSP, PL	3760	3230	2745	2305	1730	1275	920	650	0.4	1.7	4.5
225 Winchester (a, b)	55	PSP	3650	3140	2680	2270	1630	1200	875	630	0.4	1.8	4.8
243 Winchester (e)	80	PSP, PL	3500	3080	2720	2410	2180	1690	1320	1030	0.4	1.8	4.7
243 Winchester (e)	100	PP, CL, PSP	3070	2790	2540	2320	2090	1730	1430	1190	0.5	2.2	5.5
6mm Remington (c, d)	80	PSP, HP, PL	3450	3130	2750	2400	2220	1740	1340	1018	0.4	1.8	4.7
6mm Remington (c, d)	100	PCL	3190	2920	2660	2420	2260	1890	1570	1300	0.5	2.1	5.1
244 Remington (c, d)	90	PSP	3200	2850	2530	2230	2050	1630	1280	995	0.5	2.1	5.5
25-06 Remington (c, d)	87	HP	3500	3070	2680	2310	2370	1820	1390	1030	Not Available		
25-06 Remington (c, d)	120	PSP	Not Available				Not Available				Not Available		
25-20 Winchester	86	L, Lu	1460	1180	1030	940	405	265	200	170	2.6	12.5	32.0
25-20 Winchester	86	SP	1460	1180	1030	940	405	265	200	170	2.6	12.5	32.0
25-35 Winchester	117	SP, CL	2300	1910	1600	1340	1370	945	665	465	1.0	4.6	12.5
250 Savage	87	PSP, SP	3030	2660	2330	2060	1770	1370	1050	820	0.6	2.5	6.4
250 Savage	100	ST, CL, PSP	2820	2460	2140	1870	1760	1340	1020	775	0.6	2.9	7.4
256 Winchester Magnum (b)	60	OPE	2800	2070	1570	1220	1040	570	330	200	0.8	4.0	12.0
257 Roberts (a, b)	87	PSP	3200	2840	2500	2190	1980	1560	1210	925	0.5	2.2	5.7
257 Roberts	100	ST, CL	2900	2540	2210	1920	1870	1430	1080	820	0.6	2.7	7.0
257 Roberts	117	PP, CL	2650	2280	1950	1690	1820	1350	985	740	0.7	3.4	8.8
6.5 Remington Magnum (c)	100	PSPCL	3450	3070	2690	2320	2640	2090	1610	1190	Not Available		
6.5mm Remington Magnum (c)	120	PSPCL	3030	2750	2480	2230	2450	2010	1640	1330	0.5	2.3	5.7
264 Winchester Magnum	100	PSP, CL	3700	3260	2880	2550	3040	2360	1840	1440	0.4	1.6	4.2
264 Winchester Magnum	140	PP, CL	3200	2940	2700	2480	3180	2690	2270	1910	0.5	2.0	4.9

Ballistics for Standard Centerfire Rifle Ammunition Produced by the Major U. S. Manufacturers (continued)

Cartridge	Bullet Wt. Grs.	Type (g)	Velocity (fps) Muzzle	100 yds.	200 yds.	300 yds.	Energy (ft. lbs.) Muzzle	100 yds.	200 yds.	300 yds.	Mid-Range Trajectory 100 yds.	200 yds.	300 yds.
270 Winchester	100	PSP	3480	3070	2690	2340	2690	2090	1600	1215	0.4	1.8	4.8
270 Winchester (e)	130	PP, PSP	3140	2880	2630	2400	2850	2390	2000	1660	0.5	2.1	5.3
270 Winchester	130	ST, CL, BP, PP	3140	2850	2580	2320	2840	2340	1920	1550	0.5	2.1	5.3
270 Winchester (c, d)	150	CL	2800	2440	2140	1870	2610	1980	1520	1160	0.6	2.9	7.6
270 Winchester (a, b, e)	150	PP	2900	2620	2380	2160	2800	2290	1890	1550	0.6	2.5	6.3
280 Remington (c, d)	125	PCL	3190	2880	2590	2320	2820	2300	1860	1490	0.5	2.1	5.3
280 Remington (c, d)	150	PCL	2900	2670	2450	2220	2800	2370	2000	1640	0.6	2.5	6.1
280 Remington (c, d)	165	CL	2820	2510	2220	1970	2910	2310	1810	1420	0.6	2.8	7.2
284 Winchester (a, b)	125	PP	3200	2880	2590	2310	2840	2300	1860	1480	0.5	2.1	5.3
284 Winchester (a, b)	150	PP	2900	2630	2380	2160	2800	2300	1890	1550	0.6	2.5	6.3
7mm Mauser (e)	175	SP	2490	2170	1900	1680	2410	1830	1400	1100	0.8	3.7	9.5
7mm Remington Magnum	125	CL	3430	3080	2750	2450	3260	2630	2100	1660	0.6	1.8	4.7
7mm Remington Magnum (e)	150	PP, CL	3260	2970	2700	2450	3540	2940	2430	1990	0.4	2.0	4.9
7mm Remington Magnum (e)	175	PP	3070	2720	2400	2120	3660	2870	2240	1750	0.5	2.4	6.1
7mm Remington Magnum (c, d)	175	PCL	3070	2860	2660	2460	3660	3170	2740	2350	0.5	2.1	5.2
30 Carbine (e)	110	HSP, SP	1980	1540	1230	1040	950	575	370	260	1.4	7.5	21.7
30-30 Winchester (c, d)	150	CL	2410	1960	1620	1360	1930	1280	875	616	0.9	4.5	12.5
30-30 Winchester (e)	150	HP	2410	2020	1700	1430	1930	1360	960	680	0.9	4.2	11.0
30-30 Winchester (a, b)	150	PP, ST, OPE	2410	2020	1700	1430	1930	1360	960	680	0.9	4.2	11.0
30-30 Winchester (a)	170	PP, HP, CL, ST, MC	2220	1890	1630	1410	1860	1350	1000	750	1.2	4.6	12.5
30 Remington	170	ST, CL	2120	1820	1560	1350	1700	1250	920	690	1.1	5.3	14.0
30-06 Springfield	110	PSP	3370	2830	2350	1920	2770	1960	1350	900	0.5	2.2	6.0
30-06 Springfield	125	PSP	3200	2810	2480	2200	2840	2190	1710	1340	0.5	2.2	5.6
30-06 Springfield (c, d)	150	BP	2970	2710	2470	2240	2930	2440	2030	1670	0.5	2.4	6.0
30-06 Springfield (e)	150	PP	2970	2620	2300	2010	2930	2280	1760	1340	0.6	2.5	6.5
30-06 Springfield	150	ST, PCL, PSP	2970	2670	2400	2130	2930	2370	1920	1510	0.6	2.4	6.1
30-06 Springfield	180	PP, CL, PSP	2700	2330	2010	1740	2700	2170	1610	1210	0.7	3.1	8.3
30-06 Springfield (e)	180	ST, BP, PCL	2700	2470	2250	2040	2910	2440	2020	1660	0.7	2.9	7.0
30-06 Springfield	180	MCBT, MAT	2700	2520	2350	2190	2910	2540	2200	1900	0.6	2.8	6.7
30-06 Springfield	220	PP, CL	2410	2120	1870	1670	2830	2190	1710	1360	0.8	3.9	9.8

Modern Handloading

Ballistics for Standard Centerfire Rifle Ammunition Produced by the Major U. S. Manufacturers (continued)

Cartridge	Wt. Grs.	Bullet Type (g)	Velocity (fps)				Energy (ft. lbs.)				Mid-Range Trajectory		
			Muzzle	100 yds.	200 yds.	300 yds.	Muzzle	100 yds.	200 yds.	300 yds.	100 yds.	200 yds.	300 yds.
30-06 Springfield (a, b)	220	ST	2410	2180	1980	1790	2830	2320	1910	1560	0.8	3.7	9.2
30-40 Krag	180	PP, CL	2470	2120	1830	1590	2440	1790	1340	1010	0.8	3.8	9.9
30-40 Krag	180	ST, PCL	2470	2250	2040	1850	2440	2020	1660	1370	0.8	3.5	8.5
30-40 Krag	220	ST	2200	1990	1800	1630	2360	1930	1580	1300	1.0	4.4	11.0
300 Winchester Magnum	150	PP, PCL	3400	3050	2730	2430	3850	3100	2480	1970	0.4	1.9	4.8
300 Winchester Magnum	180	PP, PCL	3070	2850	2640	2440	3770	3250	2790	2380	0.5	2.1	5.3
300 Winchester Magnum (a, b)	220	ST	2720	2490	2270	2060	3620	3030	2520	2070	0.6	2.9	6.9
300 H&H Magnum (a, b)	150	ST	3190	2870	2580	2300	3390	2740	2220	1760	0.5	2.1	5.2
300 H&H Magnum	180	ST, PCL	2920	2670	2440	2220	3400	2850	2380	1970	0.6	2.4	5.8
300 H&H Magnum	220	ST, CL	2620	2370	2150	1940	3350	2740	2260	1840	0.7	3.1	7.7
300 Savage (e)	150	PP	2670	2350	2060	1800	2370	1840	1410	1080	0.7	3.2	8.0
300 Savage	150	ST, PCL	2670	2390	2130	1890	2370	1900	1510	1190	0.7	3.0	7.6
300 Savage (c, d)	150	CL	2670	2270	1930	1660	2370	1710	1240	916	0.7	3.3	9.3
300 Savage (e)	180	PP, CL	2370	2040	1760	1520	2240	1660	1240	920	0.9	4.1	10.5
300 Savage	180	ST, PCL	2370	2160	1960	1770	2240	1860	1530	1250	0.9	3.7	9.2
303 Savage (c, d)	180	CL	2140	1810	1550	1340	1830	1310	960	715	1.1	5.4	14.0
303 Savage (a, b)	190	ST	1980	1680	1440	1250	1650	1190	875	660	1.3	6.2	15.5
303 British (e)	180	PP, CL	2540	2300	2090	1900	2580	2120	1750	1440	0.7	3.3	8.2
303 British (c, d)	215	SP	2180	1900	1660	1460	2270	1720	1310	1020	1.1	4.9	12.5
308 Winchester	110	PSP	3340	2810	2340	1920	2730	1930	1349	900	0.5	2.2	6.0
308 Winchester (a, b)	125	PSP	3100	2740	2430	2160	2670	2080	1640	1300	0.5	2.3	5.9
308 Winchester (e)	150	PP	2860	2520	2210	1930	2730	2120	1630	1240	0.6	2.7	7.0
308 Winchester	150	ST, PCL	2860	2570	2300	2050	2730	2200	1760	1400	0.6	2.6	6.5
308 Winchester (e)	180	PP, CL	2610	2250	1940	1680	2720	2020	1500	1130	0.7	3.4	8.9
308 Winchester	180	ST, PCL	2610	2390	2170	1970	2720	2280	1870	1540	0.8	3.1	7.4
308 Winchester (a, b)	200	ST	2450	2210	1980	1770	2670	2170	1750	1400	0.8	3.6	9.0
32 Winchester Special (c, d, e)	170	HP, CL	2280	1920	1630	1410	1960	1390	1000	750	1.0	4.8	12.5
32 Winchester Special (a, b)	170	PP, ST	2280	1870	1560	1330	1960	1320	920	665	1.0	4.8	13.0
32 Remington (c, d)	170	CL	2120	1800	1540	1340	1700	1220	895	680	1.0	4.9	13.0
32 Remington (a, b)	170	ST	2120	1760	1460	1220	1700	1170	805	560	1.1	5.3	14.5

Ballistics for Standard Centerfire Rifle Ammunition Produced by the Major U. S. Manufacturers (continued)

Cartridge	Wt. Grs.	Bullet Type (g)	Velocity (fps) Muzzle	100 yds.	200 yds.	300 yds.	Energy (ft. lbs.) Muzzle	100 yds.	200 yds.	300 yds.	Mid-Range Trajectory 100 yds.	200 yds.	300 yds.
32-20 Winchester HV (f)	80	OPE, HP	2100	1430	1090	950	780	365	210	160	1.5	8.5	24.5
32-20 Winchester	100	SP, L, Lu	1290	1060	940	840	370	250	195	155	3.3	15.5	38.0
8mm Mauser (e)	170	PP, CL	2570	2140	1790	1520	2490	1730	1210	870	0.8	3.9	10.5
338 Winchester Magnum (a, b)	200	PP	3000	2690	2410	2170	4000	3210	2580	2090	0.5	2.4	6.0
338 Winchester Magnum (a, b)	250	ST	2700	2430	2180	1940	4050	3280	2640	2090	0.7	3.0	7.4
338 Winchester Magnum (a, b)	300	PP	2450	2160	1910	1690	4000	3110	2430	1900	0.8	3.7	9.5
348 Winchester (a)	200	ST	2530	2220	1940	1680	2840	2190	765	509	0.4	1.7	4.7
348 Winchester (c, d)	200	CL	2530	2140	1820	1570	2840	2030	1470	1090	0.8	3.8	10.0
35 Remington (c, d)	150	CL	2400	1960	1580	1280	1920	1280	835	545	0.9	4.6	13.0
35 Remington (e)	200	PP, ST, CL	2100	1710	1390	1160	1950	1300	860	605	1.2	6.0	16.5
350 Remington Magnum (c, d)	200	PCL	2710	2410	2130	1870	3260	2570	2000	1550	Not Available		
350 Remington Magnum (c, d)	250	PCL	2410	2190	1980	1790	3220	2660	2180	1780	Not Available		
351 Winchester Self-Loading	180	SP, MC	1850	1560	1310	1140	1370	975	685	520	1.5	7.8	21.5
358 Winchester (a, b)	200	ST	2530	2210	1910	1640	2840	2160	1610	1190	0.8	3.6	9.4
358 Winchester (a, b)	250	ST	2250	2010	1780	1570	2810	2230	1760	1370	1.0	4.4	11.0
375 H&H Magnum	270	PP, SP	2740	2460	2210	1990	4500	3620	2920	2370	0.7	2.9	7.1
375 H&H Magnum	300	ST	2550	2280	2040	1830	4330	3460	2770	2230	0.7	3.3	8.3
375 H&H Magnum	300	MC	2550	2180	1860	1590	4330	3160	2300	1680	0.7	3.6	9.3
38-40 Winchester	180	SP	1330	1070	960	850	705	455	370	290	3.2	15.0	36.5
44 Magnum (c, d)	240	SP	1750	1360	1110	980	1630	985	655	210	1.6	8.4	—
44 Magnum (b)	240	HSP	1750	1350	1090	950	1630	970	635	480	1.8	9.4	26.0
444 Marlin (c)	240	SP	2400	1845	1410	1125	3070	1815	1060	675	Not Available		
44-40 Winchester	200	SP	1310	1050	940	830	760	490	390	305	3.3	15.0	36.5
45-70 Government	405	SP	1320	1160	1050	990	1570	1210	990	880	2.9	13.0	32.5
458 Winchester Magnum	500	MC	2130	1910	1700	1520	5040	4050	3210	2570	1.1	4.8	12.0
458 Winchester Magnum	510	SP	2130	1840	1600	1400	5140	3830	2900	2220	1.1	5.1	13.5

(a)—Winchester only; (b)—Remington only; (c)—Peters only; (d)—Speer DWM; (e)—Cartridges also available from Federal; (f)—Not safe in Winchester 1873 rifle or handguns; (g)—HP-Hollow Point; SP-Soft Point; PSP-Pointed Soft Point; PP-Winchester Power Point; L-Lead; Lu-Lubaloy; ST-Silvertip; HSP-Hollow Soft Point; MC-Metal Case; BT-Boat Tail; MAT-Match; BP-Bronze Point; CL-Core Lokt; PCL-Pointed Core Lokt; OPE-Open Point Expanding; PL-Power-Lokt.

Modern Handloading

Ballistics for Standard Pistol and Revolver Centerfire Ammunition Produced by the Major U. S. Manufacturers

Cartridge	Grs.	Bullet Style	Muzzle Velocity	Muzzle Energy	Barrel Inches
22 Jet	40	SP	2100	390	8-3/8
221 Fireball	50	SP	2650	780	10-1/2
25 (6.35mm) Auto	50	MC	810	73	2
256 Winchester Magnum	60	HP	2350	735	8-1/2
30 (7.65mm) Luger Auto	93	MC	1220	307	4-1/2
30 (7.63mm) Mauser Auto	85	MC	1410	375	5-1/2
32 S&W Blank	No bullet		—	—	—
32 S&W Blank, BP	No bullet		—	—	—
32 Short Colt	80	Lead	745	100	4
32 Long Colt, IL	82	Lub.	755	104	4
32 Colt New Police	100	Lead	680	100	4
32 (7.65mm) Auto	71	MC	960	145	4
32 (7.65mm) Auto Pistol	77	MC	900	162	4
32 S&W	88	Lead	680	90	3
32 S&W Long	98	Lead	705	115	4
7.5 Nagant	104	Lead	722	120	4-1/2
32-20 Winchester	100	Lead	1030	271	6
32-20 Winchester	100	SP	1030	271	6
357 Magnum	158	SP	1550	845	8-3/8
357 Magnum	158	MP	1410	695	8-3/8
357 Magnum	158	Lead	1410	696	8-3/8
357 Magnum	158	JSP	1450	735	8-3/8
9mm Luger	116	MC	1165	349	4
9mm Luger Auto	124	MC	1120	345	4
38 S&W Blank	No bullet		—	—	—
38 Smith & Wesson	146	Lead	685	150	4
38 S&W	146	Lead	730	172	4
380 MK II	180	MC	620	153	5
38 Special Blank	No bullet		—	—	—
38 Special, IL	150	Lub.	1060	375	6
38 Special, IL	150	MC	1060	375	6
38 Special	158	Lead	855	256	6
38 Special	200	Lead	730	236	6
38 Special	158	MP	855	256	6
38 Special	125	SJHP	Not available		
38 Special	158	SJHP	Not available		
38 Special WC	148	Lead	770	195	6
38 Special Match, IL	148	Lead	770	195	6
38 Special Match, IL	158	Lead	855	256	6
38 Special Hi-Speed	158	Lead	1090	425	6
38 Special	158	RN	900	320	6
38 Colt New Police	150	Lead	680	154	4
38 Short Colt	128	Lead	730	150	6
38 Short Colt, Greased	130	Lub.	730	155	6
38 Long Colt	150	Lead	730	175	6
38 Super Auto	130	MC	1280	475	5
38 Auto, for Colt 38 Super	130	MC	1280	475	5
38 Auto	130	MC	1040	312	4-1/2
380 Auto	95	MC	955	192	3-3/4
38-40 Winchester	180	SP	975	380	5
41 Long Colt, IL	200	Lub.	730	230	6
41 Remington Magnum	210	Lead	1050	515	8-3/4
41 Remington Magnum	210	SP	1500	1050	8-3/8
44 S&W Special	246	Lead	755	311	6-1/2
44 Remington Magnum	240	SP	1470	1150	6-1/2

Ballistics for Standard Pistol and Revolver Centerfire Ammunition (cont.)

Cartridge	Grs.	Bullet Style	Muzzle Velocity	Muzzle Energy	Barrel Inches
44 Remington Magnum	240	Lead	1470	1150	6-1/2
44-40 Winchester	200	SP	975	420	7-1/2
45 Colt	250	Lead	860	410	5-1/2
45 Colt, IL	255	Lub., L	860	410	5-1/2
45 Auto	230	MC	850	369	5
45 ACP	230	JHP	850	370	5
45 Auto WC	185	MC	775	245	5
45 Auto MC	230	MC	850	369	5
45 Auto Match	185	MC	775	247	5
45 Auto Match, IL	210	Lead	710	235	5
45 Auto Rim	230	Lead	810	335	5-1/2

IL-Inside Lubricated; JSP-Jacketed Soft Point; WC-Wadcutter; RH-Round Nose; HP-Hollow Point; Lub-Lubricated; MC-Metal Case; SP-Soft Point; MP-Metal Point; LGC-Lead Gas Check; JHP-Jacketed Hollow Point.

Ballistics of Super Vel Cartridge Corporation Handgun Ammunition

Cartridge	Gr. Style	Bullet Style (a)	Muzzle Velocity	Muzzle Energy	Barrel Inches
380 ACP	80	JHP	1026	188	5
9mm Luger	90	JHP	1422	402	5
9mm Luger	110	SP	1325	428	5
38 Special	110	JHP/SP	1370	458	6
38 Special	147	HBWC	775	196	6
38 Special Int.	158	Lead	1110	439	6
357 Magnum	110	JHP/SP	1690	697	6
44 Magnum	180	JHP/SP	2005	1607	6
45 Auto	190	JHP	1060	743	5

(a) JHP-Jacketed Hollow Point; JSP-Jacketed Soft Point; HBWC-Hollow Base Wad Cutter.

Ballistics of Standard Shotshell Ammunition
Loaded by the Major U. S. Manufacturers

Gauge	Length Shell Ins.	Powder Equiv. Drams	Shot Ozs.	Shot Size
MAGNUM LOADS				
10	3-1/2	5	2	2, 4
12	3	4-1/2	1-7/8	BB, 2, 4
12	3	4-1/4	1-5/8	2, 4, 6
12	2-3/4	4	1-1/2	2, 4, 5, 6
16	2-3/4	3-1/2	1-1/4	2, 4, 6
20	3	3-1/4	1-1/4	2, 4, 6, 7½
20	3	Max	1-3/16	4
20	2-3/4	3	1-1/8	2, 4, 6, 7½
28	2-3/4	Max	1	6, 7½, 8, 9

Modern Handloading

Gauge	Length Shell Ins.	Powder Equiv. Drams	Shot Ozs.	Shot Size
LONG RANGE LOADS				
10	2-7/8	4-3/4	1-5/8	4
12	2-3/4	3-3/4	1-1/4	BB, 2, 4, 5, 6, 7½, 9
16	2-3/4	3-1/4	1-1/8	4, 5, 6, 7½, 9
16	2-3/4	3	1-1/8	4, 5, 6, 7½, 9
20	2-3/4	2-3/4	1	4, 5, 6, 7½, 9
28	2-3/4	2-1/4	3/4	4, 6, 7½, 9
FIELD LOADS				
12	2-3/4	3-1/4	1-1/4	7½, 8, 9
12	2-3/4	3-1/4	1-1/8	4, 5, 6, 7½, 8, 9
12	2-3/4	3	1-1/8	4, 5, 6, 8, 9
12	2-3/4	3	1	4, 5, 6, 8
16	2-3/4	2-3/4	1-1/8	4, 5, 6, 7½, 8, 9
16	2-3/4	2-1/2	1	4, 5, 6, 8, 9
20	2-3/4	2-1/2	1	4, 5, 6, 7½, 8, 9
20	2-3/4	2-1/4	7/8	4, 5, 6, 8, 9
SCATTER LOADS				
12	2-3/4	3	1-1/8	8
16	2-3/4	2-1/2	1	8
20	2-3/4	2-1/4	7/8	8
TARGET LOADS				
12	2-3/4	3	1-1/8	7½, 8, 9
12	2-3/4	2-3/4	1-1/8	7½, 8, 9
16	2-3/4	2-1/2	1	8, 9
20	2-3/4	2-1/4	7/8	8,9
28	2-3/4	2-1/4	3/4	9
410	3	Max	3/4	4, 5, 6, 7½, 9
410	2-1/2	Max	1/2	4, 5, 6, 7½, 9
SKEET & TRAP				
12	2-3/4	3	1-1/8	7½, 8, 9
12	2-3/4	2-3/4	1-1/8	7½, 8, 9
16	2-3/4	2-1/2	1	8, 9
20	2-3/4	2-1/4	7/8	8, 9
BUCKSHOT				
12	3 Mag	4-1/2	—	00 Buck—15 pellets
12	3 Mag	4-1/2	—	4 Buck—41 pellets
12	2-3/4 Mag	4	—	2 Buck—20 pellets
12	2-3/4 Mag	4	—	00 Buck—12 pellets
12	2-3/4	3-3/4	—	00 Buck— 9 pellets
12	2-3/4	3-3/4	—	0 Buck—12 pellets
12	2-3/4	3-3/4	—	1 Buck—16 pellets
12	2-3/4	3-3/4	—	4 Buck—27 pellets
16	2-3/4	3	—	1 Buck—12 pellets
20	2-3/4	2-3/4	—	3 Buck—20 pellets
RIFLED SLUGS				
12	2-3/4	3-3/4	1	Slug
16	2-3/4	3	1-7/8	Slug
20	2-3/4	2-3/4	5/8	Slug
410	2-1/2	Max	1/5	Slug

Speer/DWM Ballistics

Ammunition Manufactured by DWM in Germany and Distributed Through Speer Outlets in This Country

Cartridge	Bullet Wt. Grs.	Type (a)	Velocity (fps) Muzzle	100 yds.	200 yds.	300 yds.	Energy (ft. lbs.) Muzzle	100 yds.	200 yds.	300 yds.	Mid-Range Trajectory 100 yds.	200 yds.	300 yds.
5.6 X 35R Vierling	46	SP	2030	1500	1140		418	224	130		1.2	7.5	
5.6 x 50R (Rimmed) Mag.	50	PSP	Not Available				Not Available				Not Available		
5.6 x 52R (Savage H.P.)	71	PSP	2850	2460	2320	2200	1280	947	846	766	.3	2.3	6.5
5.6 x 61SE	77	PSP	3700	3360	3060	2790	2350	1920	1605	1345	.1	1.1	3.4
5.6 x 61R	77	PSP	3480	3140	2840	2560	2070	1690	1370	1120	.1	1.3	4.0
6.5 x 54 MS	159	SP	2170	1925	1705	1485	1660	1300	1025	810	.5	4.1	11.5
6.5 x 57 Mauser	93	PSP	3350	2930	2570	2260	2300	1760	1350	1040	.1	1.7	4.8
6.5 x 57 R	93	PSP	3350	2930	2570	2260	2300	1760	1350	1040	.1	1.7	4.8
7 x 57 Mauser	103	PSP	3330	2865	2450	2060	2550	1890	1380	977	.1	1.7	5.2
	162	TIG	2785	2480	2250	2060	2780	2200	1820	1520	.3	2.4	6.7
7 x 57 R.	103	PSP	3260	2810	2390	2000	2430	1820	1320	920	.1	1.8	5.3
	139	SP	2550	2240	1960	1720	2000	1540	1190	910	.3	2.9	8.6
	162	TIG	2710	2420	2210	2020	2640	2120	1750	1460	.3	2.4	6.9
7 x 64	103	PSP	3572	3110	2685	2283	2930	2230	1670	1190	.1	1.4	4.4
	139	SP	3000	2570	2260	1980	2780	2040	1570	1200	.2	2.2	6.4
	162	TIG	2960	2603	2375	2200	3150	2440	2030	1740	.2	2.0	6.0
	177	TIG	2880	2665	2490	2325	3270	2820	2440	2130	.2	2.0	5.6
7 x 65 R	103	PSP	3480	3010	2590	2200	2770	2100	1540	1120	.1	1.5	4.7
	139	SP	3000	2570	2260	1980	2780	2040	1570	1200	.2	2.2	6.4
	162	TIG	2887	2540	2320	2140	3000	2320	1930	1650	.2	2.2	6.3
	177	TIG	2820	2600	2420	2255	3120	2660	2300	2000	.2	2.1	5.9
7mm SE	169	ToSto	3300	3045	2825	2620	4090	3480	3010	2600	.1	1.4	3.9
7 x 75 R SE	169	ToSto	3070	2840	2630	2430	3550	3050	2620	2240	.1	1.6	4.5
30-06	180	TUG	2854	2562	2306	2077	3261	2632	2133	1726	.2	2.2	6.3
8 x 57 JS	123	SP	2968	2339	1805	1318	2415	1497	897	477	.2	2.7	8.8
	198	TIG	2732	2415	2181	1985	3276	2560	2083	1736	.3	2.5	7.1
8 x 57 JR	196	SP	2391	1991	1742	1565	2488	1736	1316	1056	.5	3.9	11.2
8 x 57 JRS	123	SP	2970	2340	1805	1318	2415	1497	897	477	.2	2.7	8.8
	196	SP	2480	2140	1870	1640	2680	2000	1510	1165	.4	3.3	9.4
8 x 60 S	198	TIG	2600	2320	2105	1930	2970	2350	1950	1620	.3	2.7	7.6
	196	SP	2585	2162	1890	1690	2905	2030	1560	1245	.4	3.2	9.2
	198	TIG	2780	2450	2205	2010	3390	2625	2130	1730	.3	2.4	6.9

Modern Handloading

Gauge	Length Shell Ins.	Powder Equiv. Drams	Shot Ozs.	Shot Size
LONG RANGE LOADS				
10	2-7/8	4-3/4	1-5/8	4
12	2-3/4	3-3/4	1-1/4	BB, 2, 4, 5, 6, 7½, 9
16	2-3/4	3-1/4	1-1/8	4, 5, 6, 7½, 9
16	2-3/4	3	1-1/8	4, 5, 6, 7½, 9
20	2-3/4	2-3/4	1	4, 5, 6, 7½, 9
28	2-3/4	2-1/4	3/4	4, 6, 7½, 9
FIELD LOADS				
12	2-3/4	3-1/4	1-1/4	7½, 8, 9
12	2-3/4	3-1/4	1-1/8	4, 5, 6, 7½, 8, 9
12	2-3/4	3	1-1/8	4, 5, 6, 8, 9
12	2-3/4	3	1	4, 5, 6, 8
16	2-3/4	2-3/4	1-1/8	4, 5, 6, 7½, 8, 9
16	2-3/4	2-1/2	1	4, 5, 6, 8, 9
20	2-3/4	2-1/2	1	4, 5, 6, 7½, 8, 9
20	2-3/4	2-1/4	7/8	4, 5, 6, 8, 9
SCATTER LOADS				
12	2-3/4	3	1-1/8	8
16	2-3/4	2-1/2	1	8
20	2-3/4	2-1/4	7/8	8
TARGET LOADS				
12	2-3/4	3	1-1/8	7½, 8, 9
12	2-3/4	2-3/4	1-1/8	7½, 8, 9
16	2-3/4	2-1/2	1	8, 9
20	2-3/4	2-1/4	7/8	8,9
28	2-3/4	2-1/4	3/4	9
410	3	Max	3/4	4, 5, 6, 7½, 9
410	2-1/2	Max	1/2	4, 5, 6, 7½, 9
SKEET & TRAP				
12	2-3/4	3	1-1/8	7½, 8, 9
12	2-3/4	2-3/4	1-1/8	7½, 8, 9
16	2-3/4	2-1/2	1	8, 9
20	2-3/4	2-1/4	7/8	8, 9
BUCKSHOT				
12	3 Mag	4-1/2	—	00 Buck—15 pellets
12	3 Mag	4-1/2	—	4 Buck—41 pellets
12	2-3/4 Mag	4	—	2 Buck—20 pellets
12	2-3/4 Mag	4	—	00 Buck—12 pellets
12	2-3/4	3-3/4	—	00 Buck— 9 pellets
12	2-3/4	3-3/4	—	0 Buck—12 pellets
12	2-3/4	3-3/4	—	1 Buck—16 pellets
12	2-3/4	3-3/4	—	4 Buck—27 pellets
16	2-3/4	3	—	1 Buck—12 pellets
20	2-3/4	2-3/4	—	3 Buck—20 pellets
RIFLED SLUGS				
12	2-3/4	3-3/4	1	Slug
16	2-3/4	3	1-7/8	Slug
20	2-3/4	2-3/4	5/8	Slug
410	2-1/2	Max	1/5	Slug

Speer/DWM Ballistics

Ammunition Manufactured by DWM in Germany and Distributed Through Speer Outlets in This Country

Cartridge	Wt. Grs.	Bullet Type (a)	Velocity (fps) Muzzle	100 yds.	200 yds.	300 yds.	Energy (ft. lbs.) Muzzle	100 yds.	200 yds.	300 yds.	Mid-Range Trajectory 100 yds.	200 yds.	300 yds.
5.6 X 35R Vierling	46	SP	2030	1500	1140	Not Available	418	224	130		1.2	7.5	Not Available
5.6 x 50R (Rimmed) Mag.	50	PSP	Not Available				Not Available				Not Available		
5.6 x 52R (Savage H.P.)	71	PSP											
5.6 x 61SE	77	PSP	2850	2460	2320	2200	1280	947	846	766	.3	2.3	6.5
5.6 x 61R	77	PSP	3700	3360	3060	2790	2350	1920	1605	1345	.1	1.1	3.4
6.5 x 54 MS	159	SP	3480	3140	2840	2560	2070	1690	1370	1120	.1	1.3	4.0
6.5 x 57 Mauser	93	PSP	2170	1925	1705	1485	1660	1300	1025	810	.5	4.1	11.5
6.5 x 57 R	93	PSP	3350	2930	2570	2260	2300	1760	1350	1040	.1	1.7	4.8
7 x 57 Mauser	103	PSP	3350	2930	2570	2260	2300	1760	1350	1040	.1	1.7	4.8
	162	TIG	3330	2865	2450	2060	2550	1890	1380	977	.1	1.7	5.2
7 x 57 R.	103	PSP	2785	2480	2250	2060	2780	2200	1820	1520	.3	2.4	6.7
	139	SP	3260	2810	2390	2000	2430	1820	1320	920	.1	1.8	5.3
	162	TIG	2550	2240	1960	1720	2000	1540	1190	910	.3	2.9	8.6
7 x 64	103	PSP	2710	2420	2210	2020	2640	2120	1750	1460	.3	2.4	6.9
	139	SP	3572	3110	2685	2283	2930	2230	1670	1190	.1	1.4	4.4
	162	TIG	3000	2570	2260	1980	2780	2040	1570	1200	.2	2.2	6.4
	177	TIG	2960	2603	2375	2200	3150	2440	2030	1740	.2	2.0	6.0
7 x 65 R	103	PSP	2880	2665	2490	2325	3270	2820	2440	2130	.2	2.0	5.6
	139	SP	3480	3010	2590	2200	2770	2100	1540	1120	.1	1.5	4.7
	162	TIG	3000	2570	2260	1980	2780	2040	1570	1200	.2	2.2	6.4
	177	TIG	2887	2540	2320	2140	3000	2320	1930	1650	.2	2.2	6.3
7mm SE	169	ToSto	2820	2600	2420	2255	3120	2660	2300	2000	.2	2.1	5.9
7 x 75 R SE	169	ToSto	3300	3045	2825	2620	4090	3480	3010	2600	.1	1.4	3.9
30-06	180	TUG	3070	2840	2630	2430	3550	3050	2620	2240	.1	1.6	4.5
8 x 57 JS	123	SP	2854	2562	2306	2077	3261	2632	2133	1726	.2	2.2	6.3
	198	TIG	2968	2339	1805	1318	2415	1497	897	477	.2	2.7	8.8
8 x 57 JR	196	SP	2732	2415	2181	1985	3276	2560	2083	1736	.3	2.5	7.1
8 x 57 JRS	123	SP	2391	1991	1742	1565	2488	1736	1316	1056	.5	3.9	11.2
	196	SP	2970	2340	1805	1318	2415	1497	897	477	.2	2.7	8.8
8 x 60 S	198	TIG	2480	2140	1870	1640	2680	2000	1510	1165	.4	3.3	9.4
	196	SP	2600	2320	2105	1930	2970	2350	1950	1620	.3	2.7	7.6
	198	TIG	2585	2162	1890	1690	2905	2030	1560	1245	.4	3.2	9.2
	198	TIG	2780	2450	2205	2010	3390	2625	2130	1770	.3	2.4	6.9

Modern Handloading

Speer/DWM Ballistics: Ammunition Manufactured by DWM in Germany and Distributed Through Speer Outlets in This Country (cont.)

Cartridge	Bullet Wt. Grs.	Type (a)	Velocity (fps) Muzzle	100 yds.	200 yds.	300 yds.	Energy (ft. lbs.) Muzzle	100 yds.	200 yds.	300 yds.	Mid-Range Trajectory 100 yds.	200 yds.	300 yds.
9.3 × 62	293	TUG	2515	2310	2150	2020	4110	3480	3010	2634	.3	2.8	7.5
9.3 × 64	293	TUG	2640	2450	2290	2145	4550	3900	3410	3000	.3	2.4	6.6
9.3 × 72 R	193	FP	1925	1600	1400	1245	1590	1090	835	666	.5	5.7	16.6
9.3 × 74 R	293	TUG	2360	2160	1998	1870	3580	3000	2560	2250	.3	3.1	8.7

(a) FP-flat point; SP-soft point; PSP-pointed soft point; TIG-Brenneke Torpedo Ideal; TUG-Brenneke Torpedo Universal; ToSto-Vom Hofe Torpedo Stop Ring; SM-Stark Mantel (strong jacket).

Ballistics for Standard Rimfire Ammunition Produced by the Major U. S. Manufacturers

Cartridge	Bullet Wt. Grs.	Type(a)	Velocity (fps) Muzzle	100 yds.	Energy (ft. lbs.) Muzzle	100 yds	Mid-Range Trajectory 100 yds.	Handgun Barrel Length	Ballistics M.V. fps	M.E. ft. lbs.
22 short	29	C, L*	1045	810	70	42	5.6	6"	865	48
22 Short Hi-Vel.	29	C, L	1125	920	81	54	4.3	6"	1035	69
22 Short HP Hi-Vel.	27	C, L	1155	920	80	51	4.2	–	–	–
22 Short	29	D	1045	–	70	–	–	–	–	(per 500)
22 Short	15	D	1710	–	97	–	–	–	–	(per 500)
22 Long Hi-Vel.	29	C, L	1240	965	99	60	3.8	6"	1095	77
22 Long Rifle	40	L*	1145	975	116	84	4.0	6"	950	80
22 Long Rifle	40	L*	1120	950	111	80	4.2	–	–	–
22 Long Rifle	40	L*	–	–	–	–	–	6¾"	1060	100
22 Long Rifle	40	C	1165	980	121	84	4.0	–	–	–
22 Long Rifle Hi-Vel.	40	C, L	1335	1045	158	97	3.3	6"	1125	112
22 Long Rifle HP (Hi-Vel.)	37	C, L	1365	1040	149	86	3.4	–	–	–
22 Long Rifle HP (Hi-Vel.)	36	C	1365	1040	149	86	3.4	–	–	–
22 Long Rifle	No	12 Shot								
22 WRF (Rem. Spl.)	45	C, L	1450	1110	210	123	2.7	–	–	–
22 WRF Mag.	40	JHP	2000	1390	355	170	1.6	6½"	1550	213
22 WRF Mag.	40	MC	2000	1390	355	170	1.6	6½"	1550	213
22 Win. Auto Inside lub.	45	C, L	1055	930	111	86	4.6	–	–	–
5mm Rem. RFM	38	PLHP	2100	1605	372	217	Not Available			

(a) C-Copper Plated; L-Lead (wax coated); D-Disintegrating; L*-Lead Lubricated; MC-Metal Case; HP-Hollow Point; JHP-Jacketed Hollow Point

Ballistics of Norma-Precision Centerfire Rifle Ammunition
Manufactured in Sweden

Cartridge	Bullet Wt. Grs.	Type (a)	Velocity (fps) Muzzle	100 yds.	200 yds.	300 yds.	Energy (ft. lbs.) Muzzle	100 yds.	200 yds.	300 yds.	Max. Height of Trajectory (in.) 100 yds.	200 yds.	300 yds.
22 Hornet	45	SPS	2690	2030	1510	1150	720	410	230	130	Not Available		
220 Swift	50	PSP	4111	3611	3133	2681	1877	1448	1090	799	.2	.9	3.0
222 Remington	50	PSP	3200	2660	2170	1750	1137	786	523	340	.0	2.0	6.2
223	55	SPP	3300	2900	2520	2160	1330	1027	776	570	.4	2.4	6.8
22-250	50	SPS	3800	3300	2810	2350	1600	1209	885	613	Not Available		
	55	SPS	3650	3200	2780	2400	1637	1251	944	704	Not Available		
243 Winchester	75	HP	3500	3070	2660	2290	2041	1570	1179	873	.0	1.4	4.1
	100	PSP	3070	2790	2540	2320	2093	1729	1433	1195	.1	1.8	5.0
6mm Remington	100	SPS	3190	2920	2660	2420	2260	1890	1570	1300	.4	2.1	5.3
250 Savage	87	PSP	3032	2685	2357	2054	1776	1393	1074	815	.0	1.9	5.8
257 Roberts	100	PSP	2822	2514	2223	1956	1769	1404	1098	850	.1	2.2	6.6
	100	PSP	2900	2588	2291	2020	1868	1488	1166	906	.1	2.1	6.2
	120	PSP	2645	2405	2177	1964	1865	1542	1263	1028	.2	2.5	7.0
6.5 Carcano	156	SPRN	2000	1810	1640	1485	1386	1135	932	764	Not Available		
6.5 Japanese	139	PSPBT	2428	2280	2130	1990	1820	1605	1401	1223	.3	2.8	7.7
	156	SPRN	2067	1871	1692	1529	1481	1213	992	810	.6	4.4	11.9
6.5 x 54 MS	139	PSPBT	2580	2420	2270	2120	2056	1808	1591	1388	.2	2.4	6.5
	156	SPRN	2461	2240	2033	1840	2098	1738	1432	1173	.3	3.0	8.2
6.5 x 55	139	PSPBT	2789	2630	2470	2320	2402	2136	1883	1662	.1	2.0	5.6
	156	SPRN	2493	2271	2062	1867	2153	1787	1473	1208	.3	2.9	7.9
270 Winchester	110	PSP	3248	2966	2694	2435	2578	2150	1773	1448	.1	1.4	4.3
	130	PSPBT	3140	2884	2639	2404	2847	2401	2011	1669	.0	1.6	4.7
	150	PSPBT	2802	2616	2436	2262	2616	2280	1977	1705	.1	2.0	5.7
7 x 57	110	PSP	3068	2792	2528	2277	2300	1904	1561	1267	.0	1.6	5.0
	150	PSPBT	2756	2539	2331	2133	2530	2148	1810	1516	.1	2.2	6.2
	175	SPRN	2490	2170	1900	1680	2410	1830	1403	1097	.4	3.3	9.0
7mm Remington Magnum	150	SPSBT	3260	2970	2700	2450	3540	2945	2435	1990	.4	2.0	4.9
	175	SPRN	3070	2720	2400	2120	3660	2870	2240	1590	.5	2.4	6.1
7 x 61 S & H (26 in.)	160	PSPBT	3100	2927	2757	2595	3415	3045	2701	2393	.0	1.5	4.3
30 U.S. Carbine	110	SPRN	1970	1595	1300	1090	948	622	413	290	.8	6.4	19.0

Ballistics of Norma-Precision Centerfire Rifle Ammunition (cont.)

Cartridge	Wt. Grs.	Type (a)	Velocity (fps) Muzzle	100 yds.	200 yds.	300 yds.	Energy (ft. lbs.) Muzzle	100 yds.	200 yds.	300 yds.	Max. Height of Trajectory (in.) 100 yds.	200 yds.	300 yds.
30-30 Winchester	150	SPFP	2410	2075	1790	1550	1934	1433	1066	799	.9	4.2	11.0
	170	SPFP	2220	1890	1630	1410	1861	1349	1003	750	.7	4.1	11.9
308 Winchester	130	PSPBT	2900	2590	2300	2030	2428	1937	1527	1190	.1	2.1	6.2
	150	PSPBT	2860	2570	2300	2050	2725	2200	1762	1400	.1	2.0	5.9
	180	PSPBT	2610	2400	2210	2020	2725	2303	1952	1631	.2	2.5	6.6
	180	SPDC	2610	2400	2210	2020	2725	2303	1952	1631	.7	3.4	8.9
7.62 Russian	180	PSPBT	2624	2415	2222	2030	2749	2326	1970	1644	.2	2.5	6.6
308 Norma Magnum	180	DC	3100	2881	2668	2464	3842	3318	2846	2427	.0	1.6	4.6
30-06	130	PSPBT	3281	2951	2636	2338	3108	2514	2006	1578	.1	1.5	4.6
	150	PS	2972	2680	2402	2141	2943	2393	1922	1527	.0	1.9	5.7
	180	PSPBT, SPDC	2700	2494	2296	2109	2914	2487	2107	1778	.1	2.3	6.4
	220	SPRN	2411	2197	1996	1809	2840	2358	1947	1599	.3	3.1	8.5
300 H & H	180	PSPBT	2920	2706	2500	2297	3409	2927	2499	2109	.0	1.9	5.3
	220	SPRN	2625	2400	2170	1986	3367	2814	2301	1927	.2	2.5	7.0
7.65 Argentine	150	PSP	2920	2630	2355	2105	2841	2304	1848	1476	.1	2.0	5.8
303 British	130	PSP	2789	2483	2195	1929	2246	1780	1391	1075	.1	2.3	6.7
	150	PSP	2720	2440	2170	1930	2465	1983	1569	1241	.1	2.2	6.5
7.7 Japanese	180	PSPBT	2540	2340	2147	1965	2579	2189	1843	1544	.2	2.7	7.3
	130	PSP	2950	2635	2340	2065	2513	2004	1581	1231	.1	2.0	5.9
	180	PSPBT	2493	2292	2101	1922	2484	2100	1765	1477	.3	2.8	7.7
8 x 57 JR	196	SPRN	2362	2045	1761	1513	2428	1820	1530	996	.4	3.7	10.6
8 x 57 JS	123	PSP	2887	2515	2170	1857	2277	1728	1286	942	.1	2.3	6.8
	159	SPRN	2723	2362	2030	1734	2618	1970	1455	1062	.2	2.6	7.9
	196	SPRN	2526	2195	1894	1627	2778	2097	1562	1152	.3	3.1	9.1
358 Winchester	200	SPS	2530	2210	1910	1640	2843	2170	1621	1195	.4	3.1	8.8
	250	SPS	2250	2010	1780	1570	2811	2243	1759	1369	.6	3.9	10.4
358 Norma Magnum	250	SPS	2790	2493	2231	2001	4322	3451	2764	2223	.2	2.4	6.6
375 H & H Magnum	300	SPS	2550	2280	2040	1830	4333	3464	2773	2231	.3	2.8	7.6
44 Magnum	240	SPFP	1750				1640				Not Available		

(a) P-Pointed; SP-Soft Point; HP-Hollow Point; FP-Flat Point; RN-Round Nose; BT-Boat Tail; MC-Metal Case; DC-Dual Core; SPS-Semi-pointed Soft Point.

Appendix

Ballistics for Weatherby Magnum Ammunition
Supplied by Weatherby's Inc. for use in Weatherby Rifles

Cartridge	Bullet Wt. Grs.	Bullet Type (a)	Velocity (fps) Muzzle	100 yds.	200 yds.	300 yds.	Energy (ft. lbs.) Muzzle	100 yds.	200 yds.	300 yds.	Mid-Range Trajectory 100 yds.	200 yds.	300 yds.
224 Weatherby Varmintmaster	50	PE	3750	3160	2625	2140	1562	1109	765	508	0.7	3.6	9.0
224 Weatherby Varmintmaster	55	PE	3650	3150	2685	2270	1627	1212	881	629	0.4	1.7	4.5
240 Weatherby	70	PE	3850	3395	2975	2585	2304	1788	1376	1038	0.3	1.5	3.9
240 Weatherby	90	PE	3500	3135	2795	2475	2444	1960	1559	1222	0.4	1.8	4.5
240 Weatherby	100	PE	3395	3115	2850	2595	2554	2150	1804	1495	0.4	1.8	4.4
257 Weatherby	87	PE	3825	3290	2835	2450	2828	2087	1553	1160	0.3	1.6	4.4
257 Weatherby	100	PE	3555	3150	2815	2500	2802	2199	1760	1338	0.4	1.7	4.4
257 Weatherby	117	SPE	3300	2900	2550	2250	2824	2184	1689	1315	0.4	2.4	6.8
270 Weatherby	100	PE	3760	3265	2825	2435	3140	2363	1773	1317	0.4	1.6	4.3
270 Weatherby	130	PE	3375	3050	2750	2480	3283	2685	2183	1776	0.4	1.8	4.5
270 Weatherby	150	PE	3245	2955	2675	2430	3501	2909	2385	1967	0.5	2.0	5.0
7mm Weatherby	139	PE	3300	2995	2715	2465	3355	2770	2275	1877	0.4	1.9	4.9
7mm Weatherby	154	PE	3160	2885	2640	2415	3406	2874	2384	1994	0.5	2.0	5.0
300 Weatherby	150	PE	3545	3195	2890	2615	4179	3393	2783	2279	0.4	1.5	3.9
300 Weatherby	180	PE	3245	2960	2705	2475	4201	3501	2925	2448	0.4	1.9	5.2
300 Weatherby	220	SPE	2905	2610	2385	2150	4123	3329	2757	2257	0.6	2.5	6.7
340 Weatherby	200	PE	3210	2905	2615	2345	4566	3748	3038	2442	0.5	2.1	5.3
340 Weatherby	210	Nosler	3165	2910	2665	2435	4660	3948	3312	2766	0.5	2.1	5.0
340 Weatherby	250	SPE	2850	2580	2325	2090	4510	3695	3000	2425	0.6	2.7	6.7
378 Weatherby	270	SPE	3180	2850	2600	2315	6051	4871	4053	3210	0.5	2.0	5.2
378 Weatherby	300	SPE	2925	2610	2380	2125	5700	4539	3774	3009	0.6	2.5	6.2
460 Weatherby	500	RN	2700	2330	2005	1730	8095	6025	4465	3320	0.7	3.3	10.0

(a) PE-Pointed Expanding; SPE-Semi-Pointed Expanding; RN-Round Nose; Nosler-Nosler Partition Controlled Expansion Bullet. All velocities taken from 26-inch barrels.

Following is a table for determining the kinetic energy of any bullet at any practical velocity.

To use this table, refer to the left-hand column and locate the velocity; then move to the immediate right and locate the energy value *per grain of bullet weight*. This is the energy possessed by each grain of weight of *any* bullet traveling at that velocity. Then, simply multiply the energy value by the weight of the bullet in grains to obtain the total kinetic energy of a bullet moving at that velocity. All you need know is the bullet's weight and its velocity—finding the energy then takes only a few seconds and should eliminate many of those arguments about which cartridge or load "shoots the hardest."

Energies of Bullets

Velocity in fps	Energy	Velocity in fps	Energy	Velocity in fps	Energy	Velocity in fps	Energy
600	.80	1010	2.26	1420	4.47	1830	7.43
610	.82	1020	2.31	1430	4.54	1840	7.51
620	.85	1030	2.35	1440	4.60	1850	7.60
630	.88	1040	2.40	1450	4.66	1860	7.68
640	.91	1050	2.45	1460	4.73	1870	7.76
650	.94	1060	2.49	1470	4.79	1880	7.84
660	.96	1070	2.54	1480	4.86	1890	7.94
670	.99	1080	2.59	1490	4.93	1900	8.01
680	1.02	1090	2.63	1500	5.00	1910	8.10
690	1.05	1100	2.68	1510	5.06	1920	8.18
700	1.08	1110	2.73	1520	5.13	1930	8.37
710	1.11	1120	2.78	1530	5.19	1940	8.35
720	1.15	1130	2.83	1540	5.26	1950	8.44
730	1.18	1140	2.88	1550	5.33	1960	8.53
740	1.21	1150	2.93	1560	5.40	1970	8.61
750	1.24	1160	2.99	1570	5.47	1980	8.70
760	1.28	1170	3.04	1580	5.54	1990	8.79
770	1.31	1180	3.09	1590	5.61	2000	8.88
780	1.34	1190	3.14	1600	5.68	2010	8.97
790	1.38	1200	3.19	1610	5.75	2020	9.06
800	1.42	1210	3.25	1620	5.82	2030	9.15
810	1.45	1220	3.30	1630	5.90	2040	9.24
820	1.49	1230	3.36	1640	5.97	2050	9.33
830	1.53	1240	3.41	1650	6.04	2060	9.42
840	1.56	1250	3.47	1660	6.12	2070	9.50
850	1.60	1260	3.52	1670	6.19	2080	9.60
860	1.64	1270	3.58	1680	6.26	2090	9.70
870	1.68	1280	3.63	1690	6.34	2100	9.80
880	1.72	1290	3.69	1700	6.41	2110	9.90
890	1.76	1300	3.75	1710	6.49	2120	9.98
900	1.79	1310	3.81	1720	6.57	2130	10.07
910	1.83	1320	3.86	1730	6.64	2140	10.17
920	1.87	1330	3.92	1740	6.72	2150	10.26
930	1.92	1340	3.98	1750	6.80	2160	10.36
940	1.96	1350	4.04	1760	6.88	2170	10.45
950	2.00	1360	4.10	1770	6.95	2180	10.55
960	2.04	1370	4.16	1780	7.03	2190	10.65
970	2.08	1380	4.22	1790	7.11	2200	10.74
980	2.13	1390	4.29	1800	7.19	2210	10.84
990	2.17	1400	4.35	1810	7.27	2220	10.94
1000	2.22	1410	4.41	1820	7.35	2230	11.04

Velocity	Energy	Velocity	Energy	Velocity	Energy	Velocity	Energy
2240	11.14	2830	17.78	3420	25.93	4010	35.71
2250	11.24	2840	17.91	3430	26.08	4020	35.89
2260	11.34	2850	18.04	3440	26.23	4030	36.07
2270	11.44	2860	18.16	3450	26.38	4040	36.25
2280	11.54	2870	18.29	3460	26.54	4050	36.43
2290	11.64	2880	18.42	3470	26.69	4060	36.61
2300	11.74	2890	18.55	3480	26.85	4070	36.79
2310	11.83	2900	18.67	3490	27.00	4080	36.97
2320	11.95	2910	18.80	3500	27.16	4090	37.15
2330	12.05	2920	18.93	3510	27.31	4100	37.33
2340	12.16	2930	19.06	3520	27.47	4110	37.51
2350	12.26	2940	19.19	3530	27.62	4120	37.70
2360	12.37	2950	19.32	3540	27.78	4130	37.88
2370	12.47	2960	19.45	3550	27.94	4140	38.06
2380	12.58	2970	19.59	3560	28.10	4150	38.25
2390	12.68	2980	19.72	3570	28.25	4160	38.43
2400	12.78	2990	19.85	3580	28.41	4170	38.62
2410	12.90	3000	20.00	3590	28.57	4180	38.80
2420	13.00	3010	20.12	3600	28.73	4190	38.99
2430	13.11	3020	20.25	3610	28.94	4200	39.18
2440	13.22	3030	20.39	3620	29.10	4210	39.36
2450	13.33	3040	20.52	3630	29.26	4220	39.55
2460	13.44	3050	20.66	3640	29.42	4230	39.74
2470	13.55	3060	20.79	3650	29.58	4240	39.92
2480	13.66	3070	20.93	3660	29.75	4250	40.11
2490	13.77	3080	21.07	3670	29.91	4260	40.30
2500	13.88	3090	21.16	3680	30.07	4270	40.49
2510	13.99	3100	21.29	3690	30.24	4280	40.68
2520	14.10	3110	21.43	3700	30.40	4290	40.87
2530	14.20	3120	21.57	3710	30.56	4300	41.06
2540	14.32	3130	21.71	3720	30.73	4310	41.25
2550	14.44	3140	21.85	3730	30.90	4320	41.45
2560	14.55	3150	21.99	3740	31.06	4330	41.64
2570	14.67	3160	22.12	3750	31.23	4340	41.83
2580	14.78	3170	22.26	3760	31.40	4350	42.02
2590	14.89	3180	22.41	3770	31.56	4360	42.22
2600	15.01	3190	22.55	3780	31.73	4370	42.41
2610	15.13	3200	22.69	3790	31.90	4380	42.61
2620	15.24	3210	22.83	3800	32.07	4390	42.80
2630	15.36	3220	22.97	3810	32.24	4400	43.00
2640	15.48	3230	23.12	3820	32.41	4410	43.19
2650	15.59	3240	23.26	3830	32.58	4420	43.39
2660	15.71	3250	23.41	3840	32.75	4430	43.58
2670	15.83	3260	23.55	3850	32.92	4440	43.78
2680	15.96	3270	23.70	3860	33.09	4450	43.98
2690	16.07	3280	23.84	3870	33.26	4460	44.18
2700	16.19	3290	23.99	3880	33.45	4470	44.38
2710	16.31	3300	24.14	3890	33.62	4480	44.58
2720	16.43	3310	24.28	3900	33.78	4490	44.77
2730	16.55	3320	24.43	3910	33.95	4500	44.97
2740	16.67	3330	24.58	3920	34.12	4510	45.17
2750	16.79	3340	24.73	3930	34.30	4520	45.37
2760	16.91	3350	24.87	3940	34.48	4530	45.58
2770	17.04	3360	25.02	3950	34.65	4540	45.78
2780	17.16	3370	25.17	3960	34.82	4550	45.98
2790	17.28	3380	25.32	3970	35.00	4560	46.18
2800	17.41	3390	25.47	3980	35.18	4570	46.38
2810	17.53	3400	25.62	3990	35.36	4580	46.59
2820	17.66	3410	25.77	4000	35.53	4590	46.79

Here are capacities of popular calibers. Capacities are measured to the base of a normally seated bullet in the instance of straight cases, and to the junction of neck and shoulder in bottle neck cases. Cubic capacity is determined by weighing amount of water case will hold, and calculating from that the volume in both cubic inches and cubic centimeters.

Case	Grains Water	Cubic Inches	Cubic Cm.	Case	Grains Water	Cubic Inches	Cubic Cm.
.22 Hornet (late)	11.4	.045	.739	303 Savage	34.3	.136	2.22
.22 K-Hornet	13.4	.053	.870	.303 British	45.9	.182	2.97
.218 Bee	14.8	.059	.960	7.65mm Arg.	52.4	.208	3.39
.222 Remington	23.8	.094	1.54	.32-40	33.4	.132	2.16
.22 Rem. Mag.	28.8	.094	1.87	.32 Win. Special	35.8	.142	2.32
.223 Remington	28.3	.112	1.84	.32 Remington	37.0	.147	2.39
.219 Wasp	27.0	.107	1.76	8x57mm Mauser	53.6	.212	3.46
.219 Zipper	33.0	.131	2.14	.338 Win. Mag.	78.6	.313	5.09
.224 Weatherby	35.8	.142	2.32	.348 Winchester	66.4	.263	4.30
.225 Winchester	38.0	.151	2.46	.358 Winchester	45.7	.182	2.96
.22-250 Remington	42.1	.167	2.73	.35 Remington	40.2	.159	2.60
.220 Swift	44.6	.177	2.89	.350 Rem. Mag.	62.2	.246	4.03
.243 Winchester	50.6	.200	3.28	.375 H & H Mag.	85.7	.340	5.53
6mm Remington	51.5	.204	3.33	.38-40 Winchester	33.8	.134	2.19
.25-20 Winchester	14.6	.058	.946	.38-55	37.5	.149	2.43
.25-35 Winchester	33.7	.134	2.18	.444 Marlin	54.0	.214	3.50
.25 Remington	35.3	.140	2.28	.45-70	48.7	.194	4.16
.250 Savage	42.0	.166	2.72	.458 Winchester	68.0	.270	4.40
257 Roberts	53.7	.213	3.46				
6.5mm Jap	44.0	.174	2.85	HANDGUNS			
6.5mm M-S	45.0	.179	2.91	.22 Remington Jet	16.9	.067	1.09
6.5x55mm	51.9	.206	3.36	.221 Rem. Fireball	20.8	.083	1.35
6.5mm Rem. Mag.	66.1	.260	4.27	.256 Winchester	19.0	.075	1.23
.264 Win. Mag.	79.8	.318	5.17	.32 S & W	3.3	.012	.214
.270 Winchester	62.9	.250	4.07	.32 S & W Long	9.5	.037	.615
280 Remington	61.9	.245	4.00	.32-20 Winchester	15.3	.061	.992
.284 Winchester	62.4	.247	4.04	.38 S & W	7.0	.028	.454
7mm Mauser	53.2	.211	3.45	.380 Auto	6.0	.024	.388
7mm Rem. Mag.	79.9	.317	5.17	9mm Luger	8.7	.035	.562
.30 Carbine	15.0	.059	.972	.38 Super Auto	10.7	.043	.691
.30-30 Winchester	35.8	.142	2.32	.38 Special	11.7	.047	.758
.30 Remington	37.0	.147	2.39	.357 Magnum	15.2	.060	.984
.300 Savage	46.3	.184	3.00	.41 Rem. Mag.	21.0	.083	1.36
.30-40 Krag	47.5	.188	3.08	.44 S & W Russian	18.8	.075	1.22
.308 Winchester	49.8	.198	3.23	.44 S & W Special	20.5	.081	1.33
.30-06 Springfield	61.3	.242	3.98	.44 Rem. Mag.	25.2	.100	1.63
.300 H & H Mag.	80.0	.318	5.18	.44-40 Winchester	32.6	.129	2.11
.308 Norma Mag.	81.2	.322	5.27	.45 Auto Rim	13.8	.053	.862
.300 Win. Mag.	83.6	.332	5.42	.45 A.C.P.	13.9	.055	.900
.300 Weatherby	91.7	.364	5.91	.45 Colt	30.3	.119	1.96

The weight of water capacity given is simply a convenience of measurement—under no circumstances should this be interpreted as loading data or a charge recommendation for *any* powder. It is only a basis for comparison of case volume and bears no practical relationship to the amount of any particular powder that might constitute a safe load.

Following are sectional densities of bullets of popular weights in standard calibers.

Bullet	Sectional Density	Bullet	Sectional Density
22 CALIBER (.222")		165 Gr.	.247
40 Gr.	.114	168 Gr.	.253
22 CALIBER (.223")		170 Gr.	.257
45 Gr.	.128	180 Gr.	.272
22 CALIBER (.224")		190 Gr.	.286
45 Gr.	.128	220 Gr.	.332
50 Gr.	.143	303 CALIBER (.312")	
53 Gr.	.151	150 Gr.	.218
55 Gr.	.157	174 Gr.	.252
60 Gr.	.171	32 CALIBER (.321")	
6mm (.243")		170 Gr.	.234
70 Gr.	.169	8mm (.323")	
75 Gr.	.181	150 Gr.	.206
87 Gr.	.210	170 Gr.	.233
100 Gr.	.241	338 CALIBER (.338")	
25 CALIBER (.257")		200 Gr.	.250
60 Gr.	.130	225 Gr.	.281
75 Gr.	.162	250 Gr.	.312
87 Gr.	.188	348 CALIBER (.348")	
100 Gr.	.216	200 Gr.	.236
117 Gr.	.253	35 CALIBER (.357")	
6.5mm (.264")		158 Gr.	.177
100 Gr.	.206	35 CALIBER (.358")	
129 Gr.	.266	200 Gr.	.224
140 Gr.	.288	250 Gr.	.280
160 Gr.	.330	275 Gr.	.308
270 CALIBER (.277")		375 CALIBER (.375")	
100 Gr.	.186	270 Gr.	.275
130 Gr.	.242	300 Gr.	.306
150 Gr.	.279	44 CALIBER (.429")	
7mm (.284")		240 Gr.	.186
120 Gr.	.212		
139 Gr.	.246	44 CALIBER (.430")	
154 Gr.	.273	265 Gr.	.204
175 Gr.	.310	45 CALIBER (.452")	
7.35mm (.300")		185 Gr.	.127
128 Gr.	.202	45 CALIBER (.454")	
30 CALIBER (.308")		250 Gr.	.173
100 Gr.	.151	45 CALIBER (.458")	
110 Gr.	.166	300 Gr.	.206
130 Gr.	.196	350 Gr.	.243
150 Gr.	.227	500 Gr.	.347

Following are tables of weights and measures that are frequently required in handloading problems that arise.

Modern Handloading

Millimeters	X	.03937	Inches
"	X	25.4	"
Centimeters	X	.3937	"
"	X	2.54	"
Meters	X	39.37	" (Act of Congress)
"	X	3.281	Feet
"	X	1.094	Yard
Square mm.	X	.0155	Square Inches
" "	X	645.1	" "
Square cm.	X	.155	" "
" "	X	6.451	" "
Cubic cm.	X	16.383	Cubic inches
" "	X	3.69	Fluid drachms
" "	X	29.57	Fluid ounces
Grams	X	15.4324	Grains (Act of Congress)
" (water)	X	29.57	Fluid ounces
Grams	X	28.35	Ounces avoirdupois
Kilograms	X	2.2046	Pounds
"	X	35.3	Ounces avoirdupois
Kilograms per sq. cm. (Atmosphere)	X	14.223	Pounds per sq. in.
Kilogrammeters	X	7.233	Foot-pounds

Measures of Weight

AVOIRDUPOIS OR COMMERCIAL WEIGHT
   1 gross or long ton equals 2240 pounds.
   1 net or short ton equals 2000 pounds
   1 pound equals 16 ounces equals 7000 grains.
   1 ounce equals 16 drachms equals 437.5 grains.

The following measures for weight are now seldom used in the United States:
   1 hundredweight equals 4 quarters equals 112 pounds (1 gross or long ton equals 20
   hundredweights); 1 quarter equals 28 pounds; 1 stone equals 14 pounds; 1 quintal
   equals 100 pounds.

TROY WEIGHT, USED FOR WEIGHING GOLD AND SILVER
   1 pound equals 12 ounces equals 5760 grains.
   1 ounce equals 20 pennyweights equals 480 grains.
   1 pennyweight equals 24 grains.
   1 carat (used in weighing diamonds) equals 3.168 grains.
   1 grain Troy equals 1 grain avoirdupois equals 1 grain apothecaries' eight.

APOTHECARIES' WEIGHT
   1 pound equals 12 ounces equals 5760 grains.
   1 ounce equals 8 drachms equals 480 grains.
   1 drachm equals 3 scriples equals 60 grains.
   1 scruple equals 20 grains.

BOOKS

Ballistics in the Seventeenth Century, by A. R. Hall. 1st J. & J. Harper ed. 1969 [from the Cambridge University Press ed. of 1952]. 186 pp., illus., with tables and diagrams.
  A profound work for advanced scholars, this is a study in the relations of science and war, with reference particularly to England.

The Bullet's Flight, from Powder to Target, by F. W. Mann. Ray Riling Arms Books Co., Phila., Pa. 1965. A reprint of the very scarce original work of 1909. Introduction by Homer S. Powley, 384 pp., illus.
  One of the best known and scholarly-developed works on basic ballistics.

Cartridge Headstamp Guide, by H. P. White and B. D. Munhall. H. P. White Laboratory, Bel Air, Md., 1963. 263 pp., illus.
  An important reference on headstamping of small arms ammo, by manufacturers in many countries. Clear illus. of 1936 headstamps of every type.

Cartridges, by H. C. Logan, Standard Public, Inc., Huntington, W. Va., 1948. 204 pp., illus. Deluxe First ed.
  "Pictorial digest of small arms ammunition," with excellent line illus. and competent text for collectors of obsolete ammunition. In very limited supply, being the scarce out-of-print and best original edition.

Cartridges for Collectors, by Fred A. Datig. Borden Publishing Co., Alhambra, Calif., Vol. I (Centerfire), 1958; Vol. II (Rimfire and Misc. Types), 1963; Vol. III (Additional Rimfire, Centerfire, and Plastic), 1967. Each of the three volumes 176 pp., well illus.
  Vol. III supplements the first two books and presents 300 additional specimens. All illus. are shown in full-scale line drawings.

Cartridges of the World, by Frank C. Barnes, John T. Amber ed., Gun Digest Co., Chicago, Ill., 1969. 378 pp. Profusely illus. Paperbound
  The second edition of a comprehensive reference for hunters, collectors, handloaders and ballisticians. Covering over 1000 cartridges, loads, components, etc., from all over the world.

Centerfire American Rifle Cartridges, 1892-1963, by Ray Bearse. A. S. Barnes & Co., S. Brunswick, N.J., 1966. 198 pp., illus.
  Identification manual covering caliber, introduction date, origin, case type, etc. Self-indexed and cross-referenced. Headstamps and line drawings are included.

Centerfire Pistol and Revolver Cartridges, by H. P. White, B. D. Munhall and Ray Bearse. A. S. Barnes, N.Y., 1967. 85 pp. plus 170 pp., illus.
  A new and revised edition covering the original Volume I, Centerfire Metric Pistol and Revolver Cartridges and Volume II, Centerfire American and British Pistol and Revolver Cartridges, by White and Mungall, formerly known as *Cartridge Identification*.

Complete Guide to Handloading, by Phil Sharpe. Funk & Wagnalls, N. Y. C., 1953 (3rd ed., 2nd rev.). 734 pp., profusely illustrated, numerous line and halftone charts, tables, lists, etc.
  The bible of handloaders ever since its first appearance in 1937, but badly dated now.

Encyclopedia of Firearms, ed. by H. L. Peterson. E. P. Dutton, N. Y. C., 1964. 367 pp., 100 pp. of illus. incl. color.
  Fine reference work on firearms, with articles by 45 top authorities covering classes of guns, manufacturers, ammunition, nomenclature, and related topics.

Guns Guns Magazine, 8150 N. Central Park Ave., Skokie, Ill. 60076.

Articles for gun collectors, hunters and shooters

Handbook for Shooters and Reloaders, by P. O. Ackley. Priv. publ., Salt Lake City, 1962-1965. Illus. Vol. I, 567 pp., Vol. II, 498 pp.

Storehouse of technical information on ammunition and its use by a noted authority, with supplemental articles by other experts. Ballistic charts plus loading data for hundreds of cartridges, standard and wildcat.

Handloader's Digest, ed. by John T. Amber. Gun Digest Publ., Northfield, Ill., 1970. 320 pp., very well illus., stiff paper covers.

This 5th edition contains the latest data on ballistics, maximum loads, new tools, equipment, reduced loads, etc., plus a fully illus. catalog section, current prices and specifications.

Home Guide to Cartridge Conversions, by Geo. C. Nonte, Jr., Stackpole Books, Harrisburg, Pa., 1967. 404 pp., illus.

A new, revised and enlarged ed. of instructions, charts and tables for making ammunition no longer available, or which has become too expensive on the commercial market.

Hornady Handbook of Cartridge Reloading. Hornady Mfg. Co., Grand Island, Nebr., 1967. 360 pp., illus.

Handloader's reference, with much detail on projectiles, ballistics, etc., on many popular U. S. and imported firearms. An excellent new work with particularly needed ballistic detail.

The Identification of Firearms and Forensic Ballistics, by G. Burrard. A. S. Barnes, New York, 1962. 217 pp., illus.

A standard, reliable, authoritative English work in the criminal-legal field of ballistics.

Interior Ballistics, How a Gun Converts Chemical Energy to Projectile Motion, by E. D. Lowry. Doubleday and Co., N. Y., 1968. 168 pp., including index and bibliography, illus. with 4 halftones and 17 line drawings.

An introduction to the history of small arms and weapons relative to the science of internal ballistics, especially for the layman and student.

Lyman Handbook No. 44. Lyman Gunsight Corp., Middlefield, Conn., 1967.

Latest edition of a favorite reference for ammunition handloaders whether novice or veteran.

Metallic Cartridges, T. J. Treadwell, compiler. The Armoury, N. Y. C., 1959. Unpaginated. 68 plates.

A reduced-size reproduction of U. S. Ordnance Memoranda No. 14, originally publ. in 1873, on regulation and experimental cartridges manufactured and tested at Frankford Arsenal, Philadelphia, Pa.

Methods in Exterior Ballistics, by F. R. Moulton. Dover Publ., N. Y. C., 1962. 257 pp.

A standard work on the mathematics of advanced theoretical and experimental exterior ballistics.

The NRA Handloader's Guide, Ashley Halsey, Jr., ed. Natl. Rifle Assn., Washington, D. C., 1969. 312 pp., illus., paperbound.

Revised edition of a reloading handbook, based on material published in *The American Rifleman.*

Pocket Manual for Shooters and Reloaders, by P. O. Ackley. Publ. by author, Salt Lake City, Utah, 1964. 176 pp., illus., spiral bound.

Good coverage on standard and wildcat cartridges and related firearms in popular calibers.

Principles and Practice of Loading Ammunition, by Lt. Col. Earl Naramore. Stackpole Books, Harrisburg, Pa., 1954. 915 text pages, 240 illustrations.

Actually two volumes in one. The first part (565 pp.) deals with ballistics and the principles of cartridge making—and the chemistry, metallurgy, and physics involved. The second part (350 pp.) is a thorough discussion of the mechanics of loading cartridges. 1967 printing.

Professional Loading of Rifle and Shotgun Cartridges . . . , by G. L. Herter. Herter's, Waseca, Minn., 1970. 430 pp., illus. Paper covers.

Loading and reloading data for everyone. Note: A new third printing, more extensive, of this book is available hard-bound. 830 pp.

Shooter's Bible Black Powder Guide, by George Nonte. Shooter's Bible, Inc., S. Hackensack, N. J., 1969. 214 pp., well illus.

Information on black powder weapons, ammunition, shooting, etc.

Shooter's Bible Reloader's Guide, 2nd ed., by R. A. Steindler. Shooter's Bible, Inc., S. Hackensack, N. J., 1968. 220 pp., fully illus.

Comprehensive coverage of technology and methods of handloading all types of small arms ammunition. This is a useful work.

Shotgun Shells: Identification, Manufacturers and Checklist for Collectors, by F. H. Steward. B. and P. Associates, St. Louis, Mo., 1969. 101 pp., illus., paper covers.

Historical data for the collector.

Shotshell Handbook, by Lyman Handbook Staff. Lyman Gunsight Corp., Middlefield, Conn., 1969. 160 pp., illus., stiff paper spiral-binding.

The first book devoted exclusively to shotshell reloading. Considers: gauge, shell length, brand, case, loads, buckshot, etc., plus excellent reference section. Some color illus.

Small Arms and Ammunition in the United States Service, 1776-1865, by B. R. Lewis. Smithsonian Inst., Washington, D. C., 1968. 338 pp. plus 52 plates.

2nd printing of a distinguished work for historians and collectors. A limited number of deluxe, signed and numbered copies (1st reprinting 1960) are available in full leather and gilt top.

Small Arms Ammunition Identification Guide. Panther Publ., Boulder, Colo., 1968. 151 pp., illus., paperbound.

Facsimile of a U. S. Army text on cartridge identification, which includes data on foreign ammunition used in Vietnam and elsewhere.

Small Arms Design and Ballistics, by Col. T. Whelen. Stackpole Books, Harrisburg, Pa., 1945. Vol. I, 352 pp., Vol. II, 314 pp., both illus.

Authoritative technical data on firearms. Vol. I covers design, function, and operation. Vol. II deals with interior and exterior ballistics.

Speer Manual for Reloading Ammunition No. 8. Speer, Inc., Lewiston, Idaho, 1970. 382 pp., illus.

A popular manual on handloading, with authoritative articles on loading, ballistics, and related subjects. Decorated paper wrappers.

Why Not Load Your Own?, by Col. T. Whelen. A. S. Barnes, New York, 1957, 4th ed., rev. 237 pp., illus.

A basic reference on handloading, describing each step, materials and equipment. Loads for popular cartridges are given.

The Winchester-Western Ammunition Handbook. Thomas Nelson & Sons, N. Y. C., 1964. 185 pp., illus.

Called the world's handiest handbook on ammunition for all types of shotguns, rifles and handguns. Full of facts, photographs, ballistics and statistics.

PERIODICAL PUBLICATIONS

Alaska Sportsman   Alaska Northwest Pub. Co., Box 4-EEE, Anchorage, Alaska 99503. Hunting and fishing articles.

American Field†   222 W. Adams St., Chicago, Ill. 60606. Field dogs and trials, occasional gun and hunting articles.

The American Rifleman (M)   National Rifle Assn., 1600 Rhode Island Ave., N.W., Wash., D. C. 20036. Firearms articles of all kinds.

The American West*   American West Publ. Co., 599 College Ave., Palo Alto, Ca. 94306.

Argosy   Popular Publ., Inc., 205 E. 42nd St., New York, N.Y. 10017.

Army (M)   Assn. of the U. S. Army, 1529 18th Ave. N.W., Wash., D. C. 20036. Occasional articles on small arms.

Australian Shooters' Journal   P.O. Box 90, Stafford, Qld., Brisbane 4053, Australia. Hunting and shooting articles.

Canadian Journal of Arms Collecting (Q)   Museums Restoration Service, P.O. Box 2037, Sta. D, Ottawa, Ont., Canada.

Deutsches Waffen Journal   Journal-Verlag Schwend GmbH, Postfach 340, Schwabisch Hall, Germany.

Antique and modern arms, their history, technical aspects, etc. German text.

Ducks Unlimited, Inc. (M)   P.O. Box 66300, Chicago, Ill. 60666.

Enforcement Journal*   Natl. Police Officers Assn., Natl. Police Academy Bldg., 1890 S. Tamiami Trail, Venice, Fla. 33595.

The Field† The Harmsworth Press Ltd., 8 Stratton St., London W.1, England.
Hunting and shooting articles.

Field & Stream Holt, Rinehart and Winston, Inc., 383 Madison Ave., New York, N.Y. 10017.
Articles on firearms plus hunting and fishing.

Fishing and Hunting Guide Fishing and Hunting Guide Ltd., P.O. Box 48, Dolton, Ill. 60419.

Fur-Fish-Game A. R. Harding Pub. Co., 2878 E. Main St., Columbus, Ohio 43209.
"Gun Rack" column by M. H. Decker.

Gunfacts Magazine Hazard Publications, Inc., Box 9335, Arlington, Va. 22209. (8 issues per yr., mo./bi-mo.)

The Gun Report World Wide Gun Report, Inc., Box 111, Aledo, Ill. 61231.
For the gun collector.

Gun Sport Gun Sport Magazine Inc., Box 116, Hughesville, Md. 20637.

Gun Week† Sidney Printing & Publishing Co., P.O. Box 150, Sidney, Ohio 45365.
Tabloid paper on guns, hunting, shooting.

Gun World Gallant Publishing Co., 130 Olinda Pl., Brea, Calif. 92621.
For the hunting, reloading and shooting enthusiast.

Guns & Ammo Petersen Pub. Co., 8490 Sunset Blvd., Los Angeles, Calif. 90069.
Guns, shooting, and technical articles.

Guns Plus Hunting* M. F. Enterprises, Inc., 222 Park Ave. So., New York, N.Y. 10003.

Guns Review Ravenhill Pub. Co. Ltd., Standard House, Bonhill St., London E.C. 2, England.
For collectors and shooters.

The Handgunner (M) U. S. Revolver Assn., 59 Alvin St., Springfield, Mass. 01104.
General handgun and competition articles.

The Handloader Magazine* Dave Wolfe Pub. Co., Box 3030, Prescott, Ariz. 86301.

Hobbies Lightner Pub. Co., 1006 S. Michigan Ave., Chicago, Ill. 60605.
Collectors departments.

International Shooting Sport* Union Internationale de Tir, 62 Wiesbaden-Klarenthal, Klarenthalerstr., Germany.
For the international target shooter.

The Journal of the Arms & Armour Society (M) F. Wilkinson (Secy.), 40 Great James St., Holborn, London W.C. 1, England.
Articles for the collector.

Law and Order Law and Order Magazine, 37 W. 38th St., New York, N.Y. 10018.
Articles on weapons for law enforcement.

The Luger Journal Robt. B. Marvin, Publ., P.O. Box 12206, Plantation, Fla. 33314.

Muzzle Blasts (M) National Muzzle Loading Rifle Assn., P.O. Box 67, Friendship, Ind. 47021.
For the black powder shooter.

National Rifle Assn. Journal (British) Natl. Rifle Assn. (Br.), Bisley Camp, Brookwood, Woking, Surrey, England.

National Sportsman's Digest National Sportsman's Club, Box 2003, Dallas, Tex. 75221.
Subs. includes membership in the Club, etc.

National Wildlife* Natl. Wildlife Fed. Inc., 1416 16th St. N.W., Washington, D. C.

New Zealand Wildlife (Q) New Zealand Deerstalkers Assoc. Inc., P.O. Box 263, Wellington, N. Z.
Hunting and shooting articles.

Ordnance* (M) American Ordnance Assn., 819 Union Trust Bldg., Wash., D. C. 20005.
Occasional articles on small arms and related subjects.

Outdoor Life Popular Science Pub. Co., 355 Lexington Ave., New York, N.Y. 10017.
Arms column by Jack O'Connor.

Outdoor World* Country Beautiful Corp., 24198 W. Bluemound Rd., Waukesha, Wis. 53186.
Conservation and wildlife articles.

Police Charles C. Thomas, publisher, 301-327 E. Lawrence Ave., Springfield, Ill. 62703.
Articles on identification.

Police Times (M)   1100 N.E. 125th St., No. Miami, Fla. 33161.
Popular Guns *   Countrywide Publications, Inc., 222 Park Ave. So., New York, N.Y. 10003.
Popular Mechanics   Hearst Corp., 224 W. 57th St., New York, N.Y. 10019.
  Hunting and shooting articles.
Precision Shooting   Precision Shooting, Inc., 8 Cline St., Dolgeville, N.Y. 13329
  Journal of the International Benchrest Shooters.
The Rifle Magazine *   Dave Wolfe Publishing Co., Box 3030, Prescott, Ariz. 86301.
  Journal of the NBRSA.
The Rifleman (Q)   National Smallbore Rifle Assoc., 113 Southwark St., London, S.E. 1, England.
  Data on British Matches and International Matches, and technical shooting articles.
Rod and Gun in Canada   Rod and Gun Pub. Corp., 1219 Hotel de Ville, Montreal 129, P.Q., Canada.
  Regular gun and shooting articles.
Saga   Gambi Public., 333 Johnson Ave., Brooklyn, N.Y. 11026.
The Shooting Industry   Publisher's Dev. Corp., 8150 N. Central Pk., Skokie, Ill. 60076.
The Shooting Times (England)†   Cordwallis Estate, Clivemont Rd., Maidenhead, Berks., England.
  Game shooting and firearms articles.
Shooting Times   Peoria Journal-Star, Inc., News Plaza, Peoria, Ill., 61601.
  Gun ads plus articles on every gun activity.
The Shotgun News‡   Snell Publishing Co., Columbus, Nebr. 68601.
  Gun ads of all kinds.
The Skeet Shooting Review   National Skeet Shooting Assn., 212 Linwood Bldg., 2608 Inwood Rd., Dallas, Tex. 75235.
  Scores, averages, skeet articles.
Sporting Goods Business   7 E. 43rd St., New York, N.Y. 10017.
  Trade journal.
The Sporting Goods Dealer   1212 No. Lindbergh Blvd., St. Louis, Mo. 63166.
  Trade journal.
Sports Afield   The Hearst Corp., 250 W. 55th St., New York, N.Y. 10019.
  Pete Brown on firearms, plus hunting and fishing articles.
Sports Age Magazine   Minneapolis, Minn.
  Trade journal.
Sports Illustrated†   Time, Inc., 540 N. Michigan Ave., Chicago, Ill. 60611.
  Articles on the current sporting scene.
Trap & Field   Review Pub. Co., 1100 Waterway Blvd., Indianapolis, Ind. 46202.
  Scores, averages, trapshooting articles.
True   Fawcett Publ., Inc., Fawcett Bldg., Greenwich, Conn. 06830.
Wildlife Review (Q)   Parliament Bldgs., Victoria, B. C., Canada.
  M Membership requirements; write for details.   † Published weekly.
  Q Published quarterly.   ‡ Published twice per month.
  * Published bi-monthly.

SHOOTING PUBLICATIONS

Write directly to the sources noted for titles listed and ask for their latest catalog.
[1] National Shooting Sports Foundation, Inc., 1075 Post Road, Riverside, Conn. 06878
[2] National Rifle Association of America, 1600 Rhode Island Ave., Washington, D. C. 20036
[3] Remington Arms Company, Dept. C—Bridgeport, Conn. 06602
[4] Olin Corporation, Conservation Dept., East Alton, Ill. 62024

A Joint Resolution—A 4-page statement by the National Police Officers Assn. and the National Shooting Sports Foundation, outlining the role of firearms in U. S. history and voicing their stand against ill-planned restrictive gun laws. Free.[1]
Basic Pistol Marksmanship—Textbook for basic pistol courses. 25c.[2]
Basic Rifle Marksmanship—Textbook for basic rifle courses. 25c ea.[2]
The Elk—125-page report on the hunting and management of this game animal, more properly called *wapiti*. Extensive biblio. $1.00.[4]
Free Films—Brochure listing outdoor movies available to sportsmen's clubs. Free.[1]

The Gun Law Problem—Information about firearms legislation. Free.[2]

How to be a Crack Shot—A 14-page booklet detailing everything necessary to becoming an outstanding shot. Free.[3]

Fundamentals of Claybird Shooting—A 39-page booklet explaining the basics of skeet and trap in non-technical terms. Many diagrams. 25c ea.[4]

Hunter Safety Instructor's Guide—How to conduct an NRA Hunter Safety Course. 25c ea.[2]

Hunting and Shooting Sportsmanship—A 4-page brochure defining the "true sportsman" and giving information on the outdoor field. Free.[1]

Junior Rifle Handbook—Information about the NRA junior program with short instruction course. (25 copies issued to each new affiliated junior club without charge.) 25c ea.[2]

NRA Hunter Safety Handbook—Textbook for students. 10c ea.[2]

National Shooting Preserve Directory—Up-to-date listing of small game preserves in the U. S. and Canada. Free.[1]

Ranger Targets—To be used in qualifying for the NRA Ranger emblem; supplied free in reasonably large quantities.[1]

Shooting's Fun for Everyone—The why, when, where, and how of riflery for boys and girls. 20 pp. 5c ea.[1]

Trap or Skeet Fundamentals—Handbooks explaining fundamentals of these two sports, complete with explicit diagrams to start beginners off right. Free.[3]

25-Foot Shooting Program—Complete information on a short range shooting program with CO_2 and pneumatic rifles and pistols. 35c.[2]

What Every Parent Should Know When a Boy or Girl Wants a Gun—Straightforward answers to the 15 questions most frequently asked by parents. 8 pp. 5c ea.[1]

The Cottontail Rabbit—56-page rundown on America's most popular hunting target. Where to find him, how to hunt him, how to help him. Bibliography included. $1.00 ea.[4]

For the Young Hunter—A 32-page booklet giving fundamental information on the sport. Single copies free, 15c each in bulk.[4]

Gray and Fox Squirrels—112-page paperbound illustrated book giving full rundown on the squirrel families named. Extensive bibliography. $1.00 ea.[4]

How to Have More Pheasant Hunting—A 16-page booklet on low cost hunting, including data on in-season stocking and how to start a small preserve. 25c.[1]

The Mallard—80-page semi-technical report on this popular duck. Life cycle, laws and management, hunting—even politics as they affect this bird—are covered. Bibliography. $1.00 ea.[4]

NRA Booklets—Ranging from 12 to 36 pages, these are articles on specific arms or arms types. Titles available are: Sightin In; The 45 Automatic; The M1 Rifle; Telescopic Sights; Metallic Sights; Duck Hunting; U. S. Cal. 30 Carbine; Remodeling the 03A3; Remodeling the 303 Lee-Enfield; Remodeling the U. S. 1917 Rifle; M1903 Springfield Rifle; Military Rifles and Civil War Small Arms, 50c ea. Gun Cabinets, Racks, Cases & Pistol Boxes, 75c. Deer Hunting, $1.00.[2]

Under the heading of "Range Plans" are 15 booklets priced from 10c to $1.00. All are described in an order form pamphlet available from the NRA.

Principles of Game Management—A 25-page booklet surveying in popular manner such subjects as hunting regulations, predator control, game refuges and habitat restoration. Single copies free, 15c each in bulk.[4]

The Ring-Necked Pheasant—Popular distillation of much of the technical literature on the "ringneck." 104-page paperbound book, appropriately illustrated. Bibliography included. $1.00 ea.[4]

Ruffed Grouse, by John Madson—108-page booklet on the life history, management and hunting of *Bonasa umbellus* in its numerous variations. Extensive biblio. $1.00[4]

Start A Gun Club—All of the basic information needed to establish a club with clay bird shooting facilities. 24 pp. 50c.[1]

Where To Shoot Muzzle Loaders in the U. S. A.—Publ. for black powder burners, and lists more than 100 muzzle loading clubs. 10c.[1]

The White-Tailed Deer—History, management, hunting—a complete survey in this 108-page paperbound book. Full bibliography. $1.00 ea.[4]

You and Your Lawmaker—A 22-page citizenship manual for sportsmen, showing how they can support or combat legislation affecting shooting and outdoor sports. 10c ea.[1]

AMMUNITION (Commercial)

Alcan Shells, (See: Smith & Wesson-Fiocchi, Inc.)
Amron Corp., 525 Progress Ave., Waukesha, Wis. 53186
Cascade Cartridge Inc., (See: Omark)
Federal Cartridge Co., 2700 Foshay Tower, Minneapolis, Minn. 55402
Frontier Cartridge Co., Inc., Box 906, Grand Island, Nebr. 68801
Omark-CCI, Inc., Box 856, Lewiston, Ida. 83501
Remington Arms Co., Bridgeport, Conn. 06602
Service Armament, 689 Bergen Blvd., Ridgefield, N. J. 07657
Smith & Wesson-Fiocchi, Inc., 3640 Seminary Rd., Alton, Ill. 62002
Speer-DWM, Box 696, Lewiston, Ida. 83501
Super-Vel Cartridge Co., Box 40, Shelbyville, Ind. 46176
Weatherby's, 2781 E. Firestone Blvd., South Gate, Calif. 90280
Winchester-Western, East Alton, Ill. 62024

AMMUNITION (Custom)

Ammodyne, Box 1589, Los Angeles, Calif. 90053
B & K Custom Rel. Serv., Rte. 1, Lake 13, Farwell, Mich. 48622
Bill Ballard, P.O. Box 656, Billings, Mont. 59103
Jerry & Betty Bird, Box 10183, Corpus Christi, Tex. 78410
Caldwell's Loading Serv., 1314 Monroe Dr., N.E., Atlanta, Ga. 30306
Russell Campbell, 219 Leisure Dr., San Antonio, Tex. 78201
Cumberland Arms, 1222 Oak Dr., Manchester, Tenn. 37355
Custom Ammo & Gunsmithing, 390 S. Main, Moab, Utah 84532
J. Dewey Gun Co., Clinton Corners, N.Y. 12514
E. W. Ellis Sport Shop, RFD 1, Box 139, Corinth, N.Y.
Ellwood Epps, 80 King St., Clinton, Ont., Canada
Steve Filipiak, 1270 So. Raleigh, Denver, Colo. 80219
H.P.K. Co., Inc., 3750 Ridge Rd., Highland, Ind.
R. H. Keeler, 1304 S. Oak, Port Angeles, Wash. 98362
KWT Inc., 710 Cooper-Foster Pk. Rd., Lorain, O. 44053 (tungsten
Dean Lincoln, 390 S. Main, Moab, Utah 84532
Pat B. McMillan, 4908 E. Indianola, Phoenix, Ariz. 85018
Mansfield Gunshop, Box 83, New Boston, N.H. 03070
Man-Tol Shells, Box 134, Bunnell, Fla. 32010
Moody's Reloading Serv., 2108 Broadway, Helena, Mont. 59601
Numrich Arms Corp., 203 Broadway, W. Hurley, N.Y. 12491
Robert Pomeroy, Morrison Ave., East Corinth, Me. 04427 (custom
Sanders Cust. Gun Serv., 2358 Tyler Lane, Louisville, Ky. 40205
Shooter's Service & Dewey, Inc., Clinton Corners, N.Y. 12514
Shot Shell Components, 365 So. Moore, Lakewood, Colo. 80226
Super Vel Cartridge Corp., Shelbyville, Ind. 46176
3-D Co., Inc., Box 4411, Lincoln, Neb. 68504
James C. Tillinghast, Box 568, Marlow, N.H. 03456
True-Blue Co., 1400 E. Palmer Ave., Glendale, Calif. 91205 (blanks)
Walmax, Inc. (See: True-Blue)
Wanda Cartridge Co., P.O. Box 45901, Houston, Tex. 77045
Whitney Cartridge Co., Box 5872, Pasadena, Calif. 91107 (shotshells)

AMMUNITION (Foreign)

Abercrombie & Fitch, Madison at 45th St., New York, N.Y. 10017
Ammodyne, Box 1859, Los Angeles, Calif. 90053 (RWS)
Canadian Ind. Ltd. (C.I.L.), Box 10, Montreal, Que., Canada
C-I-L Ammunition Inc., P.O. Box 831, Plattsburgh, N.Y. 12901
Centennial Arms Co., 3318 W. Devon Ave., Chicago, Ill. 60645
Colonial Ammunition Co., Box 8511, Auckland, New Zealand
DWM, Speer Prods. Inc., Box 641, Lewiston, Ida. 83501
Gevelot of Canada, Box 1593, Saskatoon, Sask., Canada
Hy-Score Arms Co., 200 Tillary, Brooklyn, N.Y. 11201
Paul Jaeger Inc., 211 Leedom St., Jenkintown, Pa. 19046
S. E. Lazlo, 200 Tillary, Brooklyn, N.Y. 11201
NORMA-Precision, South Lansing, N.Y. 14822
Oregon Ammo Service, Box 19341, Portland, Ore. 97219
Stoeger Arms Corp., 55 Ruta Ct., So. Hackensack, N.J. 07606
James C. Tillinghast, Box 568, Marlow, N.H. 03456

BULLET & CASE LUBRICANTS

Alpha-Molykote, Dow Corning Corp., 45 Commerce Dr., Trumbull, Conn.
Birchwood-Casey Co., Inc., 7900 Fuller Rd., Eden Prairie, Minn. 55343
Bullet Pouch, Box 4285, Long Beach, Calif. 90804 (Mirror-Lube)
Chopie Tool & Die Co., 531 Copeland, La Crosse, Wis. 54601 (Black-Solve)
Cooper-Woodward, Box 972, Riverside, Calif. 92502 (Perfect Lube)
Green Bay Bullets, 233 N. Ashland, Green Bay, Wis. 54303 (EZE-Size case
Herter's Inc., Waseca, Minn. 56902 (Perfect Lubricant)
Javelina Products, Box 337, San Bernardino, Calif. 92402 (Alox beeswax)
Jet-Aer Corp., 100 Sixth Ave., Paterson, N.J. 07524
Lenz Prod. Co., Box 1226, Sta. C., Canton, O. 44708 (Clenzoil)
Lyman Gun Sight Products, Middlefield, Conn. 06455 (Size-Ezy)
Micro Shooter's Supply, Box 213, Las Cruces, N. Mex. 88001 (Micro-Lube)
Nutec, Box 1187, Wilmington, Del. 19899 (Dry-Lube)
Pacific Tool Co., Box 4495, Lincoln, Neb. 68504
Phelps Rel. Inc., Box 4004, E. Orange, N.J. 07019

RCBS, Inc., Box 1919, Oroville, Calif. 95965
SAECO Rel. Inc., 726 Hopmeadow St., Simsbury, Conn. 06070
Scientific Lubricants Co., 3753 Lawrence Ave., Chicago, Ill. 60625
Shooters Accessory Supply (SAS), Box 250, N. Bend, Ore. 97459
Sports Distr. Co., Rte. 1, Rapid City, S.D. 57701 (Reloader No. 7)
Testing Systems, Inc., 2836 Mt. Carmel, Glenside, Pa. 19038

CHRONOGRAPHS AND PRESSURE TOOLS

A & W Eng., 6520 Rampart St., Houston, Tex. 77036 (press. tool)
Avtron, 10409 Meech Ave., Cleveland, Ohio 44105
B-Square Co., Box 11281, Ft. Worth, Tex. 76110
Chronograph Specialists, P.O. Box 5005, Santa Ana, Calif. 92704
Herter's, Waseca, Minn. 56093
Micro-Sight Co., 242 Harbor Blvd., Belmont, Calif. 94002 (Techsonic)
Oehler Research, P.O. Box 9135, Austin, Tex. 78756
Sundtek Co., P.O. Box 744, Springfield, Ore. 97477
Telepacific Electronics Co., Inc., 3335 W. Orange Ave., Anaheim, Calif.
York-Cantrell, 30241 Rosebriar, St. Clair Shores, Mich. 48082 (press. tool)

COMPONENTS—BULLETS, POWDER, PRIMERS

Accuracy Bullet Co., 2443 41st St., San Francisco, Calif. 94116
Alcan (See: Smith & Wesson-Fiocchi, Inc.)
Bahler Die Shop, Box 386, Florence, Ore. 97439 (17 cal. bullets)
Lee Baker, 4474 Yosemite Way, Los Angeles, Calif. 90065 (17 cal.
Joe J. Balickie, 409 Rose Lane, Raleigh, N.C. 27607
Ballistic Research Industries, 116 N. Main, Sebastopol, Calif. 95472
Bitterroot Bullet Co., Box 412, Lewiston, Ida. 83501
Centrix, 2116 N. 10th Ave., Tucson, Ariz. 85705
Kenneth E. Clark, 18738 Highway 99, Madera, Calif. 93637 (Bullets)
Clerke Recreation Prods., Inc., 2040 Broadway, Santa Monica, Calif. 90404
Curry Bullet Co., 4504 E. Washington Blvd., Los Angeles, Calif. 90022
Colorado Custom Bullets, Rt. 1, Box 507-B, Montrose, Colo. 81401
Division Lead, 7742 W. 61 Pl., Summit, Ill. 60502
DuPont, Explosives Dept., Wilmington, Del. 19898
Elk Mountain Shooters Supply, 2020 Road 44, Pasco, Wash. 99301
Forty Five Ranch Enterprises, Box 1080, Miami, Okla. 74354
Godfrey Reloading Supply, R.R. #1, Box 688, Brighton, Ill. 62012
Lynn Godfrey (See: Elk Mountain Shooters Supply)
G. J. Godwin, 455 Fox Lane, Orange Park, Fla. 32073 (cast bullets)
Green Bay Bullets, 233 N. Ashland, Green Bay, Wis. 54303 (lead)
H.P.K. Co., Inc., 3750 Ridge Rd., Highland, Ind. 46322 (cast bullets)
Frank A. Hemsted, Box 281, Sunland, Calif. 91040
Hercules Powder Co., 910 Market St., Wilmington, Del. 19899
Herter's Inc., Waseca, Minn. 56093
Hi-Precision Co., 109 Third Ave., N.E., Orange City, Ia. 51041
B. E. Hodgdon, Inc., 7710 W. 50th Hwy., Shawnee Mission, Kans. 66202
Hornady Mfg. Co., Box 1848, Grand Island, Neb. 68801
N. E. House Co., Middletown Rd., E. Hampton, Conn. 06424
Jurras Munition Corp., Box 140, Shelbyville, Ind. 46176
L. L. F. Die Shop, 1281 Highway 99 North, Eugene, Ore. 97402
Lee's Precision Bullets, 4474 Yosemite Way, Los Angeles, Calif. 90065
Lyman Gun Sight Products, Middlefield, Conn. 06455
Markell, Inc., 4115 Judah St., San Francisco, Calif. 94112
Meyer Bros., Wabasha, Minn. 55981 (shotgun slugs)
Michael's Antiques, Box 233, Copiague, L.I., N.Y. 11726 (Balle Blondeau)
Miller Trading Co., 20 S. Front St., Wilmington, N.C. 28401
Norma-Precision, South Lansing, N.Y. 14882
Northridge Bullet Co., P.O. Box 1208, Vista, Calif. 92083
Nosler Bullets, P.O. Box 688, Beaverton, Ore. 97005
Oregon Ammo Service, Box 19341, Portland, Ore. 97219
Robert Pomeroy, Morrison Ave., East Corinth, Me. 04427
Rainbow Prod., P.O. Box 75, Wishram, Wash. 98673 (bullets)
Remington-Peters, Bridgeport, Conn. 06602
S. W. M. Bullet Co., 1122 S. Cherry St., Port Angeles, Wash. 98362
Sanderson's, 724 W. Edgewater, Portage, Wis. 53901 (cork wad)
Sierra Bullets Inc., 421 No. Altadena Dr., Pasadena, Calif. 91107
Smith & Wesson-Fiocchi, Inc., 3640 Seminary Rd., Alton, Ill. 62002
Speedy Bullets, Box 1262, Lincoln, Neb. 68501
Speer Products, Inc., Box 896, Lewiston, Ida. 83501
C. H. Stocking, Hutchinson, Minn. 55350 (17 cal. bullet jackets)
Sullivan Arms Corp., 5204 E. 25th, Indianapolis, Ind. 46218
Super-Vel Cartr. Corp., 129 E. Franklin St., Shelbyville, Ind. 46176
Taylor Bullets, P.O. Box 21254, San Antonio, Tex. 78221
True-Blue Co., 1400 E. Palmer Ave., Glendale, Calif. 91205 (blanks)
James C. Tillinghast, Box 568, Marlow, N.H. 03456
Vitt & Boos, Sugarloaf Dr., Wilton, Conn. 06897
Walmax, Inc., 1400 E. Palmer Ave., Glendale, Calif. 91205 (blanks)
Williams Custom Guns, Rt. 3, Box 809, Cleveland, Tex. 77327 (17 cal.)
Winchester-Western, New Haven, Conn. 06504
F. Wood, Box 386, Florence, Ore. 97439 (17 cal.)
Xelex Ltd., Hawksbury, Ont., Canada (powder)

DEALERS IN COLLECTORS' CARTRIDGES

Antique Arsenal, 365 So. Moore St., Lakewood, Colo. 80226
J. A. Belton, 52 Sauve Rd., Mercier, Chateauguay Cty, Quebec, Canada
Peter Bigler, 291 Crestwood Dr., Milltown, N.J. 08850
Geo. Blakeslee, 3093 W. Monmouth, Englewood, Colo. 80110
Cameron's, 16690 W. 11th Ave., Golden, Colo. 80401
Carter Gun Works, 2211 Jefferson Pk. Ave., Charlottesville, Va. 22903
Gerry Coleman, 163 Arkell St., Hamilton, Ont., Canada
Chas. E. Duffy, Williams Lane, West Hurley, N.Y. 12419
Tom M. Dunn, 1342 So. Poplar, Casper, Wyo. 82601
Ellwood Epps, 80 King St., Clinton, Ont., Canada
Ed Howe, 2 Main St., Coopers Mills, Me. 04341
Walt Ireson, 47 Chedoke Ave., Hamilton 12, Ont., Canada
Jackson Arms, 6209 Hillcrest Ave., Dallas, Tex. 75205
Miller Bros., Rapid City, Mich. 49676
Oregon Ammo Service, Box 19341, Portland, Ore. 97219 (catalog $2.00)
Powder Horn, 3093 W. Monmouth, Englewood, Color. 80110
Martin B. Retting Inc., 11029 Washington, Culver City, Calif. 90230
Perry Spangler, 519 So. Lynch, Flint, Mich. 48503 (list 35c)
Jon Taylor House of Cartridges, 12 Cascade Bay, Brandon, Manit., Canada
Ernest Tichy, 365 S. Moore, Lakewood, Colo. 80226
James C. Tillinghast, Box 568, Marlow, N.H. 03456 (list 50c)
Wilkins Gun Shop, 1060 N. Henderson, Galesburg, Ill. 61401 (list 50c)

GUN PARTS, U. S. AND FOREIGN

American Firearms Mfg. Co., Inc., 5732 Kenwick Dr., San Antonio, Tex
Badger Shooter's Supply, Owen, Wisc. 54460
Shelley Braverman, Athens, N.Y. 12015
Philip R. Crouthamel, 817 E. Baltimore, E. Lansdowne, Pa. 19050
Charles E. Duffy, Williams Lane, West Hurley, N.Y. 12491
Federal Ordnance Inc., P.O. Box 36032, Los Angeles, Calif. 90036
Greeley Arms Co., Inc., 223 Little Falls Rd., Fairfield, N.J. 07006
Gunner's Armory, 2 Sonoma, San Francisco, Calif. 94133
H & B Gun Corp., 1228 Fort St., Lincoln Park, Mich. 48166
Hunter's Haven, Zero Prince St., Alexandria, Va. 22314
Bob Lovell, Box 401, Elmhurst, Ill. 60126
Numrich Arms Co., West Hurley, N.Y. 12491
Pacific Intl. Import Co., 2416-16th St., Sacramento, Calif. 95818
Potomac Arms Corp. (See: Hunter's Haven)
Reed & Co., Shokan, N.Y. 12481
Martin B. Retting, Inc., 11029 Washington, Culver City, Calif. 90230
Santa Barbara of America, Ltd., 930 N. Beltline Rd., #132, Irving, Tex.
Sarco , Inc., 192 Central, Stirling, N.J. 07980
R. A. Saunders, 3253 Hillcrest Dr., San Antonio, Tex. 78201 (clips)
Schmid & Ladd, 14733 Hwy. 19 So., Clearwater, Fla. 33516
Sherwood Distr. Inc., 7435 Greenbush Ave., No. Hollywood, Calif.
Clifford L. Smires, R.D., Columbus, N.J. 08022 (Mauser rifles)
Sporting Arms, Inc., 9643 Alpaca St., So. El Monte, Calif. 91735 (M-1
N. F. Strebe, 4926 Marlboro Pike, S.E., Washington, D.C. 20027
Triple-K Mfg. Co., 568-6th Ave., San Diego, Calif. 92101

GUNSMITH SCHOOLS

Colorado School of Trades, 1545 Hoyt, Denver, Colo. 80215
Lassen Junior College, 11100 Main St., Susanville, Calif. 96130
Oregon Technical Institute, Klamath Falls, Ore. 97601
Penn. Gunsmith School, 812 Ohio River Blvd., Avklon, Pittsburgh, Pa.
Trinidad State Junior College, Trinidad, Colo. 81082

GUNSMITH SUPPLIES, TOOLS, SERVICES

Adams & Nelson Co., 4125 W. Fullerton, Chicago, Ill. 60639
Alamo Heat Treating Co., Box 55345, Houston, Tex. 77055
Alley Supply Co., Box 458, Sonora, Calif. 95370
American Edelstaal, Inc., 350 Broadway, New York, N.Y. 10013
American Firearms Mfg. Co., Inc., 5732 Kenwick Dr., San Antonio, Tex.
Anderson & Co., 1203 Broadway, Yakima, Wash. 98902 (tang safe)
Armite Labs., 1845 Randolph St., Los Angeles, Calif. 90001 (pen oiler)
Atlas Arms Inc., 2952 Waukegan Rd., Niles, Ill. 60648
B-Square Inc., Box 11281, Ft. Worth, Tex. 76110
Jim Baiar, Rt. 1-B, Box 352, Columbia Falls, Mont. 59912 (hex screws)
Benrite Co., 353 Covington, San Antonio, Tex. 78220
Bonanza Sports Mfg. Co., 412 Western Ave., Faribault, Minn. 55021
Brown & Sharpe Mfg. Co., Precision Pk., No. Kingston, R.I. 02852
Bob Brownell's, Main & Third, Montezuma, Ia. 50171
W. E. Brownell, 1852 Alessandro Trail, Vista, Calif. 92083
Maynard P. Buehler, Inc., 17 Orinda Hwy., Orinda, Calif. 94563
Burgess Vibrocrafters, Inc. (BVI), Rte. 83, Grayslake, Ill. 60030
M. H. Canjar, 500 E. 45th, Denver, Colo. 80216 (triggers, etc.)
Centerline Prod., Box 14074, Denver, Colo. 80214
Chicago Wheel & Mfg. Co., 1101 W. Monroe St., Chicago, Ill. 60607
Christy Gun Works, 875-57th St., Sacramento, Calif. 95819
Clymer Mfg. Co., 14241 W. 11 Mile Rd., Oak Park, Mich. 48237 (reamers)
Colbert Industries, 10107 Adella, South Gate, Calif. 90280 (Panavise)

A. Constantine & Son, Inc., 2050 Eastchester Rd., Bronx, N.Y. 10461 (wood)
Cougar & Hunter, 66398 W. Pierson Rd., Flushing, Mich. 48433 (scope jigs)
Craft Industries, 719 No. East St., Anaheim, Calif. 92800 (Gunline tools)
Dayton-Traister Co., P.O. Box 93, Oak Harbor, Wash. 98277 (triggers)
Dem-Bart Co., 3333 N. Gove St., Tacoma, Wash. 98407 (checkering tools)
Die Supply Corp., 11700 Harvard Ave., Cleveland, Ohio 44105
Ditto Industries, 527 N. Alexandria, Los Angeles, Calif. 90004
Dixie Diamond Tool Co., Inc., 6875 S.W. 81st St., Miami, Fla. 33143
Dremel Mfg. Co., P.O. Box 518, Racine, Wis. 53401 (grinders)
Chas E. Duffy, Williams Lane, West Hurley, N.Y. 12491
Dumore Co., 1300-17th St., Racine, Wis. 53403
E-Z Tool Co., P.O. Box 3186, East 14th St. Sta., Des Moines, Ia. 50313

Edmund Scientific Co., 101 E. Glouster Pike, Barrington, N.J. 08007
F. K. Elliott, Box 785, Ramona, Calif. 92065 (reamers)
Foredom Elec. Co., Rt. 6, Bethel, Conn. 06801 (power drills)
Forster Appelt Mfg. Co., Inc., 82 E. Lanark Ave., Lanark, Ill. 61046
Keith Francis, Box 343, Talent, Ore. 97540 (reamers)
Frantz Tools, 913 Barbara Ave., Placentia, Calif. 92670
G. R. S. Corp., Box 1157, Boulder, Colo. 80302 (Gravermeister)
Gilmore Pattern Works, 1164 N. Utica, Tulsa, Okla. 74110
Gold Lode, Inc., P.O. Box 31, Addison, Ill. 60101 (gold inlay kit)
Grace Metal Prod., Box 67, Elk Rapids, Mich. 49629 (screw drivers, drifts)
Gopher Shooter's Supply, Box 246, Faribault, Minn. 55021 (screwdrivers
The Gun Case, 11035 Maplefield SE, El Monte, Calif. 91733 (triggers)
Gunline Tools, (See: Craft Ind.)
H. & M., 24062 Orchard Lake Rd., Farmington, Mich. 48024 (reamers)
Hartford Reamer Co., Box 134, Lathrup Village, Mich. 48075
Hobbi-Carve (See: St. Paul Mach.)
R. E. Hutchinson, Burbank Rd., Sutton Mass. 01527 (engine turning tool)
O. Iber Co., 626 W. Randolph, Chicago, Ill. 60606
Paul Jaeger, Inc., 211 Leedom St., Jenkinton, Pa. 19046
Kasenite Co., Inc., 3 King St., Mahwah, N.J. 07430 (surface hrdng. comp.)
Lee Mfg. Co., 237 E. Aurora St., Waterbury, Conn. 06720
Lock's Phila. Gun Exch., 6700 Rowland Ave., Philadelphia, Pa. 19149
Marker Machine Co., Box 426, Charleston, Ill. 61920
Michaels of Oregon Co., P.O. Box 13010, Portland, Ore. 97213
Viggo Miller, P.O. Box 4181, Omaha, Neb. 68104 (trigger attachment)
Miller Single Trigger Mfg. Co., Box 69, Millersburg, Pa. 17061
Frank Mittermeier, 3577 E. Tremont, N.Y., N.Y. 10465
Moderntools Corp., Box 407, Dept. GD, Woodside, N.Y. 11377
N&J Sales, Lime Kiln Rd., Northford, Conn. 06472 (screwdrivers)
Karl A. Neise, Inc., 5602 Roosevelt Ave., Woodside, N.Y. 11377
P & S Sales, P.O. Box 45095, Tulsa, Okla. 74145
Palmgren, 8383 South Chicago Ave., Chicago, Ill. 60167 (vises, etc.)
C. R. Pedersen & Son, Ludington, Mich. 49431
Ponderay Lab., 210 W. Prasch, Yakima, Wash. 98902 (epoxy glass bedding)
Redford Reamer Co., Box 6604, Redford Hts. Sta., Detroit, Mich. 48240
Richland Arms Co., 321 W. Adrian St., Blissfield, Mich. 49228
Riley's Supply Co., Box 365, Avila, Ind. 46710 (Niedner buttplates, caps)
Rob. A. Saunders (SEE: Amer. Firearms Mfg.)
Ruhr-American Corp., So. Hwy #55, Glenwood, Minn. 56334
A. G. Russell, P.O. Box 474, Fayetteville, Ark. 72701 (Arkansas stones)
Schaffner Mfg. Co., Emsworth, Pittsburgh, Pa. 15202 (polishing kits)
Schuetzen Gun Works, 1226 Prairie Rd., Colo. Springs, Colo. 80909
Shaw's, 1655 S. Euclid Ave., Anaheim, Calif. 92802
A. D. Soucy Co., Box 191, Fort Kent, Me. 04743 (ADSCO stock finish)
L. S. Starrett Co., Athol, Mass. 01331
Technological Devices, Inc., P.O. Box 3491, Stamford, Conn. 06905 '

L. B. Thompson, 568 E. School Ave., Salem, Ohio 44460 (rust bluing,
Timney Mfg. Co., 5624 Imperial Hwy., So. Gate, Calif. 90280 (triggers)
Stan de Treville, Box 2446, San Diego, Calif. 92112 (checkering patterns)
Twin City Steel Treating Co., Inc., 1114 S. 3rd, Minneapolis, Minn. 55415
R. G. Walters Co., 3235 Hancock, San Diego, Calif. 92110
Ward Mfg. Co., 500 Ford Blvd., Hamilton, Ohio 45011
Will-Burt Co., P.O. Box #1, Orrville, Ohio 44667 (vises)
Williams Gun Sight Co., 7389 Lapeer Rd., Davison, Mich. 48423
Wilson Arms Co., Box 364, Stony Creek, Branford, Conn. 06405
Wilton Tool Corp., 9525 W. Irving Pk. Rd., Schiller Park, Ill. 60176
Wisconsin Platers Supply Co., 3256-A Milwaukee, Madison, Wis. 53714
Woodcraft Supply Corp., 313 Montvale, Woburn, Mass. 01801

RELOADING TOOLS AND ACCESSORIES

A & W Eng., Inc., 6520 Rampart St., Houston, Tex. 77036 (bullet puller)
Alcan (See: Smith & Wesson-Fiocchi, Inc.)
Alpha-Molykote, Dow Corning Corp., 45 Commerce, Trumbull, Conn.
Anchor Alloys, Inc., 966 Meeker Ave., Brooklyn, N.Y. 11222 (chilled shot)
Anderson Mfg. Co., Royal, Ia. 51357 (Shotshell Trimmers)
Aurands, 229 E. 3rd St., Lewistown, Pa. 17044
Automatic Reloading Equipment, Inc., 1602 Babcock St., Costa Mesa, Cal
B-Square Eng. Co., Box 11281, Ft. Worth, Tex. 76110
Bahler Die Shop, Box 386, Florence, Ore. 97439
Bair Machine Co., Box 4407, Lincoln, Neb. 68504
Bill Ballard, P.O. Box 656, Billings, Mont. 59103
Belding & Mull, P.O. Box 428, Philipsburg, Pa. 16866

H. S. Beverage, New Gloucester, Me. 04260 (brass bullet mould)
Blackhawk East, C2274 POB, Loves Park, Ill. 61111
Bonanza Sports, Inc., 412 Western Ave., Faribault, Minn. 55021
Gene Bowlin, 3602 Hill Ave., Snyder, Tex. 79549 (arbor press)
Brown Precision Co., 5869 Indian Ave., San Jose, Calif. 95123
A. V. Bryant, East Hartford, Conn. 06424
C-H Tool & Die Corp., Box L, Owen, Wis. 54460
Camdex, Inc., 15339 W. Michaels, Detroit, Mich. 48235
Carbide Die & Mfg. Co., Box 226, Covina, Calif. 91706
C'Arco, P.O. Box 2943, San Bernardino, Calif. 92406 (Ransom rest)
Carter Gun Works, 2211 Jefferson Pk. Ave., Charlottesville, Va. 22903
Cascade Cartridge, Inc. (See: Omark)
Chellife Corp., R.D. 1, Box 260 A1, Felton, Del. 19943
Lester Coats, 416 Simpson St., No. Bend, Ore. 97459 (core cutter)
Cole's Acku-Rite Prod., P.O. Box 25, Kennedy, N.Y. 14747 (die racks)
Containter Development Corp., 424 Montgomery St., Watertown, Wis.
Cooper Engineering, 612 E. 20th, Houston, Tex. 77008
Cooper-Woodward, Box 972, Riverside, Calif. 92502 (Perfect Lube)
Design & Development Co., 1002 N. 64th St., Omaha, Neb. 68132
Clarence Detsch, 135 Larch Rd., St. Mary's, Pa. 15857 (bullet dies)
J. Dewey Gun Co., Clinton Corners, N.Y. 12514 (bullet spinner)
Division Lead Co., 7742 W. 61st Pl., Summit, Ill. 60502
Dom Enterprises, 3985 Lucas, St. Louis, Mo. 63103
Eagle Products Co., 1520 Adelia Ave., So. El Monte, Calif. 91733
W. H. English, 4411 S.W. 100th, Seattle, Wash. 98146 (Paktool)
Ellwood Epps Sptg. Goods, 80 King St., Clinton, Ont., Canada
The Fergusons, 27 W. Chestnut St., Farmingdale, N.Y. 11735
Fitz, Box 49797, Los Angeles, Calif. 90049 (Fitz Flipper)
Flambeau Plastics, 801 Lynn, Baraboo, Wis. 53913
Fordwad, Inc., 4322 W. 58th St., Cleveland, Ohio 44109
Forster-Appelt Mfg. Co., Inc., 82 E. Lanark Ave., Lanark, Ill. 61046
Full Ed'z Creel Co., 717 W. 9th St., Cheyenne, Wyo. 82001
Gene's Gun Shop, 3602 Hill Ave., Snyder, Tex. 79549 (arbor press)
Gopher Shooter's Supply, Box 246, Faribault, Minn. 55021
The Gun Clinic, 81 Kale St., Mahtomedi, Minn. 55115
H & H Sealants, Box 448, Saugerties, N.Y. 12477 (Loctite)
Hart Products, 401 Montgomery St., Nescopeck, Pa. 18635
Frank A. Hemsted, Box 281, Sunland, Calif. 91040 (swage dies)
Hensley & Gibbs, Box 10, Murphy, Ore. 97533
E. C. Herkner Co., Box 5007, Boise, Ida. 83702
Herter's Inc., RR1, Waseca, Minn. 56093
B. E. Hodgdon, Inc., 7710 W. 50 Hiway, Shawnee Mission, Kans. 66202
Hollywood Reloading, Inc., 19540 Victory, Reseda, Calif. 91335
Hulme Firearm Serv., Box 83, Millbrae, Calif. 94030 (Star case feeder)
I & I Co., 709 Twelfth St., Altoona, Pa. 16601 (multi-shellcatcher)
Independent Mach. & Gun Shop, 1416 N. Hayes, Pocatello, Ida. 83201
JASCO, Box 49751, Los Angeles, Calif. 90049
J & G Rifle Ranch, Turner, Mont. 59542 (case tumblers)
Javelina Products, Box 337, San Bernardino, Calif. 92402 (Alox beeswax)
Jay's Sports, Inc., Menominee Falls, Wis. 53051 (powder meas. stand)
Kexplore, Box 22084, Houston, Tex. 77027
Kuharsky Bros., 2425 W. 12th, Erie, Pa. 16500 (primer pocket cleaner)
Lachmiller Div. of Peng. Ind., P.O. Box 97, Parkesburg, Pa. 19365
Lee Engineering, 46 E. Jackson, Hartford, Wis. 53027
Leon's Reloading Service, 3945 N. 11 St., Lincoln, Neb. 68521
L. L. F. Die Shop, 1281 Highway 99 N., Eugene, Ore. 97402
Liberty Arms, P.O. Box 308, Montrose, Calif. 91020
Ljutic Industries, 918 N. 5th Ave., Yakima, Wash. 98902
Lock's Phila. Gun Exch., 6700 Rowland, Philadelphia, Pa. 19149
J. T. Loos, Pomfret, Conn. 06258 (primer pocket cleaner)
Lyman Gun Sight Products, Middlefield, Conn. 06455
McKillen & Heyer, Box 627, Willoughby, Ohio 44094 (case gauge)
Paul McLean, 2670 Lakeshore Blvd. W., Toronto 14, Ont., Canada (Univ Holder)
Pat B. McMillan, 4908 E. Indianola, Phoenix, Ariz. 85018
MTM Molded Prod. Co., P.O. Box 14092, Dayton, Ohio 45414
Magma Eng. Co., P.O. Box 881, Chandler, Ariz. 85224
Mayville Eng. Co., Box 267, Mayville, Wis. 53050 (shotshell loader)
Merit Gun Sight Co., P.O. Box 995, Sequim, Wash. 98382
Minnesota Shooters Supply, 1915 E. 22nd St., Minneapolis, Minn. 55404
Murdock Lead Co., Box 5298, Dallas, Tex. 75222
National Lead Co., Box 831, Perth Amboy, N.J. 08861
Normington Co., Box 156, Rathdrum, Ida. 83858 (powder baffles)
John Nuler, 12869 Dixie, Detroit, Mich. 48239 (primer sealing tool)

Ohaus Scale Corp., 29 Hanover Rd., Florham Park, N.J. 07932
Omark-CCL, Inc., Box 856, Lewiston, Ida. 83501
Pacific Tool Co., Box 4495, Lincoln, Neb. 68504
C. W. Paddock, 1589 Payne Ave., St. Paul, Minn. 55101 (cartridge boxes)
Vernon Parks, 104 Heussy, Buffalo, N.Y. 14220 (loaders bench)
Perfection Die Co., 1614 S. Choctaw, El Reno, Okla. 73036
Personal Firearms Record Book, Box 201, Park Ridge, Ill. 60068
Phelps Reloader Inc., Box 4004, E. Orange, N.J. 07019
Ferris Pindell, Connersville, Ind. 47331
Plano Gun Shop, 1521B 14th St., Plano, Tex. 75074 (powder measure)
Plum City Ballistics Range, Box 128, Plum City, Wis. 54761
Ponsness-Warren, Inc., Box 186, Rathdrum, Ida. 83858
Potter Eng. Co., 1410 Santa Ana Dr., Dunedin, Fla. 33528
Marian Powley, 19 Sugarplum Rd., Levittown, Pa. 10956
Quinetics Corp., 3740 Colony Dr., San Antonio, Tex. 78230 (kinetic bullet
RCBS, Inc., Box 1919, Oroville, Calif. 95965
Raymor Industries, 5856 So. Logan Ct., Littleton, Colo. 80120 (primer mag.)
Redco, Box 15523, Salt Lake City, Utah 84115
Redding-Hunter, Inc., 114 Starr Rd., Cortland, N.Y. 13045
Remco, 1404 Whitesboro St., Utica, N.Y. 13502 (shot caps)
Rifle Ranch, Rte. 1, Prescott, Ariz. 86301
Rochester Lead Works, Rochester, N.Y. 14608 (leadwire)
Roman Prod., Box 891, Golden, Colo. 80401
Rorschach Precision Prods., P.O. Box 1613, Irving, Tex. 75060
Rotex Mfg. Co. (See: Texan)
Ruhr-American Corp., So. East Hwy. 55, Glenwood, Minn. 56334
SAECO Rel. Inc., P.O. Box 778, Carpinteria, Calif. 93013
Savage Arms Co., Westfield, Mass. 01085
Scientific Lubricants Co., 3753 Lawrence Ave., Chicago, Ill. 60625
Shoffstalls Mfg. Co., 740 Ellis Place, E. Aurora, N.Y. 14052
Shooters Accessory Supply, Box 250, N. Bend, Ore. 97459 (SAS)
Shooters Serv. & Dewey, Inc., Clinton Corners, N.Y. 12514 (SS&D)
Sil's Gun Prod., 490 Sylvan Dr., Washington, Pa. 15301 (K-spinner)
Jerry Simmons, 713 Middlebury St., Goshen, Ind. 46526 (Pope de- & recapper)
Rob. B. Simonson, Rte. 7, 2129 Vanderbilt Rd., Kalamazoo, Mich. 49002
Smith & Wesson-Fiocchi, Inc., 3640 Seminary Rd., Alton, Ill. 62002
Sport Ammo Corp., 8407 Center Dr., Minneapolis, Minn. 55432 (mini-kit tool)
Star Machine Works, 418 10th Ave., San Diego, Calif. 92101
Strathmore Gun Spec., Box 308, Strathmore, Calif. 93267
Sullivan Arms Corp., 5204 E. 24th St., Indianapolis, Ind. 46218
Swanson Co., Inc., 2205 Long Lake Rd., St. Paul, Minn. 55112 (Safari loader)
Texan Reloaders, Inc., P.O. Box 5355, Dallas, Tex. 75222
VAMCO, Box 67, Vestal, N.Y. 13850
W. S. Vickerman, 505 W. 3rd Ave., Ellenburg, Wash. 98926
Wanda Cartr. Co., P.O. Box 45901, Houston, Tex. 77045 (plastics)
Weatherby, Inc., 2781 Firestone Blvd., South Gate, Calif. 90280
Webster Scale Mfg Co., Box 188, Sebring, Fla. 33870
Whit's Shooting Stuff, 2121 Stampede Ave., Cody, Wyo. 82414
Whitney Cartridge Co., Box 5872, Pasadena, Calif. 91107 (shotshells)
L. E. Wilson, Inc., Box 324, Cashmere, Wash. 98815
Xelex, Ltd., Hawksbury, Ont., Canada (powder)
Zenith Ent., Rt. 1, Box 52z, Del Mar, Calif. 92014
A. Zimmerman, 127 Highlant Trail, Denville, N.J. 07834 (case trimmer)

SURPLUS GUNS, PARTS AND AMMUNITION

Allied Arms Ltd., 655 Broadway, New York, N.Y. 10012
Century Arms, Inc., 3-5 Federal St., St. Albans, Vt. 05478
W. H. Craig, Box 927, Selma, Ala. 36701
Cummings Intl. Inc., 41 Riverside Ave., Yonkers, N.Y. 10701
Eastern Firearms Co., 790 S. Arroyo Pkwy., Pasadena, Calif. 91105
Fenwick's, P.O. Box 38, Weisburg Rd., Whitehall, Md. 21161
Hunter's Lodge, 200 S. Union, Alexandria, Va. 22313
Lever Arms Serv. Ltd., 771 Dunsmuir St., Vancouver 1, B.C., Can
Mars Equipment Corp., 3318 W. Devon, Chicago, Ill. 60645
National Gun Traders, 251-55 W. 22nd, Miami, Fla. 33135
Pacific Intl. Imp. Co., 2416-16th St., Sacramento, Calif. 95818
Plainfield Ordnance Co., Box 447, Dunellen, N.J. 08812
Potomac Arms Corp., Box 35, Alexandria, Va. 22313
Ruvel & Co., 3037 N. Clark St., Chicago, Ill. 60614
Service Armament Co., 689 Bergen Blvd., Ridgefield, N.J. 07657
Sherwood Distrib. Inc., 9470 Santa Monica Blvd., Beverly Hills, Cal
Z. M. Military Research Co., 9 Grand Ave., Englewood, N.J. 07631

Modern Handloading